Springer Geology

The book series Springer Geology comprises a broad portfolio of scientific books, aiming at researchers, students, and everyone interested in geology. The series includes peer-reviewed monographs, edited volumes, textbooks, and conference proceedings. It covers the entire research area of geology including, but not limited to, economic geology, mineral resources, historical geology, quantitative geology, structural geology, geomorphology, paleontology, and sedimentology.

More information about this series at http://www.springer.com/series/10172

Mark A.S. McMenamin

Deep Time Analysis

A Coherent View of the History of Life

 Springer

Mark A.S. McMenamin
Department of Geology and Geography
Mount Holyoke College
South Hadley, MA, USA

ISSN 2197-9545   ISSN 2197-9553   (electronic)
Springer Geology
ISBN 978-3-030-08947-4   ISBN 978-3-319-74256-4   (eBook)
https://doi.org/10.1007/978-3-319-74256-4

Printed on acid-free paper

This Springer imprint is published by the registered company Springer International Publishing AG part of Springer Nature
The registered company address is: Gewerbestrasse 11, 6330 Cham, Switzerland

Dedicated to Étienne Geoffroy Saint-Hilaire
(1772–1844) and the memory of
Martin D. Brasier (1947–2014)

Preface

Let us say we define *complex life* as any organism that can search its environment, seek out, locate, and utilize crystals of a single mineral type by individual selection based on distinctive characteristics. This may seem an odd definition of complex life. But as you will read in *Deep Time Analysis*, complex life appears to have done just that in the Proterozoic and at other times during earth history. The habit returns with our species in an ancient, creative use of feldspar. In addition to these curious cases, in *Deep Time Analysis* you will encounter some of the most interesting and mysterious fossils on earth. Of particular interest is a complete description of the Sonoran Ediacaran fossils, presented here for the first time.

This book aims to give you a new appreciation of the relationship between sclerotome and scleritome, a better understanding of the Ediacaran biota, and a fuller appreciation of the significance of torologous evolution. All three topics are essential to understand the development of complex life and its role on this globe.

South Hadley, MA, USA   Mark A.S. McMenamin

Acknowledgments

I thank B. Carbajal Gonzalez, B. Beckett, J. D. Beuthin, R. Boisvert, I. Burgess, F. Debrenne, K. E. Fellows, C. G. Fisher, D. Fleury, K. A. Hughes, M. Hussey, L. Orr, D. Orr, M. Gingras, D. Griffen, K. Gross, N. Hodge, C. Hoffman, M. Jercinovic, A. S. Kimberley, J. L. Kirschvink, E. A. Kishibay, J. Kingeter, W. Malik, M. Markley, D. Marshall, S. Mazur, H. Pflug, D. L. Schulte McMenamin, S. K. McMenamin, J. M. McMenamin, R. Perry, T. Ryan, C. Ruiz, E. Seilacher, A. Seilacher, D. Shepardson, H. Singleton, D. Snoeyenbos, O. Snoeyenbos West, G. Theokritoff, B. Wah, A. Werner, J. Williams, M. Williams, L. Zapata, and E. B. W. Zubrow plus two anonymous reviewers for assistance with various aspects of this research. Thanks also to Hadland Laboratories, Amherst, New Hampshire, for technical assistance. I thank Mount Holyoke College for a Faculty Fellowship during the academic year 2015–2016 and the National Science Foundation for grant EAR 857995.

Contents

Chapter 1
Seventh Law

Abstract The seven laws of morphogenetic evolution allow us to better comprehend the constraints and possibilities of evolutionary change. Parity bits of Hamming's error-correcting code procedure may be used to simulate the appearance and disappearance of the scleritome. The Seventh Law states that morphogenetic field vectors may be bundled or dilated in a geometrically regular fashion to generate compound eyes or their ectodermal equivalent. The discovery of these laws implies that we exist in a law-governed universe, where the operation of said laws implies repeated and predictable outcomes in the history of life.

Keywords Scleritome · Morphogenetic field · *Clementechiton* · *Chrysomallon* · Hamming's error correction code procedure · Parity bits · *Microdictyon*

Deep Time Analysis expands on the dynamic paleontology approach as proposed in *Dynamic Paleontology* (McMenamin 2016). *Deep Time Analysis* differs from dynamic paleontology primarily in the scope of the research questions it addresses; many of the deep time analysis projects have global implications in terms of plate tectonics, global paleoclimate, geochemistry, geochronology and paleobiogeography.

The scleritome analysis presented in *Dynamic Paleontology* examined torologous (that is, shared origin from a toroidal morphogenetic field) scleritome development and toroidal organization of animalian body plan. The point of this work was not to deny the possibility of self-assembly with regard to the surface morphology of organisms (e.g., Filippov et al. 2017), but rather to show that another factor must also be involved, namely, a surficial morphogenetic field. *Dynamic Paleontology* introduced the evolutionary concept of torologous features, that is, similar morphologies due to similar changes in morphogenetic fields (Gilbert et al. 1996; Bolker 2000) and their expression (such as the scleritome itself). When applied to shelled mollusks, the scleritome has recently been referred to as the 'secretome' (Jackson et al. 2006; Kocot et al. 2016). While focusing their research on gene expression, Kocot et al. (2016) determined that the "mantle secretome appears to regulate shell deposition and patterning and in some cases becomes part of the shell matrix." Kocot et al. (2016) seem to have stumbled upon a physical basis for torologous relationships particularly as they relate to fabrication of shell matrix. The domain shuffling associated RLCDs (repetitive low complexity domains; that is, repetitive motifs or domains in proteins) provides a splendid basis for analysis of torologous relationships particularly as they relate to cases of convergent evolution. The actual genes in use are quite variable. Kocot et al. (2016) discuss the results of Jackson et al. (2010) comparing the nacre-secreting mantle transcriptomes of a gastropod (*Haliotis asinina*) and a bivalve (*Pinctada maxima*):

> [The] majority of the secreted proteins had no similarity to sequences in public databases, and less than 15% of the secreted proteins were shared between the two species. These results indicate that the two taxa use different gene sets to construct their shells. This is in line with observations that both the crystallographic orientations of nacre tablets and their growth modes differ between these taxa, and strongly suggests that bivalve and gastropod mother-of-pearl nacre evolved to resemble each other due to homoplasy.

We see here that the genes involved differ, hence there must be morphogenetic field control governing shell formation. The gastropod (*Haliotis asinina*) and a bivalve (*Pinctada maxima*) may show a torologous relationship with regard to shell nucleation and growth. This point is underscored dramatically when one considers the *bivalve snails* of family Juliidae such as *Berthelinia*, mollusks that were considered to be clams until a live one (*Tamanovalva limax*) was found in 1959 in Japan (Kawaguti and Baba 1959) crawling on a specimen of the (oddly Ediacaran-like) green alga *Caulerpa*. These sacoglossan (algal body fluid sucking) snails have a fossil record extending back to the Paleocene (Le Renard et al. 1996).

The basic morphological patterns have been recognized for many years. The body armor of ancient fish, whose microstructure of delicate radiating tubules was first described by T. H. Huxley (Halstead 1982), is directly linked to vertebrate teeth as both develop microtubules, and both fish armor and teeth in turn have a torologous relationship to canals and punctae in tommotiid small shelly fossils and mickwitziid brachiopods, respectively (McMenamin 1992). Canals of another type of small shelly fossil (Vinther 2009), the halkieriid *Sinosachites*, are torologous to the aesthete canal system of the Ediacaran chiton *Clementechiton* (McMenamin 2016). Chen et al. (2015) emphasize what they claim to be a convergent evolutionary relationship between the foot scales of the scaly-foot gastropod (*Chrysomallon* spp.) and the girdle sclerites of chitons (*Enoplochiton niger*), however, this relationship is better considered to be a case torologous evolution. The same could be said for scale covered squid (*Lepidoteuthis grimaldii*; Escánez et al. 2017) and multiple lineages of gastropods (Chen et al. 2015) that developed multielement scleritomes by convergent evolution (Sigwart 2017).

How might we model this phenomenon, what has been called the "intrinsic skeletogenic properties of the dermis" (Cerda and Powell 2010)? To make a model of the toroidal morphogenetic field, we must borrow from the insights and techniques of computation (Hamming 1980; Feynman 1996; Stallings 2006). We will take inspiration here from the clever Hamming Code, developed by Richard Hamming of Bell Laboratories to develop error-correcting codes. Let's consider Hamming's use of intersecting circles to describe error correction on a 4-bit word, as graphically described by Stallings (2006). Seven compartments are formed by the three

Fig. 1.1 Seven
compartments formed by
the three intersecting
circles of a Venn diagram.
A 4-bit word is inscribed
in the three-petal trefoil.
The empty fields, called
parity fields, will acquire
parity bits

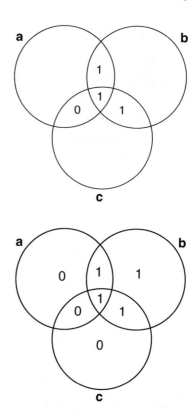

Fig. 1.2 Seven
compartments formed by
the three intersecting
circles of a Venn diagram

intersecting circles of the Venn diagram. The 4-bit word is inscribed in the three-petal trefoil as shown in Fig. 1.1. Three compartments remain, and Hamming (1980) calls these the parity bits. We will refer to the geometrical space holding the parity bit as the parity field.

The value of each parity bit is determined by the values of the 4-bit word compartments that are contained within the circumference of a given circle. This is so because each circle must have an even sum of 1 s, thus the parity bits are set to one or zero accordingly to keep the total circle sum even (Fig. 1.2). Nothing has changed within the 4-bit word trefoil. For example, in Fig. 1.2, the parity spaces have been assigned one or zero to keep the circle sums even: circle B has a value of 4, whereas circles A and C have values of 2 each, respectively. Let's introduce an error into one of the data bits, say, the lower right petal of the data trefoil changes from one to zero (Fig. 1.3). By examination of the parity bit/data bit sums, we see that there is discrepancy in the sums of circles B and C. Each has three and one as sum values, respectively, and hence they stand out as being in error because they do not have even within-circle values. In Hamming's (1980) error-correcting code procedure, only "one of the seven compartments" (Stallings 2006) is shared by the error-circles B and C. It is not shared in common by circle A (Fig. 1.3). According to Stallings (2006), the "error can therefore be corrected by changing that bit."

Fig. 1.3 Seven
compartments formed by
the three intersecting
circles of a Venn diagram.
An error has been
introduced into one of the
data bits; the lower right
petal of the data trefoil has
been changed from one to
zero

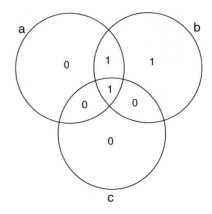

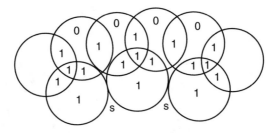

Fig. 1.4 Part of a morphogenetic field surface shown in cross section. The upper edge of the image represents the outer surface of the animal. The section through the field consists of nine interlocked circles. Three data trefoils (bold lines) are generated by the nine circles. s, point of entry of inhibitory signal molecule that maintains the value of each lozenge-shaped parity field above this point at one. The leftmost and rightmost circles are not included in the parity calculation

We will now borrow from Hamming's computational error-correcting code theory the idea of circles, divided into fields that get assigned either data or parity bits. Each circle has a "data trefoil" in the center. But instead of merely three circles, we will make a daisy chain of circles to model a portion in cross section of the morphogenetic field surface. Figure 1.4 shows a section of field consisting of nine interlocked circles. Three data trefoils are generated by the nine circles. The data values of the trefoils are all the same, namely, a value of 1 for all four fields. This will not change in the subsequent simulation run (Fig. 1.5), thus the trefoil data values may be considered here to be constant. The full circles must always obey the Hamming rule, however, such that the sum of all values within a circle remains an even number.

As per this simulation, in order for sclerites to form on the surface of the animal, a parity field must have part of its perimeter exposed to the outer surface *plus* the parity field must have a value of one. This simulation assumes that the scleritome is a primitive or original feature of early animals. This assumption will gain support from description of new Ediacaran fossils in a subsequent chapter.

Some of the upper parity fields (that is, those occurring above each trefoil) have this outer surface exposure, but two slender lozenge-shaped parity fields do not.

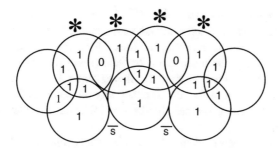

Fig. 1.5 Section of morphogenetic field enabled to grow a scleritome. The inhibitory signal molecule has been blocked at the point of entry (short bar symbol over s). As a result, the lozenge-shaped parity field changes to parity value zero. This causes the parity fields in extensive contact with the outer surface of the animal to change their parity values to one, and they are thereby enabled to grow sclerites (asterisks). The leftmost and rightmost circles are not included in the parity calculation

As shown in Fig. 1.4, with this data trefoil configuration as described above, the parity fields below the trefoils all have a parity value of one. In Fig. 1.4, the trefoil fields are outlined in bold lines. The parity fields above the trefoils may have a parity value of either one or zero. The larger, upper parity fields with an arc of the field constituting the outer surface of the organism (as seen in cross section), all have a parity value of zero. The lozenge-shaped parity fields have parity values of one. Because of this zero parity value of the larger, upper parity field, these upper parity fields will not develop sclerites, osteoderms or other skeletalized bits on the outer surface of the animal. Neither will the lower parity fields in this simulation, because they are underneath the trefoils.

Let's now make a small change the parity bit values, by changing the value of the upright lenticular field between each pair of trefoils from one to zero. This might occur as a result, say, of some type of biomolecule (perhaps a chemical signal produced by the animal), ordinarily diffusing outward from the interior of the animal, no longer reaching a particular portion of the cuticular region as it once had. This particular spot in this example is represented by the point of contact between trefoils. The slender lozenge-shaped parity fields are immediately above these points (Figs. 1.4 and 1.5). With, say, inhibition signal molecule blockage, the slender parity field changes it parity value from 1 (signal present) to 0 (signal absent). The signal molecule is presumed to have been diffusing toward the outer surface of the animal.

This simple change leads to a complete reconfiguration of the outer surface of the animal. All of the large upper parity fields must change their parity values from 0 to 1, in order to maintain the sum of data and parity values in each circle as an even number. With parity values of 1 each, the large upper parity fields are now enabled to develop sclerites, osteoderms or other skeletalized bits on the outer surface of the animal. If the signaling agent experienced downward transport, it might help maintain a skeleton-free underside to the creature (by keeping the ventral slender lozenge-shaped parity fields set at a value of zero) while the scleritome develops on the animal's upper surface due to a blocked inhibition signal (Fig. 1.5).

This model confirms McMenamin's (2015a) inference that invoking diffusing morphogen compounds as a self-contained explanation to account for morphogenesis will encounter insurmountable difficulties. There must already be a morphogenetic structure, namely, the morphogenetic field (Figs. 1.4 and 1.5), on which the signaling compounds may act as shown in this example. Figures 1.4 and 1.5 clearly demonstrate this effect, an effect that is absolutely fundamental for our understanding of biology—in some ways more fundamental than Charles Darwin's concept of natural selection.

Recognition of concentric rings of tooth elements (Smith 2011) surrounding the mouth of *Hallucigenia, Microdictyon*, priapulids, and anomalocarids is greatly in accord with the toroidal model. Also in accord is recognition of the inner, second (pharyngeal) pair of jaws in moray eels (Muranidae) and the tilapia (*Oreochromis niloticus*). Placodes, the initiation sites of follicles, teeth and feathers, are torologous in the same way. Even considering the wide spectrum of animal morphology, Honoré de Balzac observed in *La comédie humaine* that there is "in essence only a single animal" (Appel 1987; see also Edelman et al. 2016).

Torologous relationships do not imply that genes are not involved the development of the scleritome. *Scaleless (sc/sc)* recessive mutation was discovered in 1954 in a colony of chickens (*Gallus gallus domesticus*, the domesticated version of the red jungle fowl or bamboo fowl; New Hampshire breed) by researchers at the University of California at Davis (Abbott and Asmundson 1957). The naked birds are now considered particularly valuable in a warming world because they can potentially thrive in hot climates without overheating (Wells et al. 2012). Wells et al. (2012) have identified several inhibitory and activator molecular signals for feather/scale placode development (WNT/b-catenin signaling, FGF signaling, retroviral expression of BMP2 or BMP4 protein) involved in forming the patterns on the skin of chickens, but they note that "genetic evidence implicating the specific molecules acting in these multi-component pathways is as yet lacking." Wells et al.'s (2012) promising results represent "the first loss of function genetic evidence supporting a role for FGF ligand signaling in feather development, and suggests FGF20 as a novel central player in the development of vertebrate skin appendages, including hair follicles and exocrine glands."

In an early hint of an appreciation of torologous relationships among unrelated vertebrates, Etienne Geoffroy Saint-Hilaire (1807) reported the presence of vestigial teeth in mysticete (baleen whale) embryos. This key discovery was reported by Geoffroy Saint-Hilaire (1807) in the figure caption of an illustration focused on the comparative anatomy of bird skulls (Peredo et al. 2017). Geoffroy would thus have been well-conditioned to appreciate both torologous relationships and the seven laws of morphogenetic evolution as described in the first chapter here. The 'laws of evolution' constituted the primary focus of a great debate in French biology between Georges Cuvier (1769–1832) and Geoffroy Saint-Hilaire (1772–1844).

The debate pitted the laws of form (Geoffroy) against biological functionalism (Cuvier) in a contest that determined the future of biology. The conflict also had broader literary and philosophical ramifications (Appel 1987); for example, Balzac dedicated *Le Père Goriot* to Geoffroy in tribute to his labors and genius. Laws of form were considered at the time to be something that could constrain God's creative ability due to immutable laws of animal morphology (Appel 1987). Principles such as the law of homology conflicted with Cuvier's strict functionalism. In Britain, such laws of form cleared "the way for Darwin by giving greater reign to the role of law in place of the free will of the Creator" (Appel 1987). The law of homology (animals are but variations on a single theme) was seen to lend support to pantheism and anti-Catholicism, and thereby boosted the "notion of the progressive nature of humanity" (Appel 1987). That led, with more than a little irony, to the Omega Point evolutionary schema of French Jesuit and paleontologist Pierre Teilhard de Chardin (McMenamin 2015b). Laws of nature, however, can also be viewed as reflective of a higher order of Design, with laws of morphology serving as a subset of natural law and thus very much in accord with what I consider to be authentic religion. Indeed, *Cuvier's* famous law, the 'law of correlation' (claimed to allow reconstruction of the morphology and lifestyle of an entire animal from an isolated bit such as a claw or tooth; Dawson 2016), received a favorable citation in the "Palaeontology" entry in *The Catholic Encyclopedia* (Waagen 1911). In any case, laws of homology, such as the torologous relationships defined in *Dynamic Paleontology*, represent genuine aspects of nature, as seen for example in the convergent evolutionary development of elongate shoulder spike osteoderms in both the superbly well-preserved nodosaur *Borealopelta markmitchelli* (Brown et al. 2017; Greshko 2017) and aetosaurs (crocodilian relatives). Neither Geoffroy's Law of Homology nor Cuvier's Law of Correlation should be confused with John Wolfgang von Goethe's Law of Compensation, which holds that expansion of one part of an organism will necessitate reduction in another part. An example of this in *Homo sapiens* might hold that the expansion of our comparatively gigantic brains required loss of something, and the 'somethings lost' may have included our tails, thick body hair, etc. Geoffroy was a great fan of Goethe's Law of Compensation.

In the last chapter of *Dynamic Paleontology*, I advocated that we adopt a bold and fearless approach to the scientific analysis of the remote past. It is now time to extend this approach with deep time analysis.

It is possible to mistake the "ubiquity of consciousness" (Shanta 2015) with the operation of laws that deliver, through the course of evolutionary time, a series of approaches to the endpoints foreordained by the laws in question. Expressing an amusing yet widespread fallacy, the American political pundit Rush Limbaugh attacked evolutionary thinking in the wake of the 2016 Cincinnati zoo tragedy (where the gorilla Harambe was shot and killed to protect a child who had fallen into the gorilla pit). Limbaugh remarked: "A lot of people think that we all used to be gorillas, and they are looking for the missing link out there. .. they think we were originally apes… If we were the original apes, then how come Harambe is still an ape, and how come he didn't become one of us?" These remarks were subjected to widespread derision, with press sources such as the *Huffington Post* (Mazza 2016) excoriating the conservative Limbaugh for his supposed ignorance of evolutionary theory.

In fairness to Limbaugh, however, he in fact criticizes an important variant of evolutionary theory that holds that evolutionary change occurs by anagenesis. According to *Wikipedia*: "Anagenesis, also known as "phyletic transformation", is

Table 1.1 Laws of morphogenetic evolution

First law	The same forces that control macroevolution control the high precision of convergent evolution. Both processes are associated with transformations of morphogenetic fields (McMenamin 2009).
Second law	Evidence for control by morphogenetic fields is most apparent in the earliest representatives of any particular lineage of complex life (McMenamin 2009).
Third law	Higher evolutionary grades of complex life are characterized by simplification of their respective morphogenetic fields (McMenamin 2015a).
Fourth law	Sexual selection can generate prominent exceptions to the other laws of morphogenetic evolution (McMenamin 2015a).
Sbaglio's law = Fifth law	Changing the orientation of field vectors controlling the position of surface ornamentation can alter body surface form. Altering these same field vectors can *also* fundamentally alter actual body form (=*baüplan*).
Sixth law	Surface-normal field lines occur at the intersections of latitudinal and longitudinal morphogenetic field lines.
Seventh law	Field vectors may be bundled or dilated in a geometrically regular fashion to generate compound eyes or ectodermal equivalents.

when the new morphospecies is a result of rapid evolution in the ancestral form without speciation taking place, such that there are no remaining other populations of the ancestor species and the species can be considered extinct. The ancestor species is therefore superseded by the new species it morphs into." In Limbaugh's view, phyletic transformation should have transformed the great ape lineage, in its entirety, into humans, thus leading to phyletic extinction of the gorilla. This is of course an incorrect view of how evolution works, as it ignores the widespread evidence for cladogenesis or branching speciation. Limbaugh sets up and attacks a straw man (anagenesis), but his comments are not quite so unsophisticated as his detractors would maintain. Limbaugh implies that an important law of morphological change is implied by evolutionary theory, namely, that more advanced and eventually intelligent life forms will issue from less intelligent forms. In so stating he is correct at least in a general sense.

As I asked in *Dynamic Paleontology*, is it worthwhile investing in a search for laws of nature (McMenamin 2016)? The great Estonian embryologist Karl Ernst von Baer answered yes, noting that "the palm will be won by the one who can trace" the laws of animal body form back to the general laws "that direct the life of the universe" (Edelman et al. 2016). Charles Marshall (2017) writes of what he calls the Five Laws of Paleobiology. These commonsensical laws include the following: extinction happens; species do not last very long; species extinction and origination rates are in rough balance; species richness is a function of both origination and extinction, and extinction wipes out a lineage's history. Discerning the laws of nature is the driving impulse behind scientific research and investigation.

In a series of three books (McMenamin 2009, 2015a, 2016) I defined six laws of morphogenetic evolution. The following Table 1.1, includes an additional law, referred to here as the Seventh Law of Morphogenetic Evolution.

Morphogenetic field control of animal body plans is currently under intense scrutiny, as it may provide the key for understanding the evolutionary dynamics of

Fig. 1.6 *Microdictyon* cf. *M. sinicum* from the Early Cambrian Chengjiang Biota of China. Note the paired rows of dorsal trunk sclerites. Length of specimen approximately 2 cm. Photograph used here per Creative Commons BY-SA license

the otherwise inexplicable Cambrian Explosion. Seilacher (1994) advocated that if the emergent Cambrian body plans "arose largely by synergetic self-organization, their relatively sudden appearance becomes more acceptable." This represents an attempt to explain the rapid pace of the Cambrian event, but does not explain why the numerous new phyla did not appear tens or even hundreds of millions of years earlier. Some external forcing agent is surely required (McMenamin 2016).

Interestingly, external animal body form (of, say, a model scorpion) can be "reverse engineered" into what amounts to a crease pattern morphogenetic field (Lang 2011). Unfolding the evidence step-by-step, we currently seek the causes of the sudden morphogenetic field transformations that led to the origin of most animal phyla at or near the Cambrian boundary at approximately 542 million years ago.

The importance of the seventh law resides in the fact that contracting/dilating portions of the morphogenetic field, most usually in a mirror image fashion as reflected across the sagittal plane of the animal in question, can lead to major evolutionary advance by the introduction of new organs, such as compound eyes, camera eyes, or the rostral organ of the coelacanth. The geometrical placement of fenestrae (such as the anorbital fenestrae) in the skulls of diapsids is an expression of geometrically-placed morphogenetic field dilations (Edelman et al. 2016; Sheesley et al. 2014). The dorsal trunk sclerites of the bizarre Early Cambrian lobopodian *Microdictyon* show clear control by morphogenetic field influence (McMenamin 2016). These trunk sclerites (Fig. 1.6) in fact may very well be homologous to the compound eyes of arthropods and some mollusks (Dzik 2003; see also Schoenemann et al. 2009).

The morphogenetic field control of *Microdictyon* has taxonomic implications. Wotte and Sundberg (2017) reported three species of *Microdictyon*, two of them new, as *M. montezumaensis* Wotte and Sundberg 2017, *M. cuneum* Wotte and Sundberg 2017, and *M. rhomboidale* Bengtson et al. 1986. As the points of highest relief on the lobopodian sclerites ("small shelly fossils"), *M. montezumaensis* has very low nodes, *M. rhomboidale* has higher nodes, and *M. cuneum* has wedge-shaped nodes that are all inclined in the same direction. These represent varying states of entrainment of the development of nodes with either vertical or inclined morphogenetic field lines in accordance with Sbaglio's Law or the Fifth Law of Morphogenetic Evolution. Although Zhang and Aldridge (2007) were unable to suggest any *Microdictyon* species synonymies, all three of *Microdictyon* species mentioned above are probably conspecific (as *M. rhomboidale*). *M. "cuneum"* represents a sclerite from *Microdictyon rhomboidale* with a spiral cowlick in its morphogenetic field in accordance with Sbaglio's Law!

Zhang and Aldridge (2007) provided some interesting information on *Microdictyon* sclerites. The mushroom-shaped nodes of the sclerites of *Microdictyon jinshaense* appear as relict individual sclerites that have been fused together by the underlying skeletal network that binds them together. McMenamin (2016) considered cases such as this one to be examples of Williston's Law, the law of reduction of parts. Possibly related to this phenomenon is the fact that one trunk plate of *Microdictyon jinshaense* shows, in place of several nodes, an anomalously large, pointed spine projecting from the top of the sclerite (Zhang and Aldridge 2007). This elongate spine resembles a rose thorn and would seem suited to provide protection from macroscopic predators.

Of great interest is the duplicated lobopodian trunk plates of *Microdictyon jinshaense*, *Microdictyon chinense*, and their relative *Quadratapora zhenbaensis*. The sclerites or trunk plates in these examples appear as a trunk sclerite concentrically nested inside a larger sclerite. Zhang and Aldridge (2007) attribute this odd formation as an attempted, but failed, molting separation between successive plates, resulting in the two plates simply fusing together. Zhang and Aldridge (2007) criticize Dzik's (2003) interpretation of a visual role for *Microdictyon* sclerites:

> Dzik (2003) offered a radically different interpretation, believing each plate to be a visual structure with each hole in the plate occupied by a fine lens. This model may seem tenable for the relatively uncomplicated sclerites of *M. effusum*, but is less plausible for the range of sclerite morphology documented here.

Zhang and Aldridge (2007) assert that ecdysis in *Microdictyon* falsifies the visual interpretation of the trunk sclerites because light would not pass through while the new sclerite was forming underneath the previous plate. This is not a compelling critique in my view, however, as trilobites, with undoubted visual systems in the form of the ocular lobes on the cephalon, were able to thrive in spite of their vision perhaps being impaired during the molting process.

Furthermore, an undescribed species of *Microdictyon* (Figs. 1.7 and 1.8) first described in 1984 ("Genus and species indeterminate D" of McMenamin (1984)) has a much thinner sclerite type, that would be less metabolically costly to construct and would shed fairly easily during ecdysis. Figure 1.8 apparently shows mineralized remnant cuticle still adhering to the porous sclerite, indicating that the plate was shed by ecdysis at the same time as the cuticle. Wotte and Sundberg (2017) have described (as *Microdictyon* sp.) similar thin plates or sclerites.

A clear connection may be recognized between the morphology of *Microdictyon* and the morphology of the earliest animal preserved as an adult body fossil, *Clementechiton sonorensis* (Fig. 1.9). The curved bars on opposite edges of the chiton dorsal valves in *Clementechiton sonorensis* are considered here to be homologous (or better, "torologous" *sensu* McMenamin (2016) in recognition of shared origin from a toroidal morphogenetic field) to the ocular lobes of trilobites.

Fig. 1.7 *Microdictyon* sp. SEM photomicrograph, Lower Cambrian, Puerto Blanco Formation, unit 2, field sample 5.5+ of 12/1782; IGM 3614[27]; scale bar 500 microns

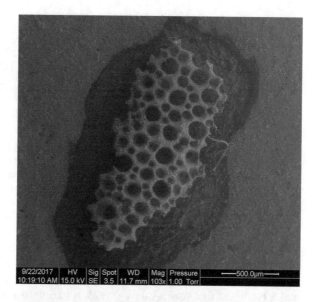

Fig. 1.8 *Microdictyon* sp. Note remaining scraps of what may represent shed (molted) cuticle of the lobopodian. SEM photomicrograph, Lower Cambrian, Puerto Blanco Formation, unit 2, field sample 5.5+ of 12/1782; IGM 3614[26]; scale bar 300 microns

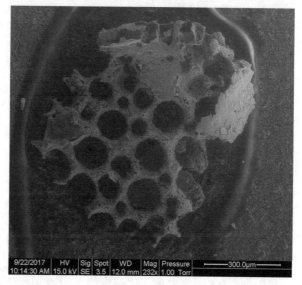

The stem-group lycophyte plants belonging to Zosterophyllopsida are, very strangely, bilaterally symmetric (McMenamin and Schulte McMenamin 1994), seemingly in accord with the Second Law of Morphogenetic Evolution where evidence for control by morphogenetic fields is most apparent in the earliest representatives of any particular lineage. The symmetrical appearance of sporangia at the ends of zosterophyll branches may represent symmetrical field dilations/herniations in accord with the Seventh Law. And finally, Harambe (a western lowland gorilla, *Gorilla gorilla gorilla*) is an excellent example of field control. The bare chest and

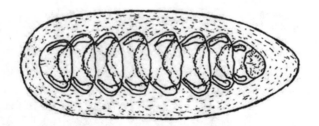

Fig. 1.9 *Clementechiton sonorensis* McMenamin and Fleury in McMenamin, 2016. This is the oldest known chiton and the oldest known body fossil of an adult animal. Its anterior end is to the left. Note the dual rows of ocular lobes on the opposite edges of each dorsal plate. These trilobite eye-like ocular lobes are considered here to be "torologous" (sensu McMenamin (2016)) to the serial trunk sclerites of *Microdictyon*. *Clementechiton* occurs in the pre-Cambrian Clemente Formation of northwestern Sonora, México. Length of specimen approximately 2 cm

very hairy arms of the male animal are a striking visual contrast, and provide evidence for sexually dimorphic field control (as a species characteristic) in the placement of body hair. The skull of *Gorilla gorilla gorilla* has a prominent sagittal crest, and a braincase that seems oddly shrunken in comparison to the rest of the skull. Morphogenetic field lines have contracted here in a case of midline field contraction, evidently to the detriment of the intellectual capacities of the species and the direct opposite of what I consider to be field line dilations (*cf.* Thompson 1917) governing the development of the skull in *Homo sapiens*. The primary biotic investment in *Gorilla gorilla gorilla* goes to great muscular strength. Human grade consciousness is thus not ubiquitous, however, the laws that have been set into place governing the course of evolution will eventually lead to its appearance when conditions are ripe.

References

Abbott UK, Asmundson VS (1957) Scaleless, an inherited ectodermal defect in the domestic fowl. J Hered 48(2):63–70

Appel TA (1987) The Cuvier-Geoffroy debate: French biology in the decades before Darwin. Oxford University Press, New York/Oxford

Bengtson S et al (1986) The Cambrian netlike fossil *Microdictyon*. In: Hoffman A, Nitecki MH (eds) Problematic fossil taxa. Oxford University Press, Oxford, pp 97–115

Bolker JA (2000) Modularity in development and why it matters to evo-devo. Am Zool 40(5):770–776

Brown CM et al (2017) An exceptionally preserved three-dimensional armored dinosaur reveals insights into coloration and Cretaceous predatorprey dynamics. Curr Biol 27:2514–2521.e3. https://doi.org/10.1016/j.cub.2017.06.071

Cerda IA, Powell JE (2010) Dermal armor histology of Saltasaurus loricatus, an upper Cretaceous sauropod dinosaur from northwest Argentina. Acta Palaeontol Pol 55(3):389–398

Chen C et al (2015) How the mollusc got its scales: convergent evolution of the molluscan scleritome. Biol J Linnean Soc 114:949–954

Dawson G (2016) Show me the bone: reconstructing prehistoric monsters in nineteenth-century Britain and America. Chicago Scholarship Online, Chicago. https://doi.org/10.7208/chicago/9780226332871.001.0001

Dzik J (2003) Early Cambrian lobopodian sclerites and associated fossils from Kazakhstan. Palaeontology 46(1):93–113

Edelman DB et al (2016) Origin of the vertebrate body plan via mechanically biased conservation of geometrically regular patterns in the structure of the blastula. Prog Biophys Mol Biol 121:212–244. https://doi.org/10.1016/j.pbiomolbio.2016.06.007

Escánez A et al (2017) New records of the scaled squid, *Lepidoteuthis grimaldii* Joubin, 1895 in the Canary Islands, eastern Atlantic Ocean. Spixiana 40(1):7–12

Feynman RP (1996) Feynman lectures on computation. Westview Press, Boulder

Filippov AÉ et al (2017) Numerical simulation of colloidal self-assembly of super-hydrophobic arachnid cerotegument structures. J Theoretical Biol 430:1–8

Gilbert SF et al (1996) Resynthesizing evolutionary and developmental biology. Dev Biol 173(2):357–372

Greshko M (2017) Turned to stone: how a Canadian mine yielded one of the best preserved dinosaurs ever found. Natl Geogr 231(6):92–105

Halstead LB (1982) The search for the past. Doubleday, Garden City/New York

Hamming RW (1980) Coding and information theory. Prentice-Hall, Upper Saddle River

Jackson DJ et al (2006) A rapidly evolving secretome builds and patterns a sea shell. BMC Biol 4:40. https://doi.org/10.1186/1741-7007-4-40

Jackson DJ et al (2010) Parallel evolution of nacre building gene sets in molluscs. Mol Biol Evol 27:591–608

Kawaguti S, Baba K (1959) A preliminary note on a two-valved sacoglossan gastropod, *Tamanovalva limax*, n. gen., n. sp., from Tamano, Japan. Biol J Okayama Univ 5(3–4):177–184

Kocot KM et al (2016) Sea shell diversity and rapidly evolving secretomes: insights into the evolution of biomineralization. Front Zool 13:23. https://doi.org/10.1186/s12983-016-0155-z

Lang RJ (2011) Origami design secrets: mathematical methods for an ancient art, 2nd edn. CRC Press, Boca Raton. http://www.langorigami.com/publication/origami-design-secrets-2nd-edition

Le Renard J et al (1996) On *Candinia* (Sacoglossa: Juliidae), a new fossil genus of bivalved gastropods. J Paleontol 70(2):230–235

Marshall CR (2017) Five palaeobiological laws needed to understand the evolution of the living biota. Nature Ecol Evol 1:0165. https://doi.org/10.1038/s41559-017-0165

Mazza E (2016) Confused Rush Limbaugh can't figure out why gorillas still exist. Huffington Post, New York. http://www.huffingtonpost.com/entry/rush-limbaugh-gorillas-evolution_us_5 74e5de6e4b0757eaeb10be0

McMenamin MAS (1984) Paleontology and stratigraphy of lower Cambrian and upper Proterozoic sediments, Caborca Region, Northwestern Sonora, Mexico. Ph.D. Dissertation, University of California at Santa Barbara. University Microfilms International, Ann Arbor

McMenamin MAS (1992) Two new species of the Cambrian genus *Mickwitzia*. J Paleontol 66(1):173–182

McMenamin MAS (2009) Paleotorus: the laws of morphogenetic evolution. Meanma Press, South Hadley

McMenamin MAS (2015a) Paramphibia: a new class of tetrapods. Meanma Press, South Hadley

McMenamin MAS (2015b) The theological treachery of partial scientific truths. New Oxford Rev 82(6):30–33

McMenamin MAS (2016) Dynamic paleontology: using quantification and other tools to decipher the history of life. Springer, Cham

McMenamin MAS, Schulte McMenamin DL (1994) Hypersea: life on land. Columbia University Press, New York

Peredo CM et al (2017) Decoupling tooth loss from the evolution of baleen in whales. Front Mar Sci 4:67. https://doi.org/10.3389/fmars.2017.00067

Saint-Hilaire G (1807) Considérations sur les pièces de la tête osseuse des animaux vertébrés, et particulièrement sur celles du crâne des oiseaux. Ann Mus Hist Nat 10:342–365

Schoenemann B et al (2009) A miniscule optimized visual system in the lower Cambrian. Lethaia 42(3):265–273

Seilacher A (1994) Candle wax shells, morphodynamics, and the Cambrian explosion. Acta Palaeontol Pol 38(3/4):273–280

Shanta BN (2015) Life and consciousness-the Vedantic view. Commun Integ Biol 8(5):e1085138

Sheesley P et al (2014) The morphogenetic mapping of the brain and the design of the nervous system. Int J Brain Sci, https://doi.org/10.1155/2014/424718

Sigwart JD (2017) Zoology: molluscs all beneath the sun, one shell, two shells, more or none. Curr Biol 27:R702–R719

Smith M (2011) *Hallucigenia* and the evolution of animal body plans. Palaeontology [Online] 7:1–9

Stallings W (2006) Computer organization and architecture, 7th edn. Prentice-Hall, Upper Saddle River

Thompson DW (1917) On growth and form. Cambridge University Press, Cambridge

Vinther J (2009) The canal system in sclerites of lower Cambrian *Sinosachites* (Halkieriidae: Sachitida): significance for the molluscan affinities of the sachitids. Palaeontology 52:689–712

Waagen L (1911) Palaeontology. In: The Catholic encyclopedia. Robert Appleton Company, New York. Retrieved January 4, 2011 from New Advent: http://www.newadvent.org/cathen/11410a.htm

Wells KL et al (2012) Genome-wide SNP scan of pooled DNA reveals nonsense mutation in FGF20 in the scaleless line of featherless chickens. BMC Genomics 13:257. http://www.biomedcentral.com/1471-2164/13/257

Wotte T, Sundberg FA (2017) Small shelly fossils from the Montezuman—Delamaran of the Great Basin in Nevada and California. J Paleontol 91(5):883–901

Zhang X-G, Aldridge RJ (2007) Development and diversification of trunk plates of the lower Cambrian lobopodians. Palaeontology 50(2):401–415

Chapter 2
Shuram Excursion

Patience obtains everything.

St. Teresa of Avila

Abstract The Shuram excursion represents the greatest negative carbon isotopic excursion in earth history, and provides an important chemostratigraphic marker horizon of global extent. The excursion is linked to the second great oxygenation event in earth history, an oxygen crisis that resulted in a transition from sulfidic oceans to a marine realm rich in sulfate. The Shuram excursion (560–550 Ma) is represented in Sonora, México by the Clemente oolite of the Clemente Formation. Ediacaran fossils (such as the Clemente biota of Unit 4 of the Clemente Formation) occur in rocks deposited below the excursion. The age of the Clemente Ediacaran biota thus falls between 550 and 560 Ma. In spite of the fact that the Sturtian glaciation apparently triggered the earliest known mass extinction on earth (the *Tindir Mass Extinction*), several lines of evidence suggest that the biosphere controlled the timing of and the onset of the Late Proterozoic glaciations, and that it also controlled the timing of the melting of the ice. Furthermore, it appears that the biosphere itself influenced the timing of the appearance of the Ediacaran biota. Whereas snowball earth events lurched suddenly from very cold (tillites) to very hot (cap carbonates) climate, the sequence going from the Gaskiers glacial event (*c.* 580 million years ago) to the Shuram was part of a wild climatic gyration where the earth went from hot (intense granite weathering at high latitudes) to cold (Gaskiers glaciation) to hot (Shuram event). The Shuram is the greatest negative carbon isotopic excursion in earth history, possibly because this is the moment in earth history when the burrowing animals assert themselves in a geochemical sense, and by remobilizing sea floor carbon, forestall a major glaciation.

Keywords Shuram excursion · Clemente Formation · Clemente oolite
· Hardgrounds · Isotope chemostratigraphy · Gaskiers glaciation · Roxbury
Conglomerate · Dropstone · Cryoconite

© Springer International Publishing AG 2018
M.A.S. McMenamin, *Deep Time Analysis*, Springer Geology,
https://doi.org/10.1007/978-3-319-74256-4_2

The Clemente oolite is one of the most intriguing single layers of strata in the entire geological record. This oolite is also known as Unit 5 of the Clemente Formation of northwestern Sonora, México. It is a highly distinctive rock unit in the field, forming prominent outcrop ledges (Figs. 2.1, 2.2 and 2.3). The rock consists of ooids, originally calcium carbonate but now altered to dolomite. Oolites are carbonate sediments consisting largely of ooids, where "individual ooids are smooth spheroids or ellipsoids of sand size in which a nucleus is encased by an aragonite or calcite cortex of approximately uniform thickness. The nuclei in ooids may be fragments of skeletal material, fecal pellets, silicate sand or silt grains, or even bits of fractured ooid" (Williams et al. 1982). Ooids and carbonate cements in Proterozoic-Cambrian sedimentary rocks are considered to be sensitive indicators "for subtle variations in seawater chemistry across the boundary" (Tucker 1989).

A photomicrograph of the Clemente oolite in thin section first appeared in print in Stewart et al. (1984). It is well worth taking a closer look at this unusual rock in thin section (Fig. 2.4). Figure 2.5 shows the short bladed cement characteristic of the oolite, plus spalling of some of the ooid laminae. Figure 2.6 shows a compound ooid, also characteristic for this rock unit. Compound ooids, or grapestones, consist of ooids that have been cemented together. Grapestones are relatively uncommon in ancient carbonate rocks (Friedman and Sanders 1978).

Figure 2.6 also shows partial silicification of the ooids. This silicification dramatically emphasizes the spherical laminae. Very interestingly, Fig. 2.7 shows development of a hardground within the Clemente oolite. A hardground, a characteristic

Fig. 2.1 Clemente Oolite, Cerro Rajón, northwestern Sonora, México. Field shot (7 March 1982) taken at the junction between the lower (pale yellow) and middle (tan-brown) parts of the unit. Length of rock hammer 35 cm (Photograph by Mark McMenamin)

feature of particular types of limestone, is a buried surface of lithified carbonate sea floor, recognizable in this case by a band of ooids that have lost much of their internal structure by recrystallization as a result of extended exposure to sea water.

Cambrian and later oolite formations are commonly encountered with hardground caps of great lateral extent (Wilson 1975). These caps may denote stratigraphic sequence boundaries. In the modern ooid shoals of the Great Bahama Bank, ooid facies are characteristic for water depths less than 3 m, whereas associated grapestone facies occur in deeper water (with less lime mud), at depths of 9–10 m (Boggs 2012). The ooid facies on the Bahama platform occur as submarine dune fields exceeding 50 km in length (Boggs 2012).

In the Proterozoic Johnnie oolite of eastern California and western Nevada, a rock unit that correlates to the Clemente oolite (see below), Corsetti (1998) reported an interesting compositional trend occurring from the Nevada or eastern exposures to the western (California) exposures of the Johnnie oolite. The percentage of rip up clasts or micritic (lime mud) intraclasts in the Johnnie oolite increases from just a few percent or less in the east to greater than 30% in the western regions of its exposures. Corsetti (1998) attributed this increase in intraclast abundance to a higher energy depositional environment, and considered this higher energy environment to represent a more offshore setting with the intraclasts serving as a proxy for agitated, high energy water. If Corsetti (1998) is correct, the offshore direction for the Johnnie oolite is to the southwest. Ironically, Corsetti (1998) inferred that this higher energy

Fig. 2.2 Clemente Oolite, northwestern Sonora, México. Field shot (7 March 1982) of oolite outcrop in the Cerro Rajón. Note darker tan-brown band in the central part of the unit. The stratigraphically upward transition to the tan-brown horizon is also seen in the previous image (Photograph by Mark McMenamin)

Fig. 2.3 Bedrock exposure
of Clement Oolite, Cerro
Rajón, northwestern
Sonora, México. Greatest
length of rock hammer
30 cm. Note young cholla
cactus (*Cylindropuntia*)
above rock hammer
(Photograph by Mark
McMenamin)

Fig. 2.4 Clemente Formation (unit 5) oolite, lower magnification thin section image. East side of
Cerro Calaveras, Sonora, México, field sample MM-82-63. The rock is dolomitic with patches of
silicification. Compound ooids and a short-bladed cement phase are seen above a hardground
surface. The hardground surface is visible here as a diagonal band that runs from near the scale bar
at the upper left to the lower right corner of the image. Scale bar in millimeters (Photograph by
Mark McMenamin)

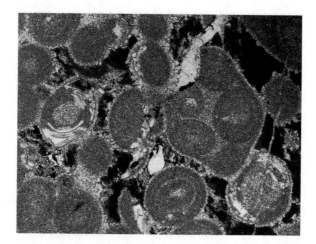

Fig. 2.5 Clemente Formation oolite in thin section, plain light view, same slide as previous image. A short bladed cement phase precipitated around the ooids. Note also spalling of ooids on left side of image. Note compound ooid at center right, and partially silicified ooid at center left. Cerro Calaveras, Sonora, México, field sample MM-82-63. Width of view 3 mm (Photograph by Mark McMenamin)

Fig. 2.6 Clemente Formation oolite in thin section of previous image, cross polarized view. Note compound ooid at center right, and partially silicified ooid at center left. Width of view 3 mm (Photograph by Mark McMenamin)

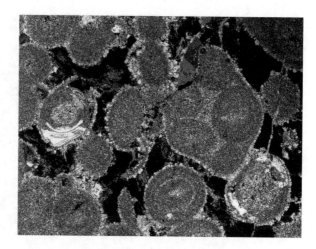

correlated with a basin high (that is, sea floor shallowing and hence exposure to greater wave energy) to the west; thus offshore shoals led to the high percentage of intraclasts in the western Johnnie Oolite exposures. A red-colored cap-rock of dolomitic composition at the top of the Johnnie Oolite, occurring in Johnson Canyon, the westernmost locality of the oolite, possibly indicates subaerial exposure (Corsetti 1998). It would be unusual to see this in the offshore direction, but is plausible if oolite shoals got exposed at the surface. This red cap rock can be comparable to the hematitic oolitic ironstones forming condensed horizons in Silurian limestone successions of Quebrada Ancha, northwestern Argentina (Stow 2006, his Fig. 12.11).

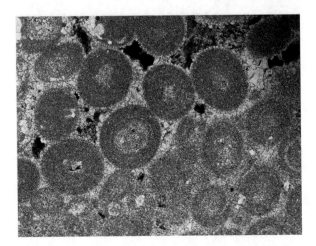

Fig. 2.7 Clemente Formation oolite in thin section with hardground. Note alignment of ooids deposited above the hardground that have settled into concave depressions at the top surface of the hardground. Imbrication of elliptical ooids beneath the hardground surface indicate water flow direction to the right. Cerro Calaveras, Sonora, México, field sample MM-82-63. Width of view 3 mm (Photograph by Mark McMenamin)

Fig. 2.8 Clemente Formation oolite in thin section with hardground, cross polarized view of previous image. Short bladed cement phase clearly visible on ooids deposited above the hardground. Cerro Calaveras, Sonora, México, field sample MM-82-63. Width of view 3 mm (Photograph by Mark McMenamin)

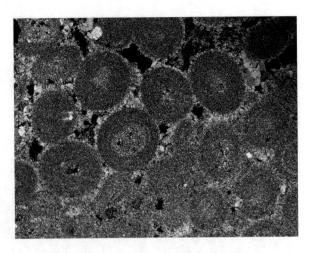

   Aligned ooids are visible immediately above the hardground surface in the Clemente oolite, still sitting where they settled into concave depressions on the hardground surface during the Late Proterozoic. Figure 2.8 shows that the short bladed cement is clearly visible coating the ooids deposited above the hardground, but has been obliterated from the ooids deposited below the hardground due to sea-floor alteration of the ooids exposed at the hardground surface. This seafloor weathering reduced the ooids of the hardground into what resembles a lime mud (micritic) mash (Figs. 2.7 and 2.8, lower right portion of both images, where ooids become less distinct due to dissolution due to hardground formation).

Fig. 2.9 Stratigraphic section of the Proterozoic to Cambrian strata of the Caborca Region of northwestern Sonora, México

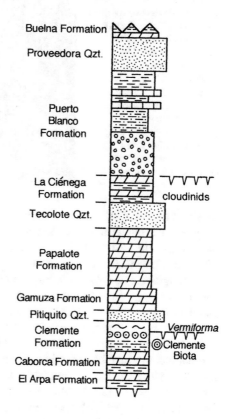

I helped name the Clemente Formation with the Great Basin stratigrapher John H. "Jack" Stewart (1928–2013) and Mexican geologist Juan Manuel Morales-Ramirez (Stewart et al. 1984). In *United States Geological Survey Professional Paper 1309*, we also named several other Proterozoic rock units, including the La Ciénega Formation and the Tecolote Quartzite, both of which occur stratigraphically upsection from (and are hence younger than) the Clemente Formation. We (Stewart et al. 1984) defined the Clemente Formation to refer to unnamed strata deposited between the Caborca Formation and the Pitiquito Quartzite (Fig. 2.9).

Due to a lamentable paucity of fossils suitable for biostratigraphy, Proterozoic strata can prove to be a challenge for anyone wishing to date them with any precision (although recent advances in geochronology have helped greatly in this matter; Hoffmann et al. 2004; Tsukui et al. 2017). Jack Stewart was first to realize that the Clemente oolite could be lithostratigraphically correlated to the Johnnie oolite (a member of the Johnnie Formation), a widespread rock unit of the southwestern United States. In *Professional Paper 1309*, we inferred that the Johnnie oolite and the Clemente oolite were once part of a single, laterally extensive yet relatively thin

Fig. 2.10 Thin section
view of Puerto Blanco
Formation Unit 1,
ilmenite-rich basalt, cross
polars. This image shows a
large radiating cluster of
clinozoisite, formed by
alteration of a calcic
plagioclase feldspar. The
rectangular black crystals
are ilmenite. Field sample
6 of 3/23/90, southern
Cerro Rajón area, Sonora,
México. Width of view
3 mm (Photomicrograph
by Mark McMenamin)

rock unit of vast areal extent (Stewart et al. 1984). This is decisively confirmed by
carbonate petrology: the short bladed cement phase, the compound ooids and the
hardground are nearly identical to similar features visible in the Johnnie oolite
(Corsetti 1998; Corsetti et al. 2006). The audacious (and correct) long distance
lithostratigraphic correlation of Stewart et al. (1984) did little, however, to resolve
the absolute age of the Clemente Formation, because both the Johnnie Formation
and the Clemente Formation have been difficult to accurately date.

When I announced the discovery of Ediacaran fossils in Sonora (McMenamin
1996, 1998), I gave an age estimate of ~600 million years for the fossils, based on
correlations to other Proterozoic stratigraphic successions in North America and
Avalonia with associated radiometric dates. Laflamme et al. (2013, their Fig. 4)
claimed an age of 542–545 Ma for the Ediacarans of Sonora, but they did not cite
any evidence in support of this assertion and they may in fact have been referring to
the tubular shelly fossils (*Sinotubulites* and other cloudinids) that occur higher in
the Sonoran stratigraphic section. Most of the Sonoran Ediacaran soft-bodied fossils discovered so far occur beneath the Clemente oolite.

The Sonoran Proterozoic stratigraphic sequence unfortunately contains few
rocks that can be radiometrically dated, and in the Clemente Formation such rock
types are entirely lacking. Considering the importance of this stratigraphic sequence
and its key Ediacaran fossils, to be described in more detail below, I made repeated
attempts to secure more confident dates for the Precambrian strata of northwestern
México. These attempts included using radiometric dating tools, stable isotope dating, and biostratigraphy.

Basaltic igneous rocks occur in both the Puerto Blanco Formation (Fig. 2.10;
Cambrian) and the La Ciénega Formation (Fig. 2.11; Late Proterozoic), and although
these stratigraphic rock units are younger than the Clemente Formation, radiometric
dates derived from these volcanic rock units could provide potentially useful constraints for the age of the Clemente Formation (Barr and Kirschvink 1983). The
Puerto Blanco Formation basalt, from the volcaniclastic unit 1 of the Puerto Blanco

Fig. 2.11 Thin section view of La Ciénega Formation igneous rock unit showing altered megacrysts. The large crystal in the image, probably originally zoned, shows pervasive alteration. Field sample 2 of 3/23/90, southern Cerro Rajón area, Sonora, México. Width of view 3 mm (Photomicrograph by Mark McMenamin)

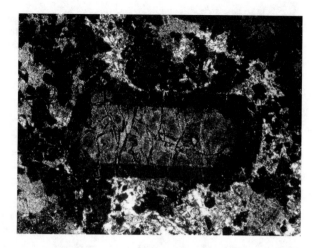

Formation (McMenamin et al. 1994), includes an ilmenite-rich metabasalt and chlorite- and clinozoisite-rich basalt (or possibly mafic andesite) with plagioclase phenocrysts up to 5 mm in diameter (Fig. 2.10). Very tiny monazite crystals were detected in these igneous rocks by Mike Jercinovic working in Mike Williams' microprobe lab at the University of Massachusetts, Amherst. Unfortunately, the monazite crystals proved to be too small to be useful for geochronology, so precise radiometric ages of the volcanic rocks remain out of reach for the time being.

During an expedition to northwestern Sonora in March 16–25, 1990, Steve Rowland of the University of Nevada, Las Vegas and I sampled the Cerro Rajón sequence and collected carbonate rocks through the entire section. Our plan was to generate a stable isotope curve for the stratigraphic section (McMenamin et al. 1992). The data (published here for the first time) are shown in Table 2.1, and Fig. 2.12 shows the carbon and oxygen stable isotope curves showing secular variation in these values spanning the Proterozoic-Cambrian boundary. At the time, using stable isotopic variations to correlate stratigraphic units and to date rock layers was a new technique, and Steve Rowland and I proceeded with caution as we began to interpret the curve shown in Fig. 2.12.

Our first concern was that the carbon and oxygen isotopic curves covaried (tracked or paralleled one another) throughout much of the section (Fig. 2.12), making us worry that it might be the case that some type of diagenetic alteration had influenced the stable isotopic data, thus rendering the curve useless as a record of secular variation in $\delta^{13}C$ and $\delta^{18}O$ near the Proterozoic-Cambrian boundary. We initially focused on a short interval at the base of the Papalote Formation (we called it "a prominent negative $\delta^{13}C$ excursion near the base of the Papalote Formation"; McMenamin et al. 1992) where the two curves diverged somewhat, suggesting (correctly as it turns out) that an actual record of secular variation in carbon and oxygen stable isotopes was present at that point.

As they grow, organisms preferentially absorb isotopically light carbon. Carbon isotopic variations in strata thus are generally interpreted in the following fashion.

Table 2.1 Carbon and oxygen isotopic data (expressed as per mil relative to PDB) and Mn/Sr ratios from the Cerro Rajón stratigraphic section. Whole rock isotopic analyses. The numbering sequence is correct as shown (the field numbering sequence included some backtracking)

Sample	$\delta^{18}O$	$\delta^{13}C$	Mn/Sr
CR-32	−13.3	−0.2	6.18
CR-31	−13.4	−1.8	5.10
CR-30	−9.0	1.4	0.77
CR-29	13.7	−0.2	3.09
CR-28	−15.5	−0.9	3.75
CR-27	−11.4	−5.0	1.10
CR-26	−6.1	−1.6	18.38
CR-25	−8.6	−1.8	10.57
CR-24	−8.5	−0.4	9.65
CR-23	−5.1	0.4	8.28
CR-22	−3.6	0.4	9.72
CR-21	−4.6	0.6	13.59
CR-20	−5.5	−0.3	8.18
CR-18	−2.8	−2.4	12.89
CR-19	−5.4	−1.6	5.60
CR-17	−5.2	−2.0	4.61
CR-16	−4.4	−1.0	6.40
CR-15	−4.7	−2.6	6.26
CR-14	−4.9	−2.9	5.41
CR-12e	−4.1	−0.2	4.55
CR-12d	−2.0	1.4	3.16
CR-12c	−2.9	−2.0	3.84
CR-12b	−2.5	−0.7	4.00
CR-12a	−3.4	2.8	3.84
CR-12	−1.4	3.3	3.61
CR-11	−2.8	2.4	6.05
CR-10	−3.4	2.7	8.47
CR-9	−2.6	4.0	6.57
CR-8	−6.9	−5.5	4.67
CR-7	−12.2	3.7	0.58
CR-6	−12.5	5.2	0.28
CR-5	−10.1	1.5	1.46
CR-4	−11.1	2.5	2.20
CR-3	−5.1	6.1	2.92
CR-2	−8.1	7.7	3.48
CR-1	−6.3	8.8	7.33

If there is a drop in $\delta^{13}C$ values, this suggests that the heavy carbon (^{13}C) is being diluted by light carbon (^{12}C), and hence the marine biosphere is not doing as well and is not preferentially absorbing as much ^{12}C as would be the case if the biosphere was flourishing. Negative isotopic excursions are often associated with mass extinction events (van de Shootbrugge et al. 2008; Suarez et al. 2017).

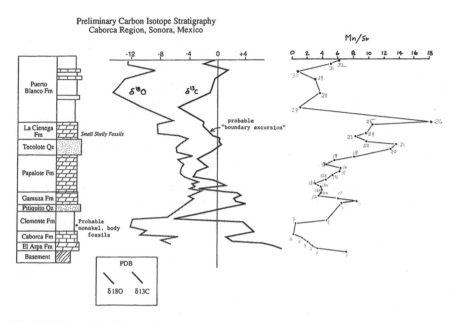

Fig. 2.12 The original Sonora Stable Isotope Curve, plotting $\delta^{18}O$, $\delta^{18}C$, and the Mn/Sr ratio. Data are listed in Table 2.1

The interpretation of isotopic curves is unfortunately not quite this simple, because a negative carbon isotopic excursion is associated with the Cambrian Explosion, the moment in earth history without precedent when life was expanding at a breakneck pace. Indeed, a consensus interpretation of any particular isotopic excursion is frequently absent, and this has curtailed the confidence that can be placed on the use of isotopic curve matching as a correlation tool, considering that the reasons for shifts in the $\delta^{13}C$ values are often not well understood. Nevertheless, the basal Cambrian boundary does seem to be marked by a sudden drop from positive to negative $\delta^{13}C$ values. This stable isotopic record, based on data from the Dvortsy section, Aldan River, Siberia can be correlated to stratigraphic sections elsewhere (Margaritz et al. 1991; Kirschvink et al. 1991; Corsetti and Hagadorn 2000). Commenting on the possible significance of this, I inferred that (McMenamin 2004a, b, c):

> [The] boundary excursion represents that moment in geological time when marine burrowing intensity crossed a threshold. At this critical point, the microbial mat seal on the seafloor may have been breached, resulting in previously immobile sediments (and their associated organic matter) becoming mobilized and injecting huge amounts of biogenic carbon into the water column. Owing to biogenic isotopic fractionation, this detrital organic matter was significantly depleted in the heavy isotope of carbon and, as it went into circulation in marine water, caused the precipitous drop or boundary excursion in the $\delta^{13}C$ value.

Thus, any particular isotopic excursion in a given local stratigraphic section may be due to secular variation in $\delta^{13}C$ values as inorganically precipitated seafloor carbonates absorb carbon from sea water. These can accurately reflect the marine isotopic signature at a given point in geological time. For this to be useful as a correlation tool, however, it is important to rule out the possibility that the isotopic value has been changed by diagenetic or metamorphic alteration of the carbonate minerals. Carbonate minerals such as aragonite and calcite are highly susceptible to diagenetic alteration; for example, the majority of Precambrian carbonate deposits have been altered from calcite to dolomite (Corsetti et al. 2006).

When Frank Corsetti completed his dissertation in 1998, he included the carbon isotopic curve (*sans* data) from the Caborca section (Corsetti 1998). As a graduate student at U. C. Santa Barbara, Corsetti had joined us on the 1990 Sonoran expedition. In his thesis, Corsetti (1998) inferred that:

> the intense negative excursion within the Clemente Formation . . . would be younger than 595 Ma . . . [and the] Clemente Oolite-Johnnie Oolite correlation appears to be corroborated by the isotopic data. The fabric of the ooids themselves, as well as the diagenetic sequence surrounding the ooids, are nearly identical as well, suggesting that the Clemente-Johnnie correlation may indeed be valid.

Thus, in spite of the Caborca samples being based on a preliminary analysis of whole rock samples (Table 2.1), Corsetti (1998) accepted the validity of our data in its entirety, noted the similarity of the negative excursion in the Clemente Oolite to that of the Johnnie Oolite (confirming the lithostratigraphic correlation of Stewart et al. 1984), and recognized the "intense" nature of the oolite isotopic excursion. It is now well-established that the Sonoran stratigraphic section contains a primary sea water signal (McMenamin et al. 1992; Sour-Tovar et al. 2007).

Figure 2.12 thus records an actual secular variation isotopic curve of great magnitude. Analysis of this curve is essential for our understanding of the geology of the late Proterozoic. Can this extreme isotopic excursion be detected in stratigraphic sections on other continents? Scrutiny of isotopic curves from the Cretaceous shows that intercontinental correlation using stable isotopes is a viable procedure (Wendler 2013):

> Because [of] the influence of early diagenetic cements . . . chemostratigraphy should not be used as a stand-alone method for trans-continental correlation, and especially minor isotopic shifts have to be interpreted with utmost care. Nevertheless, the observed consistency of the $\delta^{13}C$ correlations confirms global scale applicability of bulk sediment $\delta^{13}C$ chemostratigraphy.

Rowland and I certainly had reason to be cautious in interpreting the Sonoran isotopic curve. But the data are real and reliable in spite of being whole rock analyses. With a reliable bulk rock isotopic curve, it is now possible to provide an accurate age for the Clemente oolite. Note that this is not a "stand-alone" date but is supported by our lithostratigraphic correlations (Stewart et al. 1984) and, as discussed below, biostratigraphic analysis.

We now know that the Clemente-Johnnie oolite isotopic excursion can be identified throughout the world. It is called the Shuram excursion, and is the greatest single carbon isotopic excursion in the entire stratigraphic column. This Shuram excursion, in striking counterpoint to the initial caution Steve and I showed in identifying and interpreting it, is now being used to infer dates for high grade metamorphic rocks (Leivset Marble) in Scandinavia based on the extreme ^{13}C depletion in the Leivset Marble (Melezhik et al. 2008).

The first hint of the existence of the Shuram excursion was published by Stephen J. Burns (now at the University of Massachusetts, Amherst) and Albert Matter of the Geologisches Institut at the University of Bern, Switzerland. In a study of late Proterozoic carbonate rocks of central Oman, Burns and Matter (1990) reported that at "the Khufai/Shuram contact there is a very abrupt shift to much lighter carbon (-8 to -10 per mil) and somewhat lighter oxygen (-6 to -8 per mil) values." They published a full account of the Oman carbon and oxygen isotopes in the journal *Eclogae Geologicae Helvetiae* (Burns and Matter 1993), where they noted the excursion as "a sharp decrease, over a few meters of section, of about 8 per mil in carbon isotopic values." They seemed to scarcely be able to believe the magnitude of the stable isotopic plunge they had measured: "the magnitude of the negative carbon isotope shift in the world ocean is certainly less that that observed in the Oman sections in which the lowest values, -10 to -12 [per mil], are probably outside of the possible range for dissolved inorganic carbon in the oceans" (Burns and Matter 1993). Burns and his collaborators also studied the strontium isotopic composition of the Huqf Group carbonates (Burns et al. 1994), proposing an age of 560–540 Ma (which now appears to be accurate) for the Huqf Group of Oman. The Huqf Group includes both the Khufai and Shuram formations.

In his dissertation, Corsetti (1998) does not discuss Burns and Matter (1993) and the possible relation of their Shuram excursion to the Clemente-Johnnie oolite excursion. Corsetti was nevertheless aware of the work, as he cites Burns and Matter (1990) abstract, but he curiously omits discussion of this critical abstract in his thesis text.

We now know that the Shuram excursion and the Clemente-Johnnie oolite excursions are one and the same, an "isotopic nadir" of global extent, even though Clapham and Corsetti (2005) did not yet link it to the Shuram excursion. Now, however, it is becoming clear just what an extreme isotopic excursion the Shuram event represents. Grotzinger et al. (2011) call it the "largest-known carbon isotope excursion in Earth's history". Cui et al. (2017) call it "the most profound negative carbon isotope ($\delta^{13}C$) excursion in Earth history, the Shuram Excursion." Gong et al. (2017) refer to it as "one of the largest-known isotopic anomalies, [and] has been globally observed in Ediacaran rocks." However, the concerns of diagenetic alteration that first troubled Steve Rowland and me about the Sonoran isotopic curve continue to linger. Both Grotzinger et al. (2011) and Cui et al. (2017) note that the drop in carbon isotope values could be influenced by "alteration following the deposition of the carbonate sediments" (Grotzinger et al. 2011). Recall that Steve Rowland and I were concerned about the covarying of the carbon and oxygen isotopes in Sonora. Very interestingly, both Wang et al. (2016) and Cui et al. (2017) show that the carbon and oxygen covariation occurs worldwide.

What caused the Shuram Excursion? In other words, what hypotheses have been proposed to explain the event—what led to the deposition of all that sedimentary rock depleted in heavy carbon (^{13}C)? The first idea proposed was that environmental and/or ecological changes were responsible for the excursion (Burns and Matter 1993). The second is that the excursion was caused by massive oxidation of marine dissolved organic carbon (Cui et al. 2017). The third idea concerns mass weathering of terrestrial organic carbon. The fourth idea is oxidation of gas hydrates released from the sea floor. The fifth idea is mass oxidation of petroleum released from the subsurface. The sixth idea, which Cui et al. (2017) endorse, is that increases in atmospheric oxygen led to oxidation of surface environments that led to increased microbial sulfate reduction that led to "buildup of alkalinity in pore fluids through the anaerobic oxidation of methane" that led to precipitation in the marine carbonate sediments of "authigenic calcite (formed as early diagenetic cements and nodules) that are remarkably depleted" in heavy carbon. That is quite the concatenated chain of causality! In other words, "localized, but globally synchronous, production of ^{13}C-depleted authigenic carbonate may have led to the strong negative $\delta^{13}C$ excursions" (Cui et al. 2017). The isotope values of these calcites would then constitute the Shuram Excursion. Cui et al. (2017) stop just short of invoking what is in effect a Second Oxygen Revolution to explain the Shuram event.

This linkage to atmospheric oxygen levels is important and was discussed in the early analysis by Burns and Matter (1993):

> An increase in atmospheric oxygen has been proposed as a necessary precursor to the late Precambrian/early Cambrian radiation of animals . . . An alternative is that late Precambrian atmospheric oxygen levels were already high (see discussion in McMenamin & [Schulte] McMenamin 1990), and that the later Precambrian radiation was caused by environmental changes which upset the biological balance of the period. The sharp negative carbon isotope excursion observed in Oman may be evidence of such an environmental change.

Burns and Matter (1993) followed McMenamin and Schulte McMenamin (1990) emphasizing environmental/ecological change in the late Proterozoic, and as a result Burns and Matter (1990) championed the idea that environmental change (in the sense that "ecosystem engineering" is used today) was responsible for the Shuram excursion. I proposed a similar explanation for the isotope excursion at the base of the Cambrian as noted above (McMenamin 2004a, b, c). This represented an extension of the research program of William Beebe, who maintained that only by understanding the ecosystems where organisms live, may you understand the creatures themselves (Gould 2004; Matsen 2005).

Therefore, the ideas proposed to explain the Shuram excursion fall into one of two categories. The first category is the concept that environmental or ecological changes triggered the deficit in heavy carbon. This is plausible; recall that the basal Cambrian boundary is marked by a sudden drop from positive to negative $\delta^{13}C$ values, and this surely coincides with a major paleoecological/environmental change (the Cambrian Explosion).

The second category of explanations holds that some kind of gigantic shift in global geochemistry was responsible for the Shuram excursion. Foremost among these explanations is the idea that oxygen levels underwent a major increase at Shuram time, thus somehow causing the isotopic excursion.

Now is the time to take a closer look at the oxygen increase hypothesis for events such as the Shuram excursion at the end of the Proterozoic. A major stumbling block for the oxygen hypothesis is that, unlike the original Oxygen Revolution (~2.3 billion years ago) when reduced, soluble iron in sea water became oxidized (and insoluble), changing the color of the oceans from green to blue and forming the banded iron formations, there is no obvious big source of reduced material that could suddenly run out or get depleted, becoming fully oxidized and thus allowing massive amounts of free oxygen to accumulate in the atmosphere.

In their review of sedimentary sulfides through geologic time, Rickard et al. (2017) provide a possible solution to this conundrum by introducing the concept of sulfidic oceans. An enduring mystery in Precambrian geology is as follows: If Precambrian oceans were free of oxygen, why weren't the sediments packed with sulfide in the form of sulfide minerals such as pyrite? Canfield (1998; see also Canfield and Teske 1996; Kunzmann et al. 2017) made the case for *two* oxygen revolutions, one at 2.3 billion years ago (the original oxygen crisis) and one at the end of Proterozoic time, roughly coincident with the emergence of animals.

Canfield (1998, 2005) solved the Precambrian sulfide conundrum by arguing that the amount of sulfate was limited during the interval between the first and second oxygen revolutions. Today, after sodium and chlorine, sulfate is the most abundant substance dissolved in seawater. Dissolved sulfate is critical for microbes such as sulfate-reducing bacteria. They use the sulfate (SO_4^{2-}) as a backup oxygen supply when available free oxygen runs out in an environment such as, say, an organic-rich mud. These bacteria produce reduced sulfur as a waste product, and this accumulates in sediment as a result. This material may be oxidized by sulfur-oxidizing bacteria, who are able to use hydrogen sulfide as an energy source in the presence of oxygen, thus completing a redox exchange between the complementary types of microbes. Production of sulfide in modern marine sediments is a direct function of massive amounts of sulfate in the ocean (Rickard et al. 2017). Recall that sulfate wins third place after the halite ions sodium and chlorine in terms of amounts in sea water. Rickard et al. (2017) comment on the situation:

> This is another counterintuitive idea: significant O_2 concentrations are required for large amounts of sedimentary sulfides to be produced. In Precambrian oceans, there was limited sulfate and, therefore, the amount of sulfide in the sediments was also limited. Even so, most of the sulfide produced in sediments today is oxidized by sulfur-oxidizing microorganisms. If these organisms were absent from the Earth today, then the whole Earth would bask in the gentle pong of toxic [at least for us aerobic metazoa] H_2S.

In Canfield's (1998) view, the Proterozoic oceans that existed between the two oxygen events still had relatively low oxygen, sulfate was less common in seawater, sedimentary sulfide production was much reduced, and the oceans were sulfidic in composition. Rickard et al. (2017) argue that the geochemistry of sulfidic oceans "may have inhibited the rapid evolution of multicellular life forms." Therefore, the proposed geochemical shift is of potentially great interest to those who study the emergence of animals.

If Proterozoic marine waters were indeed sulfidic, there is a real possibility that the Shuram excursion is the signature in the sedimentary rock record of a Second Great Oxygenation event. Interestingly, Ediacaran fossil localities may show a fossil matrix that is stained deep red by iron oxidation. This terra rossa (Fig. 2.13), associated with an Ediacaran frond from the Clemente Formation (to be described in the next chapter), is indicative of highly oxidizing environmental conditions and provides support for a linkage between maximum Ediacaran biodiversity, the Shuram Excursion, and the Second Great Oxygenation event. The terra rossa shown here is interpreted as being the result of subaqueous oxidation of limestone, rather

Fig. 2.13 Terra rossa associated with Ediacaran frond fossils from the Clemente Formation of Sonora, México. The deep red color indicates highly oxidative conditions during deposition. Field sample 8 of 3/16/95; IGM 7460. Scale bar in cm (Photomicrograph by Mark McMenamin)

than a paleosol, as it is directly juxtaposed to an Ediacaran frond that is not heavily iron stained and is presumed here to have been a marine aquatic organism. According to Huang et al. (2017), sulfate levels may have taken a temporary drop because their "data of multiple S-isotopes suggest that the sulfate level in the ocean may have greatly declined as a result of the long-term oxidation of organic carbon during the 'Shuram excursion'."

Rickard et al. (2017) are imprecise regarding the date of their Second Great Oxygenation event, listing it as ~540 Ma on p. 120 and as 600 Ma in their Fig. 5. It will be important for the next chapter to obtain an accurate age for the Shuram excursion, as this will provide us a date for the Sonoran Ediacaran fossils.

By 2006, the Shuram excursion was still being dated as older than 600 Ma, with Le Guerroué (2006) saying that the event had an "onset at ~600 Ma." Le Guerroué (2006:205) proposed a correlation between the Shuram excursion and the Clemente-Johnnie oolite excursion. More precise dates for the Shuram excursion have since become available. Gong et al. (2017) place the Shuram Excursion at ca. 560 Ma. Macdonald et al. (2013) date the event at between 565 and 551 Ma, and link an isotopic excursion in the Gametrail Formation of the Mackenzie Mountains, Canada to the Shuram.

Verdel et al. (2011) place the Shuram event at between 620 and 551 Ma, and correctly note that "the occurrence of horizontal burrows and macroscopic body fossils below the Clemente oolite (McMenamin 1996) would suggest that the excursion occurred after ca. 575 Ma". I recently undertook a reevaluation of the Ediacaran fossil collection from the Clemente Formation, and as described in the next chapter was able to identify new Sonoran Ediacarans (including a beothukid, a new aculiferan and a kimberellid) that, in combination with the erniettomorph reported by McMenamin (1996), provide a firm, biostratigraphically-controlled age for the Clemente Biota at 553 Ma, with a likely range of from 555 to 551 Ma. Although there is still controversy about the age of the Shuram (e.g., Zhou et al. (2017) say that the Shuram excursion "may be significantly older than 551 Ma if it does represent a global chemostratigraphic marker"), the biostratigraphic correlation presented here (Fig. 2.14) favors the lower (~551 Ma) age for the Shuram Excursion. This biostratigraphic correlation places the Clemente Ediacaran biota squarely within the

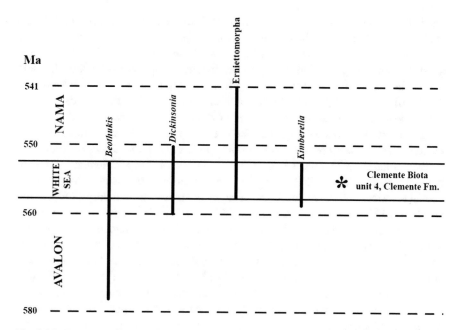

Fig. 2.14 Sonoran Ediacaran Biostratigraphy. Approximate ages in millions of years are shown at the boundaries (dashed lines) between the three Ediacaran assemblages: *Avalon, White Sea* and *Nama*. The age of the Clemente Biota (unit 4 of the Clemente Formation; its relative biostratigraphic position shown by an asterisk) is constrained by overlapping range zones of the following three groups: *Beothukis*, erniettomorphs, and *Kimberella*. The range of *Dickinsonia* is shown for comparison purposes only (the genus is not known to occur in Sonora, México)

White Sea assemblage, dated at ~560–550 Ma (Laflamme et al. 2013). This would be roughly equivalent to the Australian "*Dickinsonia costata* assemblage zone" of Richard Jenkins (Jenkins 1995).

Interestingly, the Clemente biota has elements that are also characteristic of the preceding Avalon assemblage (rangeomorph beothukids; ~580–560 Ma) and the Nama assemblage (erniettomorphs; ~545–542 Ma), although both rangeomorphs and erniettomorphs range into the White Sea assemblage. The three-fold assemblage concept has been challenged by Boag et al. (2016), who conclude that several factors lessen "the strength for either of the Avalonian or White Sea assemblages to be used as discrete faunal stages to demarcate Ediacaran biostratigraphy." The composition of the Clemente biota adds to these mounting concerns over using the Ediacaran assemblage concept in a biostratigraphic sense. These extended ranges constitute the core of the biostratigraphic correlation presented below by means of overlapping range zones. Available evidence indicates that the age of the Clemente Ediacaran biota falls between 550 and 560 Ma. The Clemente biota and all other Ediacaran biotas are younger than the Gaskiers glaciation at ~579–580 Ma.

The Gaskiers glaciation was the last Precambrian glaciation, and also the first glaciation to occur after the two great snowball earth glaciations (Pu et al. 2016). Pu et al. (2016) argue that earth "narrowly escaped a third Neoproterozoic snowball glaciation just prior to the late Ediacaran expansion of metazoan ecosystems." Tsukui et al. (2017) estimate that the Gaskiers glaciation lasted mere 1.7 million years.

What *caused*, and in a question that is nearly as important, what *ended* these great Precambrian glaciations? Suggestions that the biosphere itself controlled both the initiation of and termination of glaciation in these events (McMenamin 2004a, b) are the subject of intense continuing research. In other words, the *biosphere controlled the timing of the onset of Proterozoic glaciations, and also controlled the timing of the melting of the ice in each case.* A corollary of this is that *the biosphere itself controlled the timing of the appearance of the Ediacaran biota.*

Ever since Peter Dobson's interpretation (Dobson 1826; McMenamin 2001) of boulders of the Connecticut Valley in Connecticut and Massachusetts, USA, as glacial erratics, Ignace Venetz's rendering of the now classic theory of glaciation (Hallam 1992), and Louis Agassiz's 1837 presidential address to the Swiss Society of Natural Sciences (Neuchâtel) advocating the concept of *die Eiszeit* (Corozzi 1967), ice ages have become a widely accepted fact of the history of the earth. The severity of several of these glacial interludes has been more fully appreciated only in more recent decades (Walker 2003). With the advent of white earth or snowball earth theory, it has become increasingly clear that particular intervals of pre-Phanerozoic earth history were marked by episodes of exceptionally severe glaciation.

Evidence for these glaciations is not always unambiguous in particular regions. An extended geological debate featured hypotheses regarding the glacial origin (or otherwise) of the famous Roxbury Conglomerate of eastern Massachusetts, USA (Nance 1990). The epic nature of the debate is underscored by the proximity of the rocks in question to great institutions of higher learning such as Harvard University and MIT. As described by Skehan (2001):

> The . . . Roxbury formation consists of conglomerate, shale, sandstone, quartzite, arkose, and altered basaltic volcanic rocks. The conglomerate is often called puddingstone. Oliver Wendell Holmes (1809-1894) described the appearance of the pebbles in the fine-grained matrix as 'plums in a pudding.' His poetic theory of the origin of puddingstone is as delightful as it is imaginative.
> In Holmes's 1830 poem 'The Dorchester Giant,' the giant's unruly children:
> > . . . . *flung [the puddingstone] over to Roxbury Hills,*
> > *They flung it over the plain*
> > *And all over Milton and Dorchester too*
> > *Great lumps of pudding the giants threw . . .*

If we discount the Dorchester Giant and his offspring as causative agents for appearance of the puddingstone (Fig. 2.15), we are left with the geological debate regarding the depositional setting of the Squantum member of the Roxbury Formation. The fact that the Squantum member is sometimes referred to as the Squantum tillite emphasized the controversial nature of the origin of these rocks. It was in fact interpreted as a tillite a century ago (Skehan 2001). One difficulty with the tillite interpretation was the apparent lack of any striated cobbles in the putative till, although "rhythmically bedded siltstones above the sediment look like varves and bolster the hypothesis of a glacial origin" (Skehan 2001). The debate over a possible glacial origin has led some researchers to advocate for deposition in a glaciomarine setting, while others advocate for a depositional setting that involves the proximal parts of debris flows (Nance 1990). Debris flow deposition and deposition in a glaciomarine setting are not necessarily mutually exclusive. Nevertheless, some researchers have attempted to dismiss a glaciomarine origin (Bailey and Bland 2001; Carto 2011):

> Although glaciation has often been described to have been an important process in deposition of the [Squantum member of the Roxbury Formation and adjacent rock units] there is a marked lack of regional evidence for such processes . . . Dropstone-like outsized clasts are present in some local facies associated with conglomerates and mass flow deposits; however, they are not present in . . . fine-grained facies many reports to the contrary notwithstanding.

While working with National Science Foundation-funded geology students on a summer project focused on Late Proterozoic strata in the Boston Basin, John D. "Jack" Beuthin and I co-mentored undergraduate student Jessica Williams in her study of the fine grained sedimentary rocks (Cambridge Argillite) that occur in the same sedimentary sequence as the Squantum member of the Roxbury Formation. We jointly concluded that (Williams et al. 2008):

> Laminated Cambridge Argillite of the Boston Bay Group (Ediacaran) low-grade metasediments can be found in outcrop at Hewitt's Cove, Hingham, Massachusetts. These laminated sediments are finely graded layers 2 to 8 cm thick that are commonly interpreted in the literature as fine-grained turbidites (Stow sequences). Rhythmically laminated facies also occur that do not support a sediment-gravity flow interpretation. These rhythmites raise the possibility of sedimentation influenced by agents such as tides, waves, or wind. The Cambridge Argillite appears to consist of subtle, but definite, mosaic of laminite facies that record multiple depositional processes, and perhaps multiple water depths.

Jessica Williams, after searching diligently, unearthed a granitoid lonestone/dropstone (Williams 2008; McMenamin and Beuthin 2008) embedded in the shaly rock (laminite) of the Cambridge Argillite (Figs. 2.16 and 2.17). This was an exciting discovery, and proved to be the most important sedimentological find of

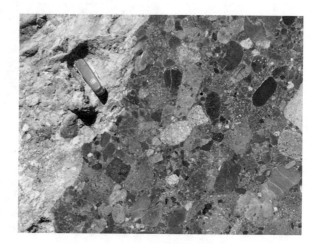

Fig. 2.15 Roxbury Formation ("puddingstone") from Boston, Massachusetts. The right side of the rock has been slabbed and polished to show individual clasts more clearly. Swiss Army knife for scale (Photograph by Mark McMenamin)

Fig. 2.16 Granitoid dropstone (lonestone) seen in thin section, from Locality F of Crosby (1894) at the Hewitt's cove site in Hingham, Massachusetts, USA. The dropstone was discovered while slabbing Cambridge Argillite at Mount Holyoke College, and was not exposed on the surface of the argillite. Note the argillite laminae draped around the clast. Scale bar in mm (Photograph by Mark McMenamin)

Fig. 2.17 The same dropstone as seen in the previous image, sectioned at a different angle to show the angular character of the clast. Scale bar in mm (Photograph by Mark McMenamin)

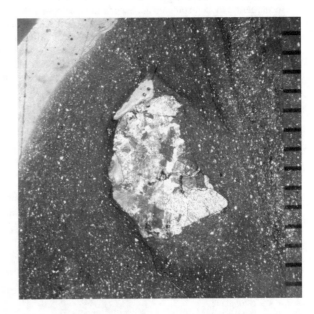

the summer. This granitoid lonestone is best interpreted as a dropstone that indicates glacial influence. Carto (2011) advocated for an alternate explanation for this lonestone, writing that "McMenamin and Beuthin (2008) stressed that its association with an intrabasinal rip-up clast was suggestive of sediment-gravity origin." This statement is incorrect as McMenamin and Beuthin (2008) were still very much considering a glacial, dropstone origin for the granitoid clast. Isolated rip up clasts can also be glacially-derived by means of the agency of grounded ice, and these by themselves do not eliminate a glacial (dropstone) interpretation for the critical Cambridge Argillite dropstone (Figs. 2.16 and 2.17).

Jack Beuthin and I concluded (McMenamin and Beuthin 2008) that the dropstone seems to best fit with the plutonic suite of the Fall River Batholith, and in particular the Fall River Granite. This granite is a medium-grained, light gray granite with few dark minerals that is approximately 600 million years in age (Skehan 2001). We concluded (McMenamin and Beuthin 2008) that evidence therefore "seems to be accumulating for glacial ice as a sedimentary agent in the Boston Basin Group." Jessica Williams's splendid undergraduate student discovery has for all intents and purposes resolved the debate regarding the origin of the Roxbury formation/Squantum tillite in favor of the hypothesis of a glacial origin for the rocks (Williams 2008; McMenamin and Beuthin 2008).

Some debates, however, seem to not wish to die, and although Passchier and Erukanure (2010) accept some glacial influence for the Squantum Member, they feel that the "glacially influenced sedimentary facies at Squantum Head are more consistent with meltwater dominated alpine glaciation or small local ice caps" based on their determination that "the chemical index of alteration (CAI) values of 61 to 75 for the non-volcanic [clasts] requires significant exposure of land surfaces to allow chemical weathering" (Passchier and Erukanure 2010). Such degree of weathering for the Squantum Member, interpreted as having been deposited during the Gaskiers glaciation, precludes snowball earth conditions at the time (Passchier and Erukanure 2010).

Although certain hypotheses, such as the possibility of extensive and thick sea ice cover at the equator during late Proterozoic glacial events, may be difficult to test, it now seems clear that continental-scale glaciers occurred on land at the equator near sea level during these glaciations. Understandable reluctance to accept these astonishing low-latitude paleomagnetic results (Cloud 1988) has given way to consensus among earth scientists that the greatest and most severe glaciations can indeed reach low latitudes.

Such a phenomenon would be remarkable all by itself, but to this must now be added breathtakingly sudden climate change from extreme glaciation (as recorded by Proterozoic tillites) to torrid conditions (as shown by the cap carbonates deposited immediately above the glacial tills). In order to properly interpret the influence of these extraordinary events on the history of the biosphere, *and the influence of the biosphere on these events*, it is best to begin by describing what is known of the earth's biota and its paleoecology before the onset of the late Proterozoic glacial events.

At 850–800 million years ago, Earth was richly populated with bacteria and eukaryotic microbes and sparsely populated with larger eukaryotic organisms. The microbes of the time took advantage of broad expanses of continental shelf to form extensive microbial mats. These mats were subsequently expressed in the rock record as organo-sedimentary structures called stromatolites (Figs. 2.18, 2.19, 2.20 and 2.21), and as wrinkly primary sedimentary structures (forming the unusual 'elephant-skin' textures or textured organic surfaces) in fine-grained siliciclastic rocks. Stromatolites at that time were abundant, well-developed (Vanyo and

Fig. 2.18 *Jacutophyton*
from the Gamuza
Formation, Cerro Rajón,
northwestern Sonora,
México. Length of rock
hammer 35 cm
(Photograph by Mark
McMenamin)

Fig. 2.19 *Jacutophyton*
from the Gamuza
Formation, Cerro Rajón,
northwestern Sonora,
México. Lens cap is 5.5 cm
in diameter (Photograph by
Mark McMenamin)

Awramik 1985; Kah et al. 2009), and capable of reaching enormous size. Some
stromatolites in eastern California and Canada attained the sizes of hills and small
mountains (Cloud et al. 1974).

The eukaryotic grade of organization has a history extending perhaps as far back
as two billion years, but the interpretation of megascopic fossils of this age (such as
Grypania) has been controversial, with some researchers contending that the fossils

Fig. 2.20 *Platella* from the Gamuza Formation, Cerro Calaveras, northwestern Sonora, México. Lens cap is 5.5 cm in diameter (Photograph by Mark McMenamin)

Fig. 2.21 *Platella* from the Gamuza Formation, northwestern Sonora, México. Detail of previous image (Photograph by Mark McMenamin)

are large bacteria or clustered bacterial colonies. Acritarchs presumed to have been formed by eukaryotes appear by about 1.8 billion years ago, but the eukaryote fossil record remains comparatively sparse (primarily acritarchs and morphologically simple macroalgae such as *Chuaria* and related forms) until approximately a billion years ago. At this time there appears to be a radiation of eukaryotes, and forms such as bangiophyte algae (Butterfield 2000) and cysts and other types of remains belonging to presumed heterotrophic and autotrophic protists (Porter 2004; Macdonald et al. 2010) appear in the fossil record. The trigger or triggers for this burst of eukaryotic radiation are not well understood, but in the same way that the evolution of whales may be linked to the closure of Tethys and an exceptionally fertile, short-lived seaway that formed between India and Asia just before the Himalayan orogeny (Gingerich et al. 1983), diversification of early eukaryotes may have been favored by life in the narrow, nutrient-rich seaways formed by oceanic closure during the amalgamation of Rodinia.

As what Cloud (1988) called the "long Umberatanan winter" descended on the planet, beginning about 750 my, the biosphere was thrust into an extended episode of global freeze. The cause of the glaciation is contentious, although there is a prevailing school of thought that invokes continental position to explain the ice ages. In its first iteration, the continental position hypothesis emphasized the possibility of elevated terrains near the equator. In wrestling with the implications of low latitude glaciation during the late Proterozoic glacial events, it was noted by some (McMenamin and Schulte McMenamin 1990) that glaciers form today at the equator in the 5200 m high Ruwenzori Mountains of Ethiopia. The elevation of these mountains is a result of crustal buoyancy associated with thermal effects of East African rifting. Preston Cloud conjectured (1988): "Could a clustering of sufficiently elevated continents at low paleolatitudes account for meteorological conditions that would lead to extensive equatorial and low-latitude glaciation?" Alternately, the continents themselves might have moved into polar regions. McMenamin and Schulte McMenamin (1990) noted two possible causes of extensive glaciation: "if parts of Rodinia were at high latitude or if global climate were severe."

In the second iteration of the continental position hypothesis as a cause of late Proterozoic glaciations, Kirschvink (1992) and subsequently Hoffman and Schrag (2002) inferred that "a preponderance of continents in middle to low latitudes created conditions favorable for snowball events." This concept of the disposition of the continents is in accord with the best current evidence for continental positions in the Late Proterozoic; in other words, continents do indeed appear to have been present at middle to low latitudes rather than at high latitudes (Donnadieu et al. 2003; Poulsen et al. 2001; Poulsen et al. 2002). This, however, brings us to a curious juncture.

At first, an explanation for the glaciation was sought in the possibility of continents being either topographically elevated or displaced to high latitudes. Failing this, explanation is now sought in the likelihood of continents occurring at middle to low latitudes, a remarkably counterintuitive explanation as these regions are considerably warmer than the poles. Poulsen et al. (2002), using a fully coupled ocean-atmosphere general circulation model, claimed to demonstrate that concentrations of continents in low latitudes actually did lower equatorial (tropical) temperatures. In none of their three model simulations (each with different continental dispositions), however, were Poulsen et al. (2002) able to generate fully white earth conditions.

One could fall back on what Cloud (1988) refers to as John Crowell's *Eclectic Hypothesis*, the concept that many different types of processes interact to cause the glaciations (Crowell 1999). Hoffman and Schrag (2002) introduced their version of this Eclectic Hypothesis: "Perhaps each event was triggered differently." I find this approach scientifically unsatisfying—it merely appeals to a diversity of causes rather than encouraging a search for necessary and sufficient causes. When considering such events as the worst glaciations on record and, say, the worst known mass extinction, appeals to many different causes may be politically safe (i.e., no one's favorite hypothesis gets left out) but scientifically unsound as they can obscure our search for what really happened. As pointed out by Cloud (1983), such attempts at explanation represent an unwelcome type of substantive uniformitarianism smacking of gradual-ism. Singular, catastrophic events of earth history will in many cases require bold explanations if we hope to further our understanding. There is nothing to be gained by cowering before the task. It seems likely that the answer to the cause of late Proterozoic glaciations, rather than being found in a series of multiple causes all of which in some way contribute to glaciation, resides in some single overarching factor.

I believe that a primary problem with current attempts to explain the late Proterozoic glaciations is that most efforts have focused too intently on a strictly physical explanation for the onset of glaciation. This tendency reaches its extreme in the case of the low-obliquity models, where a thoroughly physical mechanism (tilting Earth's rotation axis; Williams 1975) is invoked to explain the low-latitude ice. While I applaud the boldness of this hypothesis (and indeed Mars's rotation axis can tilt 60°), it unfortunately runs afoul of evidence in the terrestrial sedimentary record. For example, rhythmic sedimentation patterns (Hughes et al. 2003; Williams 1975; see especially the rhythmic varvite with dropstone from northeast of Port Augusta, South Australia (Cloud 1988)) and the morphology of sine wave columns of the Proterozoic stromatolite *Anabaria juvensis* (Vanyo and Awramik 1985) is in accordance with the hypothesis of an essentially normal earth tilt during the growth of the stromatolites. These types of both biotic and abiotic periodic structures indi-cate that the obliquity of the Proterozoic ecliptic was not significantly different from current values.

Consideration of causes of the end of glaciation are beset by similar problems that are nearly as intractable as the problems associated with consideration of the factors causing the onset of glaciation. Of particular interest are the global geo-chemical changes associated with the "cap carbonates" occurring immediately upsection from the glacial deposits. Cap carbonates are thin but very widespread limestone/dolostone rock units that are usually interpreted as evidence for sudden and massive carbonate deposition in a warm climate. Hoffman and Schrag (2002) prefer division of these units into a lower "cap dolostone" and an upper "cap lime-stone/cementstone" to reflect the dual nature of some of these deposits (particularly in Namibia and Canada).

Compounding the difficulties associated with low latitude glaciation is evidence for cap carbonate deposition immediately above the glacial deposits. This juxtaposi-tion is a major geological anomaly. Williams (1979) noted in prescient fashion the climatic implications of the juxtaposition of cap carbonates above the glacial strata:

> The sedimentological and oxygen-isotope data are consistent with relatively high formation-temperatures from the cap [carbonates]. Abrupt climatic warming—from cold glacial to at least seasonally high temperatures—at the close of the late Precambrian glacial epochs is implied.

These unusual carbonates have three signal characteristics. First, they are depleted in heavy carbon (they have a $\delta^{13}C$ value of approximately −5, indicating a lower $^{13}C/^{12}C$ ratio). As noted earlier in this chapter, a low $^{13}C/^{12}C$ ratio is usually interpreted as representing low marine biotic productivity (as indicated by a low ratio of buried organic carbon to carbonate carbon), as organisms preferentially take up the lighter isotope and leave seawater (from which the marine carbonates are precipitated) enriched in heavy carbon. Second, these deposits characteristically extend over vast map areas, having been deposited in both shallow water and relatively deep-water environments. As such they provide a useful tool for lithostratigraphic correlation. Third, they contain in their lower parts primary dolostones, a highly unusual rock type (Woods 1999) owing to the rarity of primary precipitation of the mineral dolomite (i.e., dolomite usually forms as a secondary mineral). New cap carbonate occurrences are being described as more stratigraphic sections receive scrutiny.

Kirschvink (1992) and Hoffman and Schrag (2002) relied on an abiotic geochemical model to explain the snowball earth/cap carbonate juxtaposition. Kirschvink's original hypothesis (Kirschvink 1992) for the mechanism of escape from snowball earth conditions involved the accumulation of carbon dioxide in the atmosphere as a result of volcanic input, the near cessation of photosynthesis, and the inhibition of silicate weathering via the Urey reaction. The Urey reaction consumes atmospheric carbon dioxide as calcsilicate minerals such as wollastonite undergo weathering. The weathering reaction combines wollastonite ($CaSiO_3$) with carbon dioxide and water to release the following weathering products: calcium cation, bicarbonate ion, and silica in aqueous solution.

Kirschvink (1992) attributed Late Proterozoic marine iron formations to buildup of dissolved iron in the water under the ice, followed by sudden oxidation and precipitation during the debacle (i.e., as the ice broke up). Renewed photosynthesis could then begin again to generate appreciable quantities of oxygen. Hoffman and Schrag (2002) added that cap carbonates would be an expected outcome of "extreme greenhouse conditions unique to the transient aftermath of a snowball earth." In this view the sudden leap from white earth climate to cap carbonate climate is attributable to hothouse conditions resulting from unprecedented atmospheric CO_2 buildup.

There is a problem with this scenario, however, and it concerns the speed with which the earth escapes a white earth climate. In many cases the cap carbonates are deposited directly above the glacial strata (tillites), implying a lightning fast transition from deep freeze to warm. How could this transition occur so suddenly? Certainly as the ice began to melt the planetary albedo would drop, and this process would contribute to surface warming and could even take on the characteristics of a runaway feedback. However, it seems unlikely that all or even most land surfaces were covered by ice during peak white earth conditions (recall that more land surface area is exposed as sea level drops during a glaciation), thus the albedo changes may have been incremental rather than sudden and catastrophic. The same can be said for the accumulation of atmospheric carbon dioxide. The carbon dioxide sources would include volcanic emissions and respiratory release of oxidized carbon, the latter process presumably being inhibited as much as was photosynthesis by the frigid snowball earth conditions. Enhancement of the atmospheric greenhouse could be expected to be incremental as well.

It would be possible to propose a nonlinear response of climate to greenhouse gas accumulation. In this case, once the CO_2 passed a certain threshold level, a catastrophic meltdown would occur. There is, in fact, evidence of enhanced atmospheric carbon dioxide levels as indicated by the cap carbonate deposition. But how could the climate change so quickly, especially considering the fact that the Urey reaction would kick in as a negative feedback as soon as temperatures began to rise on the continents (thus tending to limit the impact of a runaway greenhouse)?

With parts of Rodinia rifting and dispersing, the amount of relatively fresh continental material exposed to weathering along swollen rift margins may have been at an all-time high, tending to enhance the efficacy of the Urey reaction negative feedback. Such an effect might have contributed to glaciation in the first place, as the most dramatic drop in $\delta^{13}C$ and the start of the Sturtian glaciation coincide at the end of the main pulse of the supercontinental breakup, some time before 700 million years ago. This is followed by a second major glaciation, the Marinoan, at approximately 635 million years ago (Hoffmann et al. 2004). It should be noted, however, that other episodes of continental breakup (e. g., the breakup of Pangea) are not associated with glaciation. Furthermore, continental breakup is typically associated with volcanigenic release of carbon dioxide, which when occurring before a glaciation would tend to delay the onset of icehouse climate conditions.

Unless we accept as plausible the idea that deglaciation occurred solely because of abiotic accumulation of volcanigenic carbon dioxide in the atmosphere, the velocity and magnitude of global climate change at the end of the Proterozoic ice ages requires the activity of a potent "climate change catalyst." Vladimir Vernadsky (1998), who viewed life as the most important geological force, also recognized the catalytic nature of living processes and the tendency for life to expand dramatically when conditions permit. The transition from snowball earth to cap carbonate earth is so sudden that it may bear a distinctly Vernadskian biotic signature. Could the onset of glaciation have a biotic cause as well? What types of biology might lead to such rapid climate switch? If some type of microbial bloom is implicated in the climate changes, might they not leave some type of fossil evidence (even if the transition episode was short-lived)?

Onset of glaciation could be a climate response to shallow shelf areas made available by supercontinental breakup, if the biomineralizing organisms of the seafloor (a microbiota consisting of primarily filamentous microbes capable of sequestering lime mud as sea floor carbonate sediment) provided a biological pump for calcium. Climate would be cooled if these organisms took sufficient quantities of bicarbonate from seawater during their stromatolite formation and deposited it on the sea floor as limestone/dolostone. This could lead to a dangerous drawdown of atmospheric CO_2. Massive pre-ice age accumulations of carbonate are known, such as the Beck Spring Dolomite and the carbonate parts of the Crystal Spring Formation of the southern Death Valley region of California (McMenamin 2004a, b). The extensive carbonate shelf on the eastern (present day coordinates: northern) shore of the Nopah Sea (Death Valley, California) was developed by columnar stromatolites such as *Conophyton* and *Baicalia*. These shelfal sediments represent a major accumulation of carbonate (Roberts 1982). The massive to well-laminated, and in places oolitic carbonate of the Beck Spring Dolomite, upsection from the Crystal Spring

Formation, occurs immediately downsection from the Varangian/Marinoan glacial deposits of the Kingston Peak Formation. Thus the diamictites of the Kingston Peak Formation (several have been identified, the upper Wildrose Diamictite and the lower Surprise Diamictite; Prave 1999) are sandwiched between the carbonates of the Beck Spring Dolomite and the Crystal Spring Formation below, and the presumed cap carbonate (Woods 1999) of the Noonday Dolomite above. Cyclothem-like cyclic bedding in the stromatolitic Algal member of the Crystal Spring Formation (Roberts 1982) may suggest biotic influence (by drawdown of greenhouse gases) in a rapidly deteriorating global climate with continental ice already beginning to influence eustatic sea level.

The Beck Spring Dolomite of eastern California provides a fascinating glimpse into carbon isotope dynamics just before the glaciation. The lower part of the formation has carbon isotopes varying from +2 to −2 per mil $\delta^{13}C$. The middle part of the Beck Springs Formation shows strongly positive $\delta^{13}C$ values (up to +6), with values falling to as low as −4 just before the Kingston Peak glaciation (Corsetti and Kaufman 2000). This isotopic curve agrees with those from the Keele in Canada (Ice Brook Glaciation), the Ombaatije in Namibia (Ghaub Glaciation), and the Etina in Australia (Elatina Glaciation; Hoffman and Schrag 2002). The Kingston Peak, Ice Brook, Ghaub and Elatina glaciations are all considered here to be synchronous and Marinoan in age (but see Lund et al. (2003) for an alternate interpretation of glacial synchroneity. Lund et al. (2003) do, however, hold the Ice Brook and upper Kingston Peak [Wildrose] diamictites to be synchronous). The high $\delta^{13}C$ values in these sections are interpreted here to represent an active seafloor microbiota that sequestered great amounts of isotopically fractionated organic carbon in sea floor sediments, triggering a decline into icehouse conditions. The sudden drop in $\delta^{13}C$ in the Beck Spring, Keele, Ombaatije and Enorma/Trezona (Etina Group) formations, respectively, represents a secular cooling in climate regime that immediately preceded the glaciation and the associated die-off in microbe populations. Probable fungal fossils, consisting of branched filaments belonging to the species *Paleosiphonella cloudii*, occur in Beck Spring chert (Licari 1978), providing evidence for abundant organic matter which, along with the carbonate rock that entombed it, is hypothesized here as having been responsible for drawing down atmospheric carbon dioxide levels.

Gaucher (2000) attributed the drawdown of carbon dioxide to phytoplankton blooms preceding the glaciation. Gaucher et al. (2003) proposed the following scenario for slippage into an icehouse climate. First, enhanced volcanic and hydrothermal activity associated with the Rodinia breakup injects large amounts of silica, iron and other nutrients into seawater, and these accumulate in the deep sea, forming a eutrophic bottom layer in a stratified ocean. Second, fortuitous combinations of wind and water circulation trigger upwelling on a massive scale, leading to great blooms of phytoplankton and deposition of both phosphorites and banded iron formations on the sea floor below. The anoxic bottom water conditions encourage sulfate-reducing bacteria, triggering a positive shift in $\delta^{34}S$ values. Carbon removed from the atmosphere and sent to the seafloor by the phytoplankton blooms trigger the icehouse conditions. It is interesting to note that in the Arroyo del Soldado

Group of Uruguay, banded iron formations occur in positive $\delta^{13}C$ intervals (Gaucher et al. 2004). Furthermore, the highest concentrations of organic-walled microfossils occur in the iron oxide-bands of banded iron formation (Gaucher 2000). Interestingly, iron fertilization of equatorial Pacific surface waters has resulted in both phytoplankton blooms and carbon fractionation (Bidigare et al. 1999).

Note the fascinating similarity between the carbon isotopic curves of the Kingston Peak, Ice Brook, Ghaub and Elatina glaciations (or more concisely, the Marinoan glaciation) and the drop in carbon isotope values (from +6 in the Kinderhookian/Mississippian to −2 in the Morrowan/basal Pennsylvanian) followed by moderate glacial/postglacial rise in $\delta^{13}C$ in the Arrow Canyon Range of southern Nevada (Saltzman 2003). Evidence for Pennsylvanian glacio-eustacy and cyclothems begins immediately after the nadir in carbon isotope values (base of the Morrowan stage; Saltzman 2003), just as the nadir in the Marinoan sections occurs just before the first sedimentological evidence for glaciation. The Late Paleozoic glaciation was not nearly as extreme as that of the Late Proterozoic, hence the carbon isotopic results do not seem to correlate as well from continent to continent (Saltzman 2003), but nevertheless the link to glacio-eustacy is evident.

Wang et al. (2003) have also shown that carbon reservoir changes preceded ice-sheet expansion at the mid-Brunhes event during the Pleistocene glaciations. Sufficient expansion of carbon-sequestering, primary-producing organisms will be followed (after a suitable lag) by climate deterioration (and die off in primary production, thus accounting for the drop in carbon isotope values to −2 to −4 per mil from a previous high of +6–7) and then glaciation. For the Pennsylvanian, the coal forests of Hypersea (McMenamin and Schulte McMenamin 1994) acted as the sequestering agent after overcoming the challenges of nutrient acquisition, desiccation and oxidation on land. For the late Proterozoic glaciations, the carbonate-sequestering stromatolitic microbiotas acted as the climate culprit. In both cases, organic (over)productivity generated a carbon sink that eventually nudged global climate across a threshold from greenhouse to icehouse conditions. Incidentally, the telltale drop in $\delta^{13}C$ before the glaciation effectively rules out a bolide impact or other instantaneous event as a cause of the glaciation. The biomatter overproduction hypothesis advocated here may be tested by a careful comparison of the global carbon budgets before and after glaciation, using cubic kilometers of carbonate rock (or coal) deposited as proxies for variation in atmospheric carbon dioxide.

It is *well* worth considering ways in which the biota might respond to end a glaciation. In other words, we must consider how organisms might exploit a snowball earth environment and how any resulting modes of life might hasten the end of the glaciation. Ice diatoms form a thick brown layer at the base of the ice in modern polar environments. As light can pass through thick layers of ice, a dark algal layer at the bottom of a floating ice sheet could lead to considerable warming of the underside of the ice, and help to melt the ice. Although diatoms had not yet appeared to form the light-catching brown layer by the Late Proterozoic, other types of microbes likely evolved to fill extremophile niches similar to those occupied by modern ice diatoms, and might have provided a feedback mechanism for melting the ice. Exceptional populations of modern *Chlamydomonas nivalis* show optimum photosynthesis at 0° to −3 °C, although most populations do best at 10°-20 °C. Such microbes can catalyze melting of ice, both by withdrawing fresh water from the ice (and forming ice-corroding brine pockets) and by trapping heat in the brown sub-ice mat or "hyposcum."

Much of a snowball earth glaciation would occur in low and even equatorial latitudes, where light levels are intense and light penetration through ice is greatest. Organisms can thrive in such conditions; for example, Campen et al. (2003) have reported evidence for microbial metabolism within a low-latitude glacier. A Proterozoic sub-ice biota, as a result of its albedo-altering characteristics, melted the ice and was subsequently thrust into moderate climate. Ice edges seem to be a particularly favorable habitat (Leck and Persson 1996). Special phytopigments and protective phenolic compounds, such as those found in the snow alga *Chlamydomonas nivalis*, developed to cope with increased amounts of incoming radiation (Duval et al. 2000) after millions of years of life in a habitat of faint light and limited UV exposure. Clay coatings may have also played a similar role (Takazi et al. 1994).

Cryoconite is a term applied to dark-colored material that forms on glacier ice, formed primarily of wind-borne dust and microbial mat material. Cryoconite has been implicated in the acceleration of melting of Himalayan glaciers, and recent studies show that Alpine and Himalayan cryoconites consist of microbial mats formed by cyanobacteria and heterotrophic microbes (Sharp et al. 1999; Takeuchi et al. 2001; Margesin et al. 2002). Cryoconite microbes play a role in reduction of glacier albedo, and are known to colonize open areas after the ice melts. Interestingly, the distribution of individual microbes in granules of cryoconite (Takeuchi et al. 2001, their Figs. 4–6) bear a curious resemblance to the spatial distribution of Proterozoic microbes in silicified granules from the El Arpa Formation, Sonora, Mexico. The resemblance is seen both in granule morphology and arrangement of the spherical microbial mat of filamentous cyanobacteria (McMenamin 2004b). This resemblance may not be merely coincidental.

Wind-blown sediments alone, be they volcanigenic or otherwise, do not greatly alter the albedo of ice surfaces, as they are generally light tan or gray in color. But detrital cryoconites do begin to alter ice surface albedo when colonized by mat-forming microbes. Cryoconite microbial mats are known for forming dark humic acid accumulations that accelerate ice melting (Takeuchi et al. 2001). The microbes involved are able to use almost anything as a food source. The Arctic psychrophile bacterium *Polaromonas vacuolata* is able to form biofilms on a wide variety of substrates, including dichloroethane as the sole carbon source (Coleman et al. 2002) and is even able to utilize the polycyclic aromatic hydrocarbons pyrene and phenanthrene (Eriksson et al. 2002).

The ice of a snowball earth would be biotically thawed from both above (cryoconite microbial mats) and below (sub-ice algal/bacterial brown layer), thus accelerating the meltdown by amplifying the abiotic factors (particularly volcanic emission of atmospheric carbon dioxide) driving global warming. A dark, humic acid-rich cryoconite layer would tend to block light from reaching any brown layer below, but this would be compensated for by cryoconite-induced thinning of the ice sheet above the brown layer. Furthermore, light entering the ice laterally from some distance away could become trapped in the ice between the upper and lower dark layers, further contributing to meltdown. All of this would be accelerated by the high light intensities at low latitudes.

It would be to the immediate advantage of cryoconite microbes and brown layer microbes to expand their range and influence, thus melting more ice and expanding the areal extent of the band-shaped zone between the ice margin and the limit of light penetration through the ice. There would be but few limits to growth, especially for a cold-adapted biota, as the chemical composition of the oceans of the time may have resembled "the nutrient media typically used for growing cyanobacteria in pure culture" (Duval et al. 2000) and a steep nutricline may not yet have been re-established in the oceans. In Vernadskian fashion, uncontrolled growth of the sub-ice microorganisms might well have catalyzed rapid collapse of the ice sheets, heating them, as it were, from above, below and within (internally) by passive solar gain. The organisms would also begin penetrating into the ice itself by means of brine pockets and cryoconite holes (Wharton et al. 1985).

The oceans could thus become "prematurely" free of ice; that is, clear of ice more quickly than otherwise and allowing the resumption of the normal operation of bio-geochemical cycles. The accumulated levels of atmospheric carbon dioxide would then represent a highly metastable situation, and carbon dioxide would literally collapse out of the atmosphere as the Urey reaction kicked in with full force in an anomalously hot climate, causing the "freeze-fry" transition. Carbon stripped from the atmosphere and sequestered as dissolved calcium carbonate would be delivered in massive amounts to the ocean basins. Without the influence of life, the Proterozoic glaciations might have lasted considerably longer than was actually the case.

The atmospheric component of meltdown had a significant biotic component as well. Leck et al. (2004) demonstrated that atmospheric particulates generated by microbes associated with high Arctic (89°N, 1°E) pack ice and microbe activity in nearby open ocean areas have potentially important climate implications. Leck et al. (2004) hypothesized that the particulates (curious virus-like particles that act as centers for the condensation of dimethyl sulfide oxidation products) influence low altitude cloud formation and are thus important in the melting of Arctic pack ice. The particulate centers seem to be mostly associated with the thin surface layers of open water, whereas the dimethyl sulfide gas was mostly formed by microbes living around the edges of the pack ice (Leck et al. 2004; Leck and Persson 1996).

The odd radiating crystal textures and unusual dolostones of the cap carbonates are evidence for precipitation as a result of what could be called oceanic supersaturation with respect to calcium carbonate. The depletion in heavy carbon (low $\delta^{13}C$ values) of the cap carbonates is also considered evidence for a chemical oceanographic origin (Kennedy et al. 2001), although the source of this carbon is not entirely clear. Atmospheric carbon and carbon in methane from gas hydrates undergoing decomposition as global climate warmed have been suggested as sources. Newly described gas hydrate chimneys could have profound effects on the stability of gas hydrates (Wood et al. 2002).

Some type of runaway biotic feedback must be responsible for the rapid transition from snowball conditions to tropical cap carbonate deposition. Conventional abiotic geochemical systems simply cannot respond with the requisite speed. Caldeira and Kasting (1992) estimated that it would require an atmospheric carbon dioxide content (pCO$_2$) of 0.12 bar to rescue the earth from hard snowball conditions. This estimate was assumed by Higgins and Schrag (2003) in the initiation of their climate model at values of 0.04, 0.08 and 0.12 bar, respectively. I suspect that these pCO$_2$ values are far too low. Note that the greenhouse effect is not going to work very well on a pearly white planet because the incoming solar radiation will simply bounce off the ice rather than being transformed into infrared (heat) radiation to be trapped by what greenhouse gases may exist in the atmosphere at the time. Joseph Kirschvink (personal communication) has calculated that the atmosphere requires 0.6 bar CO$_2$ to end hard snowball earth conditions, and that it takes 70 million years to thaw the earth once this level of atmospheric carbon dioxide has been attained. The rock record suggests that the transition out of glacial conditions took less (and perhaps much less) than a million years (judging from the sharp formational contact at the base of the cap carbonates), in which case Kirschvink's estimate is more than seventy times the amount of time allowed by the rock record to accomplish the transition. Snowball earth thus provides a challenge for anyone wishing to rely exclusively on abiotic mechanisms to explain climate change. Abiotic geochemistry alone is inadequate to explain the rapid deglaciation (or for that matter, the onset of glaciation).

The rapid onset of Cenozoic glaciation in Antarctica (the ice cap begins to form in the Early Oligocene, 33.5 million years ago) is thought to have been a result of declining atmospheric carbon dioxide (DeConto and Pollard 2003). The reason for this decline is often attributed to falling amounts of carbon dioxide from mantle sources as the breakup of Pangea and the tectonic dispersion of its fragments began to slow down. Life processes, such as the changing productivity of the southern oceans and transitions from forest to peatland, possibly on Antarctica itself (R. DeConto, personal communication), may also have played an important role in helping to cool climate by pulling carbon dioxide out of the atmosphere. Even though peatlands only account for about 1% of the land surface area today, peatlands (mires) are thought to contain about a third of the organic carbon stored in soil (Gorham 1991; Turunen et al. 2002).

Life processes, for Vernadsky the most important geological force (Vernadsky 1998), can certainly impact global climate. Such influence, however, must always be considered in Vernadskian terms. The Vernadskian "pressure of life" will expand explosively at unexpected times and places, as with the Cambrian Explosion that ended the Proterozoic (McMenamin 2004a). Microbes flourished even under glacial conditions, and prokaryotic and eukaryotic microbes (primarily cyanobacteria and vase-shaped microfossils) do not suffer mass extinctions due to the glaciations (Corsetti et al. 2003). Explosive blooms of *Bavlinella faveolata* are known to occur globally in post-Varangerian deposits (Gaucher et al. 2003). The inferred growth of the cryoconites and the beneath-the-ice microbe brown layer, hypothesized here to have assisted the collapse of the Neoproterozoic ice sheets, provides further examples of the opportunistic potential of life.

Yet another problem involves the calcium required to balance the massive accumulation of carbon in the cap carbonates. Hoffman and Schrag (2002) calculate that to maintain the marine calcium saturation needed to sustain the cap carbonate deposition, a five-fold increase in marine calcium ion concentration is required. The question arises of how this much calcium was delivered to the sea. Glacial flour (rock powder) and frost-shattered silicate rock would rapidly weather via the Urey reaction, thus putting calcium in solution (while at the same time drawing down atmospheric carbon dioxide). But Higgins and Schrag (2003) and Hoffman (written communication) argue that oceanic supersaturation was achieved by rapid weathering of carbonates. Marine shelves were extensively exposed to weathering as deglaciation began, and some of these terranes consisted of exposed carbonate rock. Carbonate weathering proceeds hundreds of times faster than silicate weathering, and this would have been especially enhanced by high pCO_2 levels which would have boosted levels of carbonic acid in rainwater. But the question remains—was there sufficient carbonate rock exposed to rapid weathering to maintain supersaturation and sustain the cap carbonate precipitation? The five-fold increase in concentration required is huge. Such a hypothetical erosive unconformity (eroded downward into the carbonates) should be expressed in the rock record by a karst zone of unprecedented proportions. Intriguingly, Neoproterozoic karsts are known from several sequences, including the Sierras Bayas Group, Argentina (Barrio et al. 1991) and the Arroyo del Soldado Group, Uruguay, at the contact between the Polanco and Barriga Negra Formations (Gaucher 2000). However, whether the extent of this karst formation can account for the missing calcium is unknown, and furthermore, the post-glacial erosion would first be directed to the blanket of till covering the proximal parts of the exposed continental shelves. This blanket would presumably be dominated by siliciclastic debris, and it would tend to protect the shelfal carbonates from the acid rain.

Part of the problem here, once again, is over-reliance on physical science models that do not sufficiently account for the influence of the biota. We should not underestimate the influence of the Vernadsky's (1998) 'pressure of life'. Runaway expansions of the biota eventually run out of space on a spherical planet, forcing a new and eventually stable climate equilibrium. Perhaps the two levels of stasis in $\delta^{13}C$ levels in Namibia before the Ghaub Glaciation (at about +6 per mil before the big drop in values, and at about −4 per mil after the drop and shortly before deposition of the glacial strata; Hoffman and Schrag 2002; Hoffmann et al. 2004) represent biotic response (in the direction of reestablishing climate equilibrium) in the face of declining levels of a critical greenhouse gas.

Note that many of the cap dolostones are distinguished by impressive evidence for life. Some of the largest stromatolites known (Cloud et al. 1974; Hoffman and Schrag 2002 and references therein) occur within the cap dolostone. Carbonate bioherms formed of tubular stromatolites ("stromatolitic pipe rock") occur in the dolostone part (Keilberg member) of the Maieberg Formation in Namibia and represent biotic trapping of carbonate on a tremendous scale (Hoffman and Schrag 2002). As the sediment trapping and binding activities of filamentous mat-forming microbes are well-known, it may be safe to assume that such a great (five-fold) supersaturation with respect to calcium concentration was not required to form the cap carbonates. The filamentous seafloor microbes trapped and bound lime mud, and this action provided a biological pump for calcium carbonate, taking it from seawater (with elevated but not necessarily supersaturated concentrations of calcium) and sequestering it to the sea floor as cap carbonate. Indeed, the mountainous sizes of some of these stromatolites indicate that the stromatolite carbonate pump was working overtime, indicating a strong biotic (again more Vernadskian than Gaian; McMenamin 2004b) signature to cap carbonate formation.

The next generation of general circulation models should (in addition to improving their treatment of cloud cover in light of the results of Leck et al. (2004)) consider parameters designed to describe the surge aspect of biotic influence on global climate. Another good example of the phenomenon already mentioned is the expansion of Hypersea in the Devonian (McMenamin and Schulte McMenamin 1994) and associated climate instability later in the Paleozoic (another glaciation; Saltzman 2003). Certainly our species could be considered responsible for a similar type of geologically sudden perturbation, only this time via addition rather than subtraction of carbon dioxide from the atmosphere.

The intensity of snowball earth conditions during the Marinoan and Sturtian ice ages needs to be better understood if we hope to present a more complete description of extreme climate dynamics and their influence on Neoproterozoic sedimentation (Miller et al. 2003). For example, carbonate sediments (peloids, oncolites and marine cements) occurring within the glacial strata and presumably precipitated directly from glacial seawater of ancient Australia and Namibia have relatively high $\delta^{13}C$ values. These high values suggest to some that the global biological carbon pump during the glaciation was functioning normally, an interpretation that argues against ice-covered oceans (Kennedy et al. 2001). Paul F. Hoffman and Adam C. Maloof (written communication) argue that these high values would be consistent with a detrital origin for the intra-glacial diamictite carbonates, in which case the carbonates would have a heavy carbon signature inherited from more ancient carbonates. This would also avoid the difficulty of having to deposit newly formed carbonate in mid-glaciation. Or perhaps the intra-glacial sediment carbonates are indeed primary, but result from an unusual ecosystem at the base of the ice or on the seafloor below the ice.

Strange and innovative biology seems to be associated with the sub-ice biota. The hyposcum habitat would have been a suitable proving ground for a tenebrous proto-Garden of Ediacara ecology. Many paleontologists have assigned the Ediacaran fossils to conventional animal phyla such as the Cnidaria, Echinodermata or Annelida, but the arguments supporting these assignments have not convinced everyone. This is primarily so because of the bizarre morphology of the fossils. The strange shapes of these creatures have inspired comparisons to metacellular algae (McMenamin 1998) and even fungi (Peterson et al. 2003). Until recently, skeptics (with regard to the animal affinity hypothesis) pointed out that not a single uniquely animalian trait had been identified on any of the thousands of Ediacaran fossils collected so far. I will directly address this problem in the next chapter.

It seems reasonable to infer that an extremophile biota could exert a major influence on white earth conditions, and that white earth conditions in turn could have a significant impact on the evolution of ecosystems. Likewise, the perplexing cap carbonate deposition bears a signature of biotic influence on a massive scale, comparable to the great initial global oxidation event represented by the early Proterozoic banded iron formations (Johnson et al. 2003). The first inference could be tested by scalable experiments on modern ice microbes associated with floating, grounded or anchored ice. Certainly the extreme fluctuation in carbon isotope values of Late Proterozoic time implies substantive changes in the global and local biogeochemistry of microorganisms. Similarly, the role of microorganisms in the formation of cap carbonates could be tested by a series of experiments involving rates of artificial stromatolite growth under a variety of simulated marine chemistries, including of course marine water supersaturation with regard to the calcium ion.

Ever since Mawson (1949) and Harland and Rudwick (1964) drew attention to the severity of the late Proterozoic glaciations, there have been concerns about the "bizarre potential, and possible past states of Earth's climate" (Fairchild 2001). In particular, there are fears that alteration of the northward course of the Gulf Stream and shutdown of North Atlantic bottom water flow could threaten the mild climate of northwestern Europe. We need to combine analysis of ocean dynamics with analysis of biotic feedback mechanisms to arrive at an understanding of the constraints (or lack thereof) associated with extreme climatic fluctuations of the past. For a century and a half our species has been trying to retrodict (as opposed to predict) evolutionary changes and predict the course of future change as a function of carbon dioxide content of the atmosphere (Koene 1856). It may not matter much for bacteria, however, as there is evidence suggesting that Proterozoic pre-glacial bacterial communities were much the same as the syn-glacial bacterial communities (the oscillatoracean cyanobacterium *Cephalophytarion grande* passes through the Marinoan glaciation unharmed; McMenamin 2004b). Not so for eukaryotic organisms, where catastrophic extinctions may have occurred in the pre-Sturtian marine protistans (Runnegar 2000). The Tindir microbiota from the Yukon-Alaska border region (Allison and Hilgert 1986) is estimated to be 812–717 million years in age (Macdonald et al. 2010; Cohen and Knoll 2012). This curious and unique microbiota consists of previously unknown types of mineralized scales. Their appearance has been linked to early micropredation (Porter 2011). The current composition of these intricate and beautiful scale fossils is phosphatic but might originally have been siliceous (like diatoms and radiolarians) "if the apatite in the fossils is diagenetic in origin" (Cohen and Knoll 2012). These forms all apparently vanish by the end of the Sturtian glaciation, hence they are apparently victims of the earliest known mass extinction on earth, what I refer to here as the *Tindir Mass Extinction*. This extinction event completely wipes out a veritable 'lost world' of Proterozoic marine protists, and apparently represents the same mass extinction inferred by Vidal and Knoll (1982) and McMenamin (1990) to have been caused by the snowball earth glaciations.

It seems likely that the biosphere, simply by exercising its Vernadskian prerogative—the pressure of life—strongly influences when the earth goes into glaciation, as well as when it comes out of glaciation. Massive carbonate accumulation and giant stromatolites, combined with hypothesized "massive phytoplankton blooms" (Gaucher et al. 2003) led to a significant drop in the carbon dioxide content of the atmosphere, and pushed the global climate regime across a threshold to runaway cooling. Gaucher et al. (2003) link the onset of glaciation to "enhanced bioproductivity driven by high nutrient availability" due to the increased hydrothermal activity associated with the tectonic breakup of Rodinia. The biota, in Gaucher's scheme (Gaucher et al. 2003), is not of course the only factor leading to glaciation, but rather a critical piece in a combination of events leading from rifting, to upwelling, to enhanced bioproductivity, and finally to glaciation. A component of the modern biosphere in the guise of *Homo sapiens* appears to be having a major and perhaps soon to be catastrophic (Rothman 2017) influence on the stability of polar ice sheets. If so, we may be merely following a trajectory blazed by our Proterozoic microbial forebears.

Cryoconites and hyperscums developed during the ice age and contributed to melting the ice, and would have been most effective at doing so when atmospheric carbon dioxide levels reached high levels. Heinrich layers (thin till deposits representative of pulses of glacial advance) in late Proterozoic glacial stratigraphic sequences may represent episodic ice collapses caused by cryoconite/hyperscum microbial blooms. A propensity to carbonate deposition was present before the glaciation (biotic dominated—giant stromatolites and microbial laminites; abiotic dominated—oolites of the Beck Spring Dolomite) as well as after the deposition of the cap carbonates (biotic dominated—cloudinid patch reefs; *Jacutophyton* stromatolites (Kah et al. 2009); abiotic dominated—oolite of the Johnnie Formation). Substrate disturbance by burrowing metazoa after the ice ages may have disrupted the microbial mat-influenced component of Ediacaran carbonate sequestration. Post-glacial stromatolites tend to have porous, clotted and thrombolitic textures and would therefore be less effective at keeping carbon compounds (as carbonate and organic carbon) out of marine circulation. The question must now be asked: Did burrowing metazoa inhibit the ice-age inducing giant stromatolites, thus saving the world from a continuous killer freeze? One need not postulate "purpose" on the part of the organisms themselves to invoke such an occurrence (indeed the Vernadskian viewpoint forbids doing so); but in the same way that certain shapes of organisms seemed destined to appear by convergent evolution (Conway Morris 2003), certain states of the global biosphere-climate system may be more-or-less inevitable. This might imply a certain predictability to the succession of such states, allowing us to postulate biotic triggers for both the onset of glaciation and the subsequent elimination of low-latitude ice.

A valid criticism of the biospheric control model presented here might be articulated as follows: If the biosphere can so easily tip the global carbon budget scales and cause glaciation, then why were there not more glaciations in the Proterozoic? The Katangan Copperbelt of central Africa has two diamictites (the Petit Conglomérat and the Grand Conglomérat) that may represent earlier Proterozoic glaciations (Jackson et al. 2003) deposited in a nascent rift basin in the early stages of Rodinia breakup. The latter of these, the Grand Conglomérat of the Kundelungu Supergroup, is dated at approximately 940 million years and is overlain by banded shale (?possible varvite) and a dolostone (unit Ki.1.3 of Jackson et al. 2003) that may be its corresponding cap carbonate. We may be seeing here another case of the hypothesized link (Gaucher et al. 2003) between rifting, nutrient influx and carbon drawdown in the initial pulse of supercontinental breakup. The alternation between the thicker, older Grand Conglomérat and the thinner Petit Conglomérat (Unit Ks.1.1) of the Upper Kundelungu Supergroup is reminiscent of the alternation between the thicker, older Surprise Diamictite of the Kingston Peak formation and the younger, thinner Wildrose Diamictite of the upper Kingston Peak in California. In both cases thick dolostone sequences (Roan Supergroup and Beck Spring Dolomite, respectively) immediately precede the first, and thicker, diamictites. A case can be made that we are seeing in both the Kingston Peak and Kundelungu a record of extreme climate oscillation driven by Vernadskian overconsumption of atmospheric carbon dioxide. This inference accords nicely with that of Hoffman and Schrag (2002) that a preponderance of continents in mid-low latitudes "created conditions favorable for snowball events," for indeed these would be the continental positions most favorable for expansion of the biota in well lit, shallow shelfal seas.

As part of her undergraduate independent project, Mount Holyoke student Izzy Burgess (2017) and I concluded that Jessica Williams' lonestone/dropstone clast (Figs. 2.16 and 2.17) more likely belongs to the Deadham Granite (the source of Plymouth Rock) than to the Fall River Granite, although both igneous intrusive rock units are of approximately the same age and both belong to the Avalonia terrane. McMenamin and Beuthin (2008) advocated for a Fall River identification due to the fact that in the clasts "alkali feldspar is pink perthite, and its sodic feldspar is a highly altered saussurite rimmed by an inclusion-free one of sodium-rich plagioclase." The Deadham Granite (Chute 1969; Burgess 2017) is characterized by "[m]assive medium- to coarse grained pink granite and quartz monzonite. Contains 40 to 60 percent microcline microperthite, 10 to 30 percent zoned plagioclase with albite-oligoclase rims and more calcic altered cores, 20 to 25 percent quartz, and 3 to 5 percent biotite. Accessory minerals are magnetite, monazite, zircon, sphene, and pyrite. Cores of the plagioclase grains are much altered to epidote and sericite, and part or all of the biotite is altered to chlorite." Regardless of the exact source of this granitoid rock, it is clear that it is highly altered by weathering. Izzy Burgess and I infer that this weathering took place before the granitoid clasts were incorporated into the Squantum Member or the Cambridge Argillite, respectively.

Proceeding from the work of Passchier and Erukanure (2010), and considering the granitoid rock weathering results of Momeni et al. (2015), Burgess (2017) concluded that the degree of weathering observed in both the Jessica Williams dropstone (Figs. 2.18, 2.19, 2.20 and 2.21) and a granitoid clast from the puddingstone itself (Fig. 2.22) was high (W3 grade of weathering, plus presumably high chemical index of alteration [CAI] level). Burgess (2017) realized that immediately before the Gaskiers glaciation, Avalonia was at high southerly latitudes (south polar region; Li et al. 2013). This poses a paleoclimate conundrum. How could such intense weathering occur at such high latitudes?

The snowball earth glaciations (Sturtian, Marinoan) are each associated with a tillite-cap carbonate couplet that led Hoffman and Schrag (2002) to postulate the

Fig. 2.22 Granitoid clast
in the Roxbury
Conglomerate seen in thin
section, from Hewitt's
Cove, Massachusetts, USA
(Slide 7 of 6/15/2007B).
Feldspars in the clast show
alteration due to
weathering. The clast may
be derived from the
Dedham Granodiorite, a
610 million year old rock
unit that is the source of
Plymouth Rock. Scale bar
in mm

'freeze-fry' scenario. The Gaskiers glaciation is sometimes thought of as a 'failed' snowball earth event: "Newly emergent, burrowing metazoa of the [l]ate Proterozoic eventually halted the development of ice-age inducing conditions, and may have prevented even worse glaciations by releasing hydrocarbons sequestered in seafloor sediment" (McMenamin 2004b). But in light of Burgess's (2017) preliminary results the work of Passchier and Erukanure (2010), it appears that there may have been an episode of sharply elevated global temperatures that allowed granite weathering usually associated with hot, warm climates of the tropical and temperate latitudes to occur at high latitude. Burgess (2017) noted that since Avalonia was not on the equator, but rather close to the south polar regions (Hoffman and Li 2009), then there must have been "high global temperatures for the Dedham Granite to be weathered to the W3 level." Burgess (2017) refers to this as the 'fry freeze' scenario.

Pronounced global warming (as seen by alteration of Avalonian granite clasts at high southerly latitudes), probably caused by release of carbon dioxide and methane due to marine animal burrowing, triggered rapid weathering of silicates (especially granitoids), resulting in drawdown of atmospheric carbon dioxide as a result of the Urey reaction and related weathering reactions. This triggered a glaciation at 579–580 million years ago (Gaskiers), but a developing snowball earth event was forestalled by another increase in burrowing activity, one that released even more carbon dioxide by mobilizing more seafloor organic matter. This fry-freeze-fry alternation is likely due to fluctuations in the biosphere, and indicates the sensitive dependence of climate on interactions between the marine biosphere (especially burrowing animals) and the carbon cycle. The rate at which carbon is added to the marine realm is plausibly linked to the magnitude of perturbations to the earth's carbon cycle (Rothman 2017). Whereas the snowball earth events went suddenly from very cold (tillites) to very hot (cap carbonates) climate, the Gaskiers glacial event evidently went from hot (granite weathering at high latitudes) to cold (significant glaciation) to hot (Shuram event). The Shuram is the greatest negative carbon isotopic excursion in earth history for a good reason; this is the moment in earth history when the burrowing animals really assert themselves in a global geochemical sense, and thus stop a glacial event from becoming much worse.

The implication of this finding is clear. Extreme states of climate, whether fry to freeze or freeze to fry, may be juxtaposed with fast transitions between the successive extreme climate states. Rapid onset of extreme states, as, for instance, may have occurred during the end-Cretaceous mass extinction (Petersen et al. 2016), could magnify the killing efficiency of mass extinction events. As my daughter Sarah has shown (McMenamin et al. 2008), rapid shifts can eliminate long standing favorable habitats for many species, thus triggering extinctions (Dietl and Flessa 2017).

Every crisis, however, brings an opportunity. Jacobs and Speck (2017) articulated a new perspective on alternation of climate states:

> Neoproterozoic snowball Earth events seemingly anticipate the evolution of animals, while warm events in the Phanerozoic associate with episodes of extinction. This pattern implies a cold 'cradle' for animal diversity, followed by episodic warm 'graves,' and this pattern appears to be a function of temperature on oxygen . . . both snowball and greenhouse earths are expected to have stratified oceans suggesting that partially glaciated worlds should have advantages in effective oxygenation and the evolution of complex multicellular life. This is consistent with an inter-snowball, or post-snowball 'Gaskiers', initiation and radiation of Metazoa.

Jacobs and Speck (2017) infer that the Gaskiers event was *just the right amount* of glaciation to enhance oxygenation and thus encourage Metazoan diversification. If they are correct in this inference, then a potentially great crisis for the biosphere (a potentially extreme snowball earth glaciation) was transformed by complex life itself into an unprecedented opportunity for earth's biota.

References

Allison CW, Hilgert JW (1986) Scale microfossils from the early Cambrian of northwest Canada. J Paleontol 60:973–1015

Bailey RH, Bland BH (2001) Recent developments in the study of the Boston Bay Group. In: West DP, Bailey RH (eds) Guidebook for geological field trips in New England. Geological Society of America Annual Meeting, Boston, pp U1–U23

Barr TD, Kirschvink JL (1983) The paleoposition of North America in the early Paleozoic: new data from the Caborca sequence in Sonora, Mexico. Eos 64(45):689–690

Barrio CA et al (1991) El contacto entre la Formación Loma Negra (Grupo Sierras Bayas) y la Formación Cerro Negro, un ejemplo de paleokarst, Olavarría, Provincia de Buenos Aires. Rev Asoc Geol Argent 46:69–76

Bidigare R et al (1999) Iron-stimulated changes in carbon isotopic fractionation by phytoplankton in equatorial Pacific waters. Paleoceanography 14:589–595

Boag T et al (2016) Ediacaran distributions in space and time: testing assemblage concepts of earliest macroscopic body fossils. Paleobiology 42(4):574–594

Boggs S (2012) Principles of sedimentology and stratigraphy, 5th edn. Prentice Hall, Boston

Burgess I (2017) Flipped fry freeze. Independent Study Project (supervised by Mark McMenamin), Mount Holyoke College Department of Geology and Geography, pp 1–10

Butterfield NJ (2000) *Bangiomorpha pubescens* n. gen., n. sp.: implications for the evolution of sex, multicellularity, and the Mesoproterozoic/Neoproterozoic radiation of eukaryotes. Paleobiology 26:386–400

Burns SJ, Matter A (1990) Carbon and oxygen isotope stratigraphy of latest Precambrian to Cambrian(?) carbonates of central Oman. Geol Soc Am Abstr Progr 22(7):190
Burns SJ, Matter A (1993) Carbon isotopic record of the latest Proterozoic from Oman. Eclogae Geol Helv 86(2):595–607
Burns SJ et al (1994) The strontium isotopic composition of carbonates from the late Precambrian (~560-540 Ma) Huqf Group of Oman. Chem Geol 111(1–4):269–282
Caldeira K, Kasting JF (1992) Susceptibility of the early earth to irreversible glaciation caused by carbon dioxide clouds. Nature 359:226–228
Campen RK et al (2003) Evidence of microbial consortia metabolizing within a low-latitude mountain glacier. Geology 31:231–234
Canfield DE (1998) A new model for Proterozoic ocean chemistry. Nature 396:450–453
Canfield DE (2005) The early history of atmospheric oxygen: homage to Robert A. Garrels. Annu Rev Earth Planet Sci 33:1–36
Canfield DE, Teske A (1996) Late Proterozoic rise in atmospheric oxygen concentration inferred from phylogenetic and sulphur-isotope studies. Nature 382:127–132
Carto SL (2011) Sedimentology of the Squantum 'tillite', Boston Basin, USA: modern analogues and implications for the paleoclimate during the Gaskiers glaciation (c. 580 Ma). Ph.D. Dissertation, University of Toronto
Chute NE (1969) Bedrock geologic map of the Blue Hills quadrangle, Norfolk, Suffolk, and Plymouth Counties, Massachusetts. U S Geol Surv Quadrangle 796:1
Clapham ME, Corsetti FA (2005) Deep valley incision in the terminal Neoproterozoic (Ediacaran) Johnnie Formation, eastern California, USA: tectonically or glacially driven? Precambrian Res 141:154–164
Cloud PE (1983) Banded iron formation—a gradualist's dilemma. In: Trendall AF, Morris RC (eds) Iron-formation: facts and problems. Elsevier, Amsterdam, pp 401–416
Cloud PE (1988) Oasis in space. Norton, New York
Cloud PE et al (1974) Giant stromatolites and associated vertical tubes from the upper Proterozoic Noonday Dolomite, Death Valley region, eastern California. Geol Soc Am Bull 85:1869–1882
Cohen PA, Knoll AH (2012) Scale microfossils from the mid-Neoproterozoic Fifteenmile Group, Yukon Territory. J Paleontol 86(5):775–800
Coleman NV et al (2002) Biodegradation of cis-dichloroethene as the sole carbon source by a beta-proteobacterium. Appl Environ Microbiol 68:2726–2730
Conway Morris S (2003) Life's solution. Cambridge University Press, Cambridge
Corozzi AV (ed) (1967) Studies on glaciers preceded by the discourse of Neuchâtel by Louis Agassiz. Hafner, New York
Corsetti FA (1998) Regional correlation, age constraints, and geologic history of the Neoproterozoic-Cambrian strata, southern Great Basin, USA: Integrated carbon isotope stratigraphy, biostratigraphy, and lithostratigraphy. Ph.D. Dissertation, University of California at Santa Barbara
Corsetti FA et al (2003) A complex microbiota from snowball earth times: microfossils from the Neoproterozoic Kingston Peak Formation, Death Valley, USA. Proc Natl Acad Sci USA 100:4399–4404
Corsetti FA et al (2006) Trends in oolite dolomitization across the Neoproterozoic-Cambrian boundary: a case study from Death Valley, California. Sed Geol 191:135–150
Corsetti FA, Hagadorn JW (2000) Precambrian-Cambrian transition: Death Valley, United States. Geology 28(4):299–302
Corsetti FA, Kaufman AJ (2000) High resolution chemostratigraphy of the Neoproterozoic Beck Spring Dolomite, Great Basin, USA. Geol Soc Am Abstr 32:144
Crosby WO (1894) Geology of the Boston Basin, Hingham. Occasional Papers of the Boston Society of Natural History 4:179–288
Crowell JC (1999) Pre-Mesozoic ice ages: their bearing on understanding the climate system. Geological Society of America, Boulder
Cui H et al (2017) Was the Ediacaran Shuram excursion a globally synchronized early diagenetic event? Insights from methane-derived authigenic carbonates in the uppermost Doushantuo Formation, South China. Chem Geol 450:59–80

DeConto RM, Pollard D (2003) Rapid Cenozoic glaciation of Antarctica induced by declining atmospheric CO_2. Nature 421:254–249

Dietl GP, Flessa KW (2017) Conservation paleobiology: science and practice. Univ Chicago Press, Chicago

Dobson P (1826) Remarks on bowlders [*sic*]. Am J Sci Ser 1(10):217–218

Donnadieu Y et al (2003) Is there a conflict between the Neoproterozoic glacial deposits and the snowball earth interpretation?: an improved understanding with numerical modeling. Earth Planet Sci Lett 208:101–112

Duval B et al (2000) Phenolic compounds and antioxidant properties in the snow alga *Chlamydomonas nivalis* after exposure to UV light. J Appl Phycol 11:559–566

Friedman GM, Sanders JE (1978) Principles of sedimentology. Wiley, New York

Eriksson M et al (2002) Bacterial growth and biofilm production on pyrene. FEMS Microbiol Ecol 40:21–27

Fairchild IJ (2001) Encapsulating climate catastrophe: snowball earth. Geoscientist 11:4–5

Gaucher C (2000) Sedimentology, palaeontology and stratigraphy of the Arroyo del Soldado Group (Vendian to Cambrian, Uruguay). Beringeria 26:1–120

Gaucher C et al (2003) Integrated correlation of the Vendian to Cambrian Arroyo del Soldado and Corumbá Groups (Uruguay and Brazil): palaeogeographic, palaeoclimatic and palaeobiologic implications. Precambrian Res 120:241–278

Gaucher C et al (2004) Chemostratigraphy of the Lower Arroyo del Soldado Group (Vendian, Uruguay) and palaeoclimatic implications. Gondwana Res 7(3):715–730

Gingerich PD et al (1983) Origin of whales in epicontental remnant seas: new evidence from the early Eocene of Pakistan. Science 220:403–406

Gong Z et al (2017) Rock magnetic cyclostratigraphy of the Doushantuo Formation, south China and its implication for the duration of the Shuram carbon isotope excursion. Precambrian Res 289:62–74

Gorham E (1991) Northern peatlands: role in the carbon cycle and probable responses to global warming. Ecol Appl 1:182–195

Gould CG (2004) The remarkable life of William Beebe. Island Press, Washington, DC

Grotzinger JP et al (2011) Enigmatic origin of the largest-known carbon isotope excursion in Earth's history. Nat Geosci 4:285–292

Hallam A (1992) Great geological controversies, 2nd edn. Oxford University Press, New York

Harland WB, Rudwick MJS (1964) The great infra-Cambrian ice age. Sci Am 211:28–36

Higgins JA, Schrag DP (2003) The aftermath of a snowball earth. Geochem Geophys Geosyst 4(3). https://doi.org/10.1029/2002GC000403

Hoffman PF, Li Z-X (2009) A palaeogeographic context for Neoproterozoic glaciation. Pal Pal Pal 277:158–172

Hoffman PF, Schrag DP (2002) The snowball earth hypothesis: testing the limits of global change. Terra Nova 14:129–115

Hoffmann K-H et al (2004) U-Pb zircon date from the Neoproterozoic Ghaub Formation, Namibia: constraints on Marinoan glaciation. Geology 32(9):817–820

Huang J et al (2017) Multiple sulfur isotopic records associated with the 'Shuram excursion' from South China. Geol Soc Am Abstr Progr 49(6). https://doi.org/10.1130/abs/2017AM-306932

Hughes GB et al (2003) Modern spectral climate patterns in rhythmically deposited argillites of the Gowganda Formation (early Proterozoic), southern Ontario, Canada. Earth Planet Sci Lett 207:12–23

Jackson M et al (2003) Neoproterozoic allochthonous salt tectonics during the Lufilian orogeny in the Katangan copperbelt, central Africa. Geol Soc Am Bull 115:314–330

Jacobs DK, Speck HP (2017) Cold cradles and warm graves—how temperature constrains oxygen impacting diversity. Geol Soc Am Abstr Progr 49(6). https://doi.org/10.1130/abs/2017AM-308455

Li Z-X et al (2013) Neoproterozoic glaciations in a revised global palaeogeography from the breakup of Rodinia to the assembly of Gondwanaland. Sediment Geol 294:219–232

Jenkins RJF (1995) The problems and potential of using animal fossils and trace fossils in terminal Proterozoic biostratigraphy. Precambrian Res 73:51–69

Johnson CM et al (2003) Ancient geochemical cycling in the earth as inferred from Fe isotope studies of banded iron formations from the Transvaal craton. Contrib Mineral Petrol 144:523–558

Kah LC et al (2009) Reinterpreting a Proterozoic enigma: *Conophyton-Jacutophyton* stromatolites of the Mesoproterozoic Atar Group, Mauritania. Int Assoc Sedimentol Spec Publ 41:277–295

Kennedy MJ et al (2001) Are Proterozoic cap carbonates and isotopic excursions a record of gas hydrate destabilization following earth's coldest intervals? Geology 29:443–446

Kirschvink JL (1992) Late Proterozoic low-latitude global glaciation: the snowball earth. In: Schopf JW, Klein C (eds) The Proterozoic biosphere. Cambridge University Press, Cambridge, MA, pp 51–52

Kirschvink JL et al (1991) The Precambrian/Cambrian boundary: magnetostratigraphy and carbon isotopes resolve correlation problems between Siberia, Morocco, and South China. Eos 1(4):69–91

Koene CJ (1856) Popular lectures: concerning the creation from the formation of the earth to the extinction of the human species, or insights into the natural history of air and its miasmas in connection with acid factories and complaints of those who suffer from their pollution. P. Larcier, Brussels

Kunzmann M et al (2017) Bacterial sulfur disproportionation constrains timing of Neoproterozoic oxygenation. Geology 45(3):207–210

Laflamme M et al (2013) The end of the Ediacara biota: extinction, biotic replacement or Cheshire Cat? Gondwana Res 23:558–573

Leck CM, Persson C (1996) The central Arctic Ocean as a source of dimethyl sulfide-seasonal variability in relation to biological activity. Tellus 48:156–177

Leck CM et al (2004) Can marine micro-organisms influence melting of the Arctic pack ice? Eos 85:25–32

Le Guerroué E (2006) Sedimentology and chemostratigraphy of the Ediacaran Shuram Formation, Nafum Group, Oman. Ph.D. Dissertation, Swiss Federal Institute of Technology Zürich

Licari GR (1978) Biogeology of the late pre-Phanerozoic Beck Spring dolomite of eastern California. J Paleontol 52:767–792

Lund K et al (2003) SHRIMP U-Pb geochronology of Neoproterozoic Windermere Supergroup, central Idaho: implications for rifting of western Laurentia and synchroneity of Sturtian glacial deposits. Geol Soc Am Bull 115:349–372

Macdonald FA et al (2010) Early Neoproterozoic scale microfossils in the lower Tindir Group of Alaska and the Yukon Territory. Geology 38:143–146

Margesin R et al (2002) Characterization of heterotrophic microorganisms in alpine glacier cryoconite. Arct Antarct Alp Res 34:88–93

Margaritz M et al (1991) Precambrian/Cambrian boundary problem: carbon isotope correlations for Vendian and Tommotian time between Siberia and Morocco. Geology 19:847–850

Macdonald FA et al (2013) The stratigraphic relationship between the Shuram carbon isotope excursion, the oxygenation of Neoproterozoic oceans, and the first appearance of the Ediacara biota and bilaterian trace fossils in northwestern Canada. Chem Geol 362:250–272

Matsen B (2005) Descent: the heroic discovery of the abyss. Vintage Books, New York

Mawson D (1949) The late Precambrian ice-age and glacial record of the Bibliando dome. J Proc R Soc NSW 82:150–174

McMenamin MAS (1990) 2.13.1 mass extinction: events: Vendian. In: Briggs DEG, Crowther PR (eds) Palaeobiology: a synthesis. Blackwell Scientific Publications, Oxford, pp 179–181

McMenamin MAS et al (1992) Vendian body fossils (?) and isotope stratigraphy from the Caborca area, Sonora, Mexico. North American Paleontological Convention 5:206

McMenamin MAS et al (1994) Upper Precambrian-Cambrian faunal sequence, Sonora, Mexico and lower Cambrian fossils from New Jersey, United States. In: Landing E (ed) Festschrift Honoring Donald W. Fisher, New York State Mus Bull 481:213–227

McMenamin MAS (1996) Ediacaran biota from Sonora, Mexico. Proc Natl Acad Sci 93:4990–4993

McMenamin MAS (1998) The garden of Ediacara: discovering the first complex life. Columbia Univ Press, New York

McMenamin MAS (2001) Review of McDonald, NG, The Connecticut Valley in the age of dinosaurs: a guide to the geologic literature. Isis 92:134–135

McMenamin MAS (2004a) Climate, paleoecology and abrupt change during the Late Proterozoic: a consideration of causes and effects. In: Jenkins GS et al (eds) The extreme Proterozoic: geology, geochemistry, and climate. American Geophysical Union, Washington, DC, pp 215–229

McMenamin MAS (2004b) Gaia and glaciation: Lipalian (Vendian) environmental crisis. In: Schneider SH et al (eds) Scientists debate Gaia: the next century. MIT Press, Cambridge, MA, pp 115–127

McMenamin MAS (2004c) Vendian and Ediacaran. In: Selley RC et al (eds) Encyclopedia of geology. Elsevier, Oxford, pp 371–381

McMenamin MAS, Beuthin JD (2008) Fine clastics of the Boston Bay Group: new data and interpretations concerning depositional processes and environments. In: de Wet AP (ed) Keck Geology Consortium, 21st Keck Research Symposium in geology, short contributions, April 2008. Franklin and Marshall College, Lancaster, pp 209–212

McMenamin MAS, Schulte McMenamin DL (1990) The emergence of animals: the Cambrian breakthrough. Columbia Univ Press, New York

McMenamin MAS, Schulte McMenamin DL (1994) Hypersea: life on land. Columbia Univ Press, New York

McMenamin SK et al (2008) Climatic change and wetland desiccation cause amphibian decline in Yellowstone National Park. Proc Natl Acad Sci USA 105(44):16988–16993

Melezhik VA et al (2008) The Shuram-Wonoka event recorded in a high-grade metamorphic terrane: insight from the Scandinavian Caledonides. Geol Mag 145(2):161–172

Miller NR et al (2003) Significance of the Tambien Group (Tigrai, n. Ethiopia) for snowball earth events in the Arabian-Nubian shield. Precambrian Res 121:263–283

Momeni AA et al (2015) New engineering geological weathering classifications for granitoid rocks. Eng Geol 185:43–51

Nance RD (1990) Late Precambrian-early Paleozoic arc-platform transitions in the Avalon terrane of the northern Appalachians: review and implications. Geol Soc Am Spec Pap 245:1–11

Passchier S, Erukanure E (2010) Palaeoenvironments and weathering regime of the Neoproterozoic Squantum 'tillite', Boston Basin: no evidence of a snowball earth. Sedimentology 57:1526–1544

Petersen SV et al (2016) End-cretaceous extinction in Antarctica linked to both Deccan volcanism and meteorite impact via climate change. Nat Commun 7:12079. https://doi.org/10.1038/ncomms12079

Peterson KJ et al (2003) A fungal analog for Newfoundland Ediacaran fossils? Integr Comp Biol 43:127–136

Porter SM (2004) The fossil record of early eukaryotic diversification. Paleontol Soc Pap 10:35–50

Porter SM (2011) The rise of predators. Geology 39(6):607–608

Poulsen CJ et al (2001) Impact of ocean dynamics on the simulation of the Neoproterozoic 'snowball earth'. Geophys Res Lett 28:1575–1578

Poulsen CJ et al (2002) Testing paleogeographic controls on a Neoproterozoic snowball earth. Geophys Res Lett 29(11). https://doi.org/10.1029/2001GL014352

Prave AR (1999) Two diamictites, two cap carbonates, two $\delta^{13}C$ excursions, two rifts: the Neoproterozoic Kingston Peak Formation, Death Valley, California. Geology 27:339–342

Pu JP et al (2016) Dodging snowballs: geochronology of the Gaskiers glaciation and the first appearance of the Ediacaran biota. Geology 44(11):955–958

Rickard D et al (2017) Sedimentary sulfides. Elements 13(2):117–122

Roberts MT (1982) Depositional environments and tectonic setting of the Crystal Spring Formation, Death Valley region, California. In: Cooper JD et al (eds) Geology of selected areas in the San Bernardino Mountains, western Mojave Desert, and southern Great Basin, California, Death Valley Publishing Company, Shoshone, California, pp 143–154

Rothman DH (2017) Thresholds of catastrophe in the earth system. Sci Adv 3(9). https://doi.org/10.1126/sciadv.1700906

Runnegar B (2000) Loophole for snowball earth. Nature 405:403–404

Saltzman MR (2003) Late Paleozoic ice age: oceanic gateway or pCO$_2$? Geology 31:151–154

Sharp M et al (1999) Widespread bacterial populations at glacier beds and their relationship to rock weathering and carbon cycling. Geology 27:107–110

Skehan JW (2001) Roadside geology of Massachusetts. Mountain Press, Missoula, Montana

Sour-Tovar F et al (2007) Ediacaran and Cambrian index fossils from Sonora, Mexico. Palaeontology 50(1):169–175

Stewart JH et al (1984) Upper Proterozoic and Cambrian rocks in the Caborca region, Sonora, Mexico-physical stratigraphy, biostratigraphy, Paleocurrent studies and regional relations. U S Geol Surv Prof Pap 1309:1–36

Stow DAV (2006) Sedimentary rocks in the field-A color guide. Academic Press, Burlington

Suarez CA et al (2017) A chronostratigraphic assessment of the Moenave Formation, USA using C-isotope chemostratigraphy and detrital zircon geochronology: implication for the terrestrial end Triassic extinction. Earth Planet Sci Lett 475(1):83–93

Takazi K et al (1994) Clay aerosols and Arctic ice algae. Clay Clay Miner 42:402–408

Takeuchi N et al (2001) Structure, formation, and darkening process of albedo-reducing material (cryoconite) on a Himalayan glacier: a granular algal mat growing on the glacier. Arct Antarct Alp Res 33:115–122

Tsukui K et al (2017) Developing an enhanced chronology for the terminal Ediacaran-Cambrian transition on a global scale. Geol Soc Am Abstr Progr 49(6). https://doi.org/10.1130/abs/2017AM-308028

Tucker ME (1989) Carbon isotopes and Precambrian-Cambrian boundary geology, South Australia: ocean basin formation, seawater chemistry and organic evolution. Terra Nova 1:573–582

Turunen J et al (2002) Estimating carbon accumulation rates of undrained mires in Finland—application to boreal and subarctic regions. The Holocene 12:79–90

van de Shootbrugge et al (2008) Carbon cycle perturbation and stabilization in the wake of the Triassic-Jurassic boundary mass-extinction event. Geochem Geophys Geosyst 9:Q04028. https://doi.org/10.1029/2997GC001914

Vanyo JP, Awramik SM (1985) Stromatolites and earth-sun-moon dynamics. Precambrian Res 29:121–142

Verdel C et al (2011) The Shuram and subsequent Ediacaran carbon isotope excursions from southwest Laurentia, and implications for environmental stability during the metazoan radiation. Geol Soc Am Bull 123(7/8):1539–1559

Vernadsky V (1998) The biosphere. Copernicus, New York

Vidal G, Knoll AH (1982) Radiations and extinctions of plankton in the late Proterozoic and early Cambrian. Nature 297:57–60

Wang P et al (2003) Carbon reservoir changes preceded major ice-sheet expansion at the mid-Brunhes event. Geology 33:239–242

Wang X et al (2016) Paired carbonate and organic carbon isotope variations of the Ediacaran Doushantuo Formation from an upper slope section at Siduping, South China. Precambrian Res 273:53–66

Walker G (2003) Snowball earth: the story of the great global catastrophe that spawned life as we know it. Crown Books, New York

Wendler I (2013) A critical evaluation of carbon isotope stratigraphy and biostratigraphic implications for late Cretaceous global correlation. Earth-Sci Rev 126:116–146

Wharton RA et al (1985) Cryoconite holes on glaciers. Bioscience 35:440–503

Williams GE (1975) Late Precambrian glacial climate and the earth's obliquity. Geol Mag 112:441–465

Williams GE (1979) Sedimentology, stable-isotope geochemistry and palaeoenvironment of dolostones capping late Precambrian glacial sequences in Australia. J Geol Soc Aust 26:377–386

Williams H et al (1982) Petrography: an introduction to the study of rocks in thin sections. Freeman, New York

Williams J (2008) Laminites and dropstones in the Cambridge Argillite (Ediacaran), Hewitt's Cove, Hingham, Massachusetts. In: de Wet AP (ed) Keck Geology Consortium, 21st Keck Research Symposium in geology, short contributions, April 2008. Franklin and Marshall College, Lancaster, pp 234–237

Williams J et al (2008) Laminites in the Cambridge Argillite (Ediacaran), Hewitt's Cove, Hingham, Massachusetts. Geol Soc Am Abstr Progr 40(1):69

Wilson JL (1975) Carbonate facies in geologic history. Springer, New York

Wood WT et al (2002) Decreased stability of methane hydrates in marine sediments owing to phase-boundary roughness. Nature 420:656–660

Woods KN (1999) Investigating the nature of the dolomite in a possible Neoproterozoic cap carbonate: the Noonday Formation, Death Valley, CA. Geol Soc Am Abstr Progr 31:486

Zhou C et al (2017) The stratigraphic complexity of the middle Ediacaran carbon isotopic record in the Yangtze Gorges area, South China, and its implications for the age and chemostratigraphic significance of the Shuram excursion. Precambrian Res 288:23–38

Chapter 3
Clemente Biota

One of my correspondents expressed concern that I was perpetuating a dubious interpretation of Kimberella as a mollusk; for another, Kimberella-as-mollusk is crucial to the interpretation of early bilaterian evolution...Perhaps by the time you read this, things will be clearer.

Peter Godfrey-Smith Other Minds (2016)

Abstract A diverse fauna of Ediacarans from the Clemente Formation of northwestern Sonora, México, includes *Pteridinium* cf. *P. simplex*, a recumbent sand frond *Beothukis* cf. *B. mistakensis*, the kimberellomorph *Kimberella* cf. *K. quadrata*, the solzid kimberellomorph *Zirabagtaria ovata* n. gen. n. sp., the praecambridiid *Palankiras palmeri* n. gen. n. sp., *Vendamonia truncata* n. gen. n. sp., and the aculiferans *Clementechiton sonorensis* McMenamin and Fleury, 2016 and *Korifogrammia clementensis* n. gen. n. sp. The Clemente biota provides new data regarding the Ediacaran Cuticle Paradox, which holds that in spite of their apparent simplicity, the Ediacaran cuticle in fact hosted a highly complex morphogenetic field. In a corollary of Williston's Law, this cuticle underwent successive simplification at the end of the Proterozoic.

Keywords Clemente Formation · Proterozoic · *Pteridinium* · *Beothukis* · *Kimberella* · *Zirabagtaria* · *Palankiras* · *Vendamonia* · *Korifogrammia* · Aculiferans

The feeding strategies of Ediacaran creatures (the problem is the same whether they are considered to be early animals or non-animals) are quite cryptic even after decades of research. Detailed study of the morphology of the fossils has also failed to reveal satisfying answers. Fortunately, geochemistry is beginning to provide helpful clues to the mysteries of Ediacaran paleontology.

Proposals that the Avalonian Ediacaran biota lived in dense stands in pitch black, deep water make no sense. There is no light and limited particulate food available at such depths. How then were the creatures able to survive? Even osmotrophy would be a challenge with so many individuals in close proximity.

Cui et al. (2017) argue that a new conceptual model (globally synchronized calcite mineralization) "provides a non-actualistic interpretation for the largest negative $\delta^{13}C$ excursion in earth history, and suggests a unique geochemical transition in seawater during the Ediacaran Period." Such claims may provide a clue to Ediacaran food sources. If sulfate was becoming abundant in sea water for the first time, it may have provided Ediacarans with a non-actualistic trophic strategy. In other words, Ediacarans may have been able to utilize the newly available sulfate for a type of enhanced osmotrophy where anoxic environments (not all of them necessarily in deep water) could, for the first time, provide nutrients such as fixed carbon and reduced sulfur compounds that could be metabolized by Ediacarans (and their symbionts) by utilizing oxygen derived from sulfate. This could potentially combine with excesses of glycolate dehydrogenase released into sea water, also a result of increasing ambient oxygen levels (McMenamin 1993). Indeed, Hoyal Cuthill and Conway Morris (2017) conclude that the "growth of rangeomorph branch internodes declined as their relative surface area decreased. This suggests that frond size and shape were directly responsive to nutrient uptake."

A recent test, however, of the hypothesis of chemosymbiosis in the Ediacaran holdfast *Aspidella* failed to show any isotopic evidence for chemosymbiosis. Taking advantage of a rare occurrence of Ediacaran fossil preservation in carbonate rocks (Khatyspyt Formation, Olenek uplift, Siberia), Bykova et al. (2017) tested the fossils for isotopic evidence for chemosymbiosis with microbes such as methylotrophs, methanogens or sulfide-oxidizing bacteria. Their results showed that there was no difference in the $\delta^{13}C_{org}$ isotopic signature between the *Aspidellas* and their surrounding matrix, thus eliminating the hypothesis of chemosymbiosis for the holdfasts.

Bykova et al. (2017) noted that photosymbiosis was still a possibility for these organisms. The fronds connected to the holdfasts were not preserved in the Khatyspyt Formation. Bykova et al. (2017) reasoned that if the *Aspidellas* had had photosymbionts, then the symbionts would have lived in the upper or frondose part of its body, whereas putative chemosymbionts would reside in the holdfast where they could "explore redox boundaries." Bykova et al.'s (2017) results were not in accord with the concept that the *Aspidella* holdfasts hosted chemosymbiont microbes "that would impart a distinct $\delta^{13}C_{org}$ signature." Bykova et al. (2017) could not eliminate the possibility that *Aspidella* hosted photosymbionts—such symbiotic microbes would generate organic carbon using the Rubisco enzyme that could not be distinguished from carbon derived from other primary producers and sequestered in matrix sediments. Thus photosymbiosis remains a possibility for the Khatyspyt *Aspidellas*, whereas chemosymbiosis may be ruled out at least for the lower (holdfast) part of the creatures. This does not, however, rule out the possibility of Ediacaran sulfate osmotrophy as noted above. Bykova et al. (2017) do note that the Khatyspyt *Aspidellas may* have lived under oxygenated conditions.

Fig. 3.1 *Yorgia* reconstruction (Artwork by Aleksey Nagovitsyn, adapted after Fedonkin (2003). Used here per Creative Commons BY-SA 3.0 license)

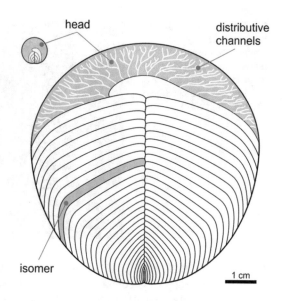

head

distributive channels

isomer

1 cm

The systematic placement of many Ediacarans remains unsettled. Even the iconic *Dickinsonia* has never received a proper phylum-level assignment nor convincing stem taxon link to better known groups. This organism, even though it reaches up to a meter in diameter, continues to defy all attempts at formulating a satisfying paleontological and taxonomic interpretation.

Evans et al. (2017) determined that growth in *Dickinsonia* was "highly regulated to maintain an ovoid shape via terminal addition and predictable expansion of modules [the isomers or 'semisegments' of Ivantsov (2007)]," and concluded that *Dickinsonia* "likely belongs in the Eumetazoa." This interpretation receives some support from Hoekzema et al. (2017), who conclude that *Dickinsonia* (oddly) added new segments not at the small segment pole ("anti-deltoidal end") but at the large segment pole ("deltoidal end"), where the deltoidal region (a narrow triangle at this tip of the organism) split in two to form additional 'segmental' units. This is a curious growth pattern for an animal to say the least. Nevertheless, Hoekzema et al. (2017) infer that *Dickinsonia* belongs in "the Eumetazoa plus Placozoa total group."

This is an excellent finding if true, but we must first examine several issues with the Evans et al. (2017) *Dickinsonia* paper. First, Evans et al. (2017) claim bilateral symmetry in *Dickinsonia's* 'quilt', not glide symmetry. However, closely related forms such as *Yorgia* (Yorgiidae and Dickinsoniidae are allied families; Dzik and Martyshyn 2015) unquestionably do develop glide symmetry, with isomers offset along the midline (Fig. 3.1). Thus it is entirely possible that *Dickinsonia* could, at least on occasion, grow with offset in its tubular modules along its midline, in other words, with glide symmetry.

Second, Evans et al. (2017), in support of their placement of *Dickinsonia* in the Eumetazoa, claim that there is no midline running through *Dickinsonia's* AMU (anterior most unit), the structure in the place where its 'head' ought to be. However,

Fig. 3.2 *Yorgia*. Specimen
from the Arkhangelsk
Regional Museum
(Photograph by Aleksey
Nagovitsyn, used here per
GNU Free Documentation
Version 1.3 license)

Evans et al.'s (2017) own fig. 2C seems to contradict this interpretation, as an isomer groove runs through the center of the AMU. In any case, the midline or isomer groove runs all the way to the anterior edge of *Yorgia* as seen here in Fig. 3.2. Thus, Ediacaran morphology continues to confound the sharpest researchers. If *Dickinsonia* and related forms such as *Yorgia* and *Podolimirus* are indeed eumetazoans, they are very strange animals indeed.

Evans et al.'s (2017) difficulties with *Dickinsonia* represent a reworking of the classic paleontological problem centering around the question of symmetry in the Ediacaran body plan. In what was arguably Martin F. Glaessner and Mary Wade's most contentious paper (Glaessner and Wade 1971), they deformed (either as a thought experiment or graphically, see their fig. 4) the Ediacarans with glide symmetry such as *Vendia* and *Praecambridium* in order to render these forms bilaterally symmetric by fiat. Glaessner and Wade (1971) were forcing the morphology to accord with their expectations. In the caption for their fig. 4 showing an intentionally deformed *Vendia sokolovi*, Glaessner and Wade (1971) state that five "segments are shown in this interpretation of the fossil as bilaterally symmetrical, with a slight distortion." Glide symmetry in a putative arthropod ancestor was evidently too jarring a concept to be tolerable for Glaessner and Wade (1971). It did not help that the subject of their paper, *Praecambridium sigillum*, in some cases *also* developed axial offset: "The apparent staggering of the lateral 'outgrowths' from the median ridge of *Vendia* is also found in some specimens of *Praecambridium* where this arrangement can be seen to be the result of a slight, oblique distortion" (Glaessner and Wade 1971). The taphonomic distortion argument largely collapses, as noted above, when one considers the unquestioned glide symmetry of *Yorgia* and *Podolimirus* as noted above.

In fairness to Martin Glaessner and Mary Wade, both great and inspiring paleontologists, they do cite in Glaessner and Wade (1971) a key, brilliant comment made early on in Ediacaran studies by the Russian paleontologist B. S. Sokolov (1965, p. 80): "*Vendia*…[is] an impression of an externally trilobitomorph organism which however could be related to the rangeids [i.e., frondose Ediacarans such as *Rangea*]." Demonstrating how a single well-crafted sentence can spawn an entire research program, Sokolov's (1965) insightful comment inspired Seilacher (1989) to develop his vendozoa/vendobionta concept that united Ediacaran fronds and trilobitomorphs. McMenamin (1998), influenced in turn by Seilacher, argued that "externally trilobitomorph" Ediacarans might represent a case of convergent evolution and hence may not be closely related to trilobites at all. Adding to the confusion, I later changed my mind about this and now consider *Spriggina* to be a trilobitomorph ecdysozoan (McMenamin 2003b), largely because its homologies with trilobites are too numerous and too specific to be accounted for by convergent evolution.

In a study of putative rheotaxis in the Ediacaran *Parvancorina*, Paterson et al. (2017) claim to show current orientation for *Parvancorina*. However, Paterson et al. (2017) conclude that the pointed posterior end of the creature would be the end pointing into the current, a less than stable configuration considering the morphology of *Parvancorina's* anchor-shaped dorsal ridge. The data in their figure 2C (Parv Bed) does suggest some sort of orientation to the *Parvancorinas* (anterior end to the east), but there is a lot of scatter in the data. Their figure 3 (bed assemblage data MM3) shows orientation of *Parvancorina* that is at a 41° angle to the inferred current direction, with many of the *Parvancorinas* oriented at an even higher value, approximately 115°, to the inferred current direction (as inferred from folds in *Dickinsonia* and 'mop structures'; Paterson et al. 2017). The *Parvancorina* orientations in their Figures 2 and 3 are thus quite different. It seems far more reasonable to infer from these data that these *Parvancorinas* were engaging in phototaxis (or possibly chemotaxis; McMenamin 1998) rather than rheotaxis.

Darroch et al. (2017) inferred facultative mobility in *Parvancorina*. However, their interpretation of the computational fluid dynamics results is unconvincing. Darroch et al. (2017:3) state that the "characteristic flow patterns seen around *Parvancorina* largely disappear in the null models." However, if you inspect their fig. 1, the computer simulation of water flow in their fig. 1b (anterior end of *Parvancorina* oriented into current) is broadly identical to the null model (their fig. 1e) if you observe the outline of the low velocity blue areas. There are chevron-shaped dark blue areas behind the *Parvancorina* 'anchor' in their fig. 1b, and these might plausibly serve as catchment areas for filter feeding. However, recall that Paterson et al. (2017) suggest that the *Parvancorina* was oriented with its posterior pointed up-current. Darroch et al. (2017) show this *Parvancorina* orientation in their fig. 1d, but in this case the low velocity blue area is far to the right, away from and beyond the body of the *Parvancorina*, and thus this low velocity zone would be useless for filter feeding.

Paterson et al. (2017) and Darroch et al. (2017) mention the possibility of both suspension feeding and deposit feeding as possibilities for *Parvancorina*, but neither of these trophic strategies seems plausible. In the case of suspension feeding, there is no structure visible such as an ambulacrum in *Parvancorina* to assist the process. In any case, the field data regarding orientation and the computational fluid dynamics modeling results taken together go against an inference of suspension feeding. Also, there is no filter feeding organ such as a lophophore preserved in *Parvancorina*. And with regard to deposit feeding, there are no deposit feeding trace fossils directly associated with *Parvancorina*. These should occur if *Parvancorina* was indeed a deposit feeder.

Curiously, neither Paterson et al. (2017) nor Darroch et al. (2017) consider the possibility of light orientation for photosymbiosis in *Parvancorina* in spite of the fact that the genus is known to have inhabited a photic zone depositional environment. Phototactic behavior could very nicely explain the change in orientation with regard to ambient current direction between Paterson et al.'s (2017) *Parvancorinas* of Parv Bed and those of MM3. Photosymbiosis is not even mentioned in these papers; it should have been considered as a possibility.

In a successful conclusion of a research objective initiated in the early 1980s, I discovered a new Ediacaran fossil locality in Sonora, México (McMenamin 1996). For a variety of reasons, not the least of which were the difficulties encountered in locating the fossils in the first place (McMenamin 1998), this biota has never before been fully described. Advances in the paleontological study of Ediacarans and associated biogenic structures (Seilacher et al. 2003; McCall 2006; Laflamme et al. 2013), plus advances in geochemistry and geochronology discussed in the previous chapter, now allow me to complete the descriptions and more fully discuss the significance of the critical Clemente Formation fossils.

The Altar Desert region of Sonora, México is known for its abundant fossils of Cambrian and Proterozoic organisms, occurring in strata deposited atop granites ranging in age from 1.71 to 1.1 billion years old (Lang et al. 2005). The oldest known fossil nest has recently been described from algal-archaeocyath limestones of the Puerto Blanco Formation (McMenamin 2016). The Sonoran Proterozoic-Cambrian rock sequence at Cerro Rajón has produced two different Ediacaran biotas. The youngest of these occurs in the La Ciénega Formation, is latest Proterozoic in age, and may be correlated with the Nama Assemblage. The tripartite division of Ediacaran biotas into the Nama (~544–542 Ma), the White Sea (~555–550 Ma), and the Avalonian (~580–560 Ma; Laflamme et al. 2013) was challenged by Grazhdankin (2004) as being more reflective of facies distributions of the fossils rather a geochronological time separation of the assemblages, and in accord with Grazhdankin's (2004) conclusions we will see below that the Clemente biota contains elements of all three assemblages (Fig. 2.14). Nevertheless, the threefold division is useful for discussion and will be used here in a roughly geochronological sense.

Fossils of probable latest Ediacaran age occur in the La Ciénega Formation, including a rare, soft-bodied discoidal form (*Protolyellia*-like; field sample 2 of 3/22/90; McMenamin 2001) plus *Sinotubulites* and other calcareous tubular shelly fossils. Among the oldest shelly fossils on earth, these constituted a late Proterozoic "worm world" (550–541 Ma) that is seen by some researchers as transitional to the Cambrian explosion (Schiffbauer et al. 2016). This worm world fauna (Zhuravlev et al. 2012) from Sonora includes the genera *Sinotubulites*, *Cloudina*, *Chenmengella*, and *Qinella*. It now seems clear that those who attempted to lump all of these taxa into the single genus *Cloudina* were oversimplifying matters (e.g., Sour-Tovar et al. 2007). Interestingly, *Qinella* was recently discovered in Lower Cambrian strata in

Fig. 3.3 Eophyton toolmarks from the La Ciénega Formation, Sonora, México. Field sample MM-82-49a; IGM 3619. Width of view 18 mm (Photograph by Mark McMenamin)

Death Valley, California (McMenamin 2016). This case indicates that calcareous tubular fossils are characteristic of, but by no means restricted to, Proterozoic strata, and thus they should not necessarily be used as index fossils for the Proterozoic.

Before introducing the Ediacarans of the Clemente Formation, now is the time to clear up some misconceptions about several Proterozoic dubiofossils from Sonora. I had earlier argued that a structure from a shaly interval in the La Ciénega Formation ("*Rusophycus multilineatus* with *Planolites*"; Fig. 3.3) represented early ichnofossil evidence for predation (McMenamin 2003a, 2016). Sour-Tovar et al. (2007) challenged this interpretation, arguing that this structure was in fact an "inorganic structure," and they were largely correct in their inference. Thanks to recently published research by Enrico Savazzi (2015), the La Ciénega Formation structure may now be properly identified as an Eophyton drag marking caused by current drag of an organism with a holdfast and a 'sail.' A dense holdfast scored parallel reciprocal grooves into the substrate as its sail (such as a frond, tentacles, buoyant seaweed, etc.) was dragged along by current. The La Ciénega Formation structure shows the characteristic 'capped end' associated with a short drag track. The structure, particularly in combination with a capped end, is similar to Eophyton furrows with capped ends illustrated by Savazzi (2015, his Figure 7), thus interpretation of the Mexican hyporelief structure now seems secure. The putative "Planolites" prey ichnofossil (McMenamin 2003a) is in fact a slightly wavy major groove associated with the 'cap' of the Eophyton structure. Although it probably has at least a partially organic origin, Sour-Tovar et al. (2007) were mostly correct to interpret this Eophyton marking as an inorganic structure.

Such are the perils of Precambrian paleontology. A large number of working hypotheses are required to help us converge on the correct interpretation or, in other words, the right answer. This has reached a successful conclusion in the case of Eophyton of Sonora. There is no dishonor in proposing a plausible hypothesis that is subsequently falsified by new data or better interpretations, as was the case here.

Fig. 3.4 Eophyton toolmarks from the Clemente Formation, Sonora, México. Note deep red staining of matrix, possibly indication high levels of ambient free oxygen during deposition of the sediment that formed this rock. Field sample 6 of 3/16/95. IGM 7462. Scale bar in mm (Photograph by Mark McMenamin)

A word of explanation is nevertheless in order, and this will help to clear up another misconception.

I had based my identification of the La Ciénega Formation *Rusophycus multilineatus* on an illustration published in Alpert (1976). Alpert (1976, his Plate 1, Figure 2) reported *Monomorphichnus multilineatus* from the Deep Spring Formation of eastern California. Both *Monomorphichnus* and *Rusophycus* are generally considered to be trilobite ichnofossils, and both the La Ciénega Formation of México and the Deep Spring Formation of California are very ancient, that is, too deep in the stratigraphic section, to be producing trilobitoid trace fossils. The problem is outlined by Jensen and Mens (2001), with the "laterally repeated ridges" representing *Monomorphichnus* scratch marks formed by the right and left side walking legs of the trilobite, respectively:

> In sections worldwide, arthropod-type trace fossils, such as *Rusophycus*, generally appear significantly down-section of the first trilobite body fossils. In particular *Monomorphichnus*, sets of occasionally laterally repeated ridges occur close to the Precambrian-Cambrian boundary in many sections (Crimes 1987).

I first described the La Ciénega Formation structure in my dissertation (McMenamin 1984) as follows:

> This is the oldest trilobitoid arthropod trace fossil known from the Caborca region and was found in a green platy to fissile siltstone in which other trace fossils are rare. This trace is interpreted here as the rusophycid form of *Monomorphichnus multilineatus* Alpert (1976), hence the shared trivial designation. The ridges (interpreted as scratches) in the La Ciénega Formation trace are most distinct and widest in the center and thin towards the edge, as in *M. multilineatus* from the upper member of the Deep Spring Formation (especially Pl. 1, fig. 2 in Alpert 1976).

This pattern of ridges or grooves is a better match to the (centrally located) major groove and lateral minor groove arrangement of Savazzi (2015). Nevertheless, the linkage that I had made between the La Ciénega Formation and Deep Spring Formation structures was in fact correct. They can now be shown to both be

Fig. 3.5 Detrital microcline feldspar with tartan twinning in thin section (center of image) surrounded by mostly straight extinction monocrystalline quartz grains. Sandstone, Unit 4 of the Clemente Formation, Sonora, México. Greatest dimension of feldspar grain 960 microns. Specimen with *Palaeophycus tubularis*; field sample 7 of 3/16/95 (Photograph by Mark McMenamin)

Eophyton drag structures. Close inspection of Alpert's (1976, his Plate 1, Fig. 2) image confirms this, as a subsidiary set of Eophyton parallel grooves comes in to the main set at a fairly high angle (just below the center of Alpert's photograph).

This same arrangement, at nearly the same angle, is seen in the "Eophyton toolmarks from the early [*sic*] Cambrian Mickwitzia Sandstone of Sweden" (Savazzi 2015, his Fig. 6E), and also, as discussed below, Eophyton from the Clemente Formation (Fig. 3.4). We may conclude that Eophyton toolmarks are characteristic for both the La Ciénega Formation and the Deep Spring Formation, and Alpert's (1976) ichnotaxa list must be amended accordingly. These Eophytons are probably at least partly biogenic structures, but the identity of the maker of the drag markings remains problematic. Whether the name Eophyton should be italicized depends on whether one considers it to be a primarily inorganic structure caused by physical process (no italics) or whether one considers it to be a primarily biogenic structure, in other words, a true trace fossil or ichnofossil (in which case italics are required). Seilacher (1994) emphasizes Eophyton's "essentially physical nature," calls it a "pseudofossil," and remarks with his characteristic enthusiasm that it is "so common in the Lower Cambrian of the Baltic region…we have the rare case that a mechanoglyph becomes a stratigraphic index!"

It should be noted that the "drag of a heavy sand skeleton" interpretation (Seilacher 1994) for Eophyton is far from completely satisfactory. Even Savazzi (2015) laments "that some of the available specimens are difficult to interpret…[I] have decided to apply caution in interpreting instances of furrows that display unusual characteristics…a degree of uncertainty remains." In discussing a striking Eophyton slab from the Mickwitzia Sandstone of Lugnås (Riksmuseet, Stockholm), Seilacher (1994) describes how a pair of Eophyton drag marks move in parallel to make a parabolic turn that creates a pattern resembling the "golden arches" of a McDonald's restaurant! Current drag of a stalked organism does seem to be the best interpretation of Eophyton structures at present, but as Savazzi (2015) implies, new evidence could unexpectedly revise this interpretation.

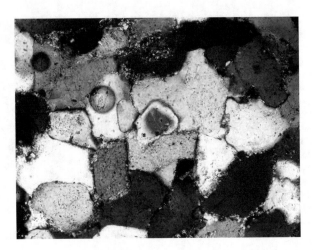

Fig. 3.6 Detrital accessory mineral, possibly zircon, as seen in thin section, surrounded by mostly straight extinction monocrystalline quartz. Sandstone, Unit 4 of the Clemente Formation, Sonora, México. Greatest dimension of accessory mineral as seen in thin section 360 microns. Specimen with *Palaeophycus tubularis*; field sample 7 of 3/16/95 (Photograph by Mark McMenamin)

Fig. 3.7 Detrital accessory mineral, possibly tourmaline, as seen in thin section in cross-polarized light. Note thin, colorless overgrowth (*cf.* Pettijohn et al. (1972), p. 388), surrounded by mostly straight extinction monocrystalline quartz. Note also the roughly triangular cross section of the presumed tourmaline crystal. Sandstone, Unit 4 of the Clemente Formation, Sonora, México. Greatest dimension of accessory mineral as seen in thin section 460 microns. Specimen with *Palaeophycus tubularis*; field sample 7 of 3/16/95 (Photograph by Mark McMenamin)

Ediacarans of the Clemente Formation occur in a shallow marine setting, characterized by muds, siltstones and sandstones. Small scale cross beds visible in thin section are delineated by opaque oxide layers. The mostly subangular monocrystalline quartz grains (most with straight extinction; some grains show wavy extinction

Fig. 3.8 Very fine sandstone in thin section seen with cross polars. Unit 4 of the Clemente Formation, Sonora, México. Subangular quartz (φ = 3.5; field sample 4 of 3/16/95); microcline feldspar in center of photomicrograph; flat white mica flake on lower right edge of photomicrograph. This rock constitutes the matrix of the Ediacaran holdfast *Sekwia* sp. (IGM 7454). Width of view 1.2 mm (Photograph by Mark McMenamin)

Fig. 3.9 Cathodoluminescence in fine sandstone in thin section. Unit 4 of the Clemente Formation, Sonora, México. Matrix of *Sekwia* sp. (4 of 3/16/95; IGM 7454). Note blue and orange luminescing grains. Width of view approximately 1.5 mm (Photograph by Mark McMenamin)

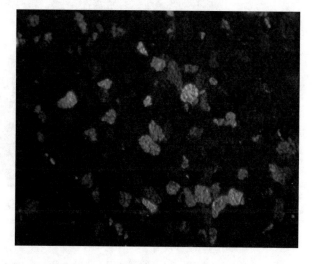

under cross polars [metamorphic source], plus rare polycrystalline grains) are associated with scattered feldspar grains, some with microcline tartan twinning (Fig. 3.5; this suggests continental crustal granitoid source), and some white mica flakes (these are often associated with the opaque layers). Prominent syntaxial overgrowths are common on quartz grains. The rare polycrystalline quartz grains may have very diffuse boundaries between the constituent subgrains. Rare accessory minerals (Fig. 3.6; possibly zircon, and; Fig. 3.7; possibly tourmaline) occur in the sandstone (field sample 7 of 3/16/95).

Fig. 3.10 Siltstone (6 of
3/16/95) in thin section
seen with cross polars.
Unit 4 of the Clemente
Formation, Sonora,
México. Width of view
1.2 mm (Photograph by
Mark McMenamin)

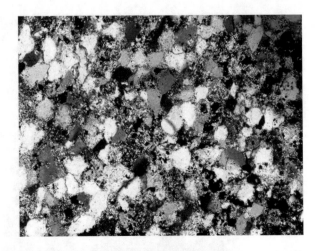

Fig. 3.11 Cathodoluminescence in siltstone seen in thin section (6 of 3/16/95). Unit 4 of the
Clemente Formation, Sonora, México. Note blue and orange luminescing grains. Width of view
approximately 1.5 mm (Photograph by Mark McMenamin)

The sediment enclosing the fossils ranges from mostly monocrystalline quartz
(Figs. 3.8 and 3.9; very fine sand, $\phi = 3.5$; IGM 7454; 4 of 3/16/95) to approximately 50% subangular monocrystalline quartz with 50% cherty-clayey matrix
(Figs. 3.10 and 3.11; 6 of 3/16/95; small scale grading visible on the thin section).
Blue-luminescing grains and some orange-luminescing grains appear in thin section
under cathodoluminesence (Figs. 3.9 and 3.11). The matrix of the fragmentary erni-
ettid (field sample 5 of 3/16/95; McMenamin 1996) has some very fine sand sized
cryptocrystalline (cherty) lithic fragments and little to no feldspar. The rock does
not fizz with dilute hydrochloric acid, so its carbonate content must be low.

Low to high-angle cross beds 10–25 cm thick (medium grained sand, $\phi = 1.5$–1.0; field sample 7 of 3/16/95) and Eophyton drag mark/tool mark structures (Fig. 3.4; parallel hyporelief grooves with "roughly constant reciprocal spacing", Savazzi 2015) indicate moderate to energetic current velocities. Starved ripples stabilized by microbial mats occur in the Clemente Formation 15–20 meters above the Clemente oolite (McMenamin 2016; field sample 8 of 3/16/95). I had earlier interpreted the Eophyton structures as the arthropod ichnofossil *Monomorphichnus* (McMenamin 2011; field specimens 6 of 3/16/95 (Fig. 3.4) and 7′ of 3/16/95), but it now seems clear that we are dealing with a grooved drag marking as per Savazzi (2015). Other types of drag marks (not Eophyton) also occur in the Clemente Formation (Stewart et al. 1984). This revision of the Clemente Formation structures from *Monomorphichnus* to Eophyton actually reinforces the biostratigraphic placement proposed here for the Clemente Biota, as *Monomorphichnus* is considered characteristic for Paleozoic marine strata and Eophyton could have a potentially much longer range extending far back in to the Proterozoic.

Finally, mention must be made of the enigmatic trace fossil *Vermiforma*. As I noted in *Dynamic Paleontology* (McMenamin 2016), *Vermiforma* "is known with confidence from only one site" in the Carolina Slate Belt. Specimens from the Clemente Formation were originally interpreted as "probable metazoan traces" (McMenamin et al. 1983). These were later interpreted as pseudofossils (McMenamin 1984; McMenamin and Schulte McMenamin 1990), but I returned to the ichnofossil interpretation in 1996, concluding that these were *Vermiforma* ichnofossils (McMenamin 1996, 1998). These are indeed ichnofossils, but their identification as *Vermiforma* is quite provisional at this point so they will be referred to here as *?Vermiforma*.

Ediacaran body fossils known to occur in the Clemente Formation (Table 3.1) include *Aspidella* sp. (Fig. 3.12), *Sekwia* sp. (Fig. 3.13), several species of erniettomorphs, and *Clementechiton sonorensis* (Fig. 1.2), plus new forms described below.

Aspidella sp. and *Sekwia* represent holdfasts of Ediacaran fronds (the upper parts of the fronds were not preserved). *Aspidella* is both the first-to-be named Ediacaran fossil (Billings 1872) and the most abundant Ediacaran body fossil, however, the taxon itself is likely a "garbage can" taxon that includes holdfasts and other circular impressions formed by a variety of different Ediacaran organisms (MacGabhann 2007; Bykova et al. 2017) rather than representing a single biological genus as implied by Gehling et al. (2000). The Sonoran *Aspidella* sp. is very similar to an *Aspidella* specimen identified from the White Sea biota of Russia (Ivantsov and Leonov 2008).

Of the two different species of Sonoran erniettomorphs, the first (erniettomorph I) develops pointed tubular module ends (and hence a serrate margin; McMenamin 1996, Fig. 2C therein). The second (erniettomorph II) develops swollen tubular module ends (Fig. 3.14). As both specimens are incomplete, they remain here in open nomenclature.

Table 3.1 The Clemente biota: Proterozoic fossils and ichnofossils of the Clemente Formation

Taxon/IGM number	Interpretation	Illustrations	Previous name(s)
Aspidella sp. IGM 7455	Holdfast for large Ediacaran frond	Figure 3.12	Cf. *Cyclomedusa plana, Evandavia aureola*
Sekwia sp. IGM 7454	Holdfast for small Ediacaran frond	Figure 3.13	*Beltanelloides sorichevae*
Erniettomorph 1 IGM 5001	Erniettomorph with pointed tube terminations; external mold of tubular modules	McMenamin (1996, his Fig. 2C)	Erniettid
Erniettomorph 2 IGM 7456	Erniettomorph with swollen tubular module ends; radiating tubes filled with very fine sand	Figure 3.14	Hiemaloriid Ediacaran
Pteridinium cf. *P. simplex* IGM 7460	Erniettomorph frond with broad medial zone (*cf.* Glaessner and Wade 1966, p. 617)	Figures 3.15 and 3.16	Chiton (McMenamin 2011)
Beothukis cf. *B. mistakensis* IGM 4493-4494	Rangeomorph frond; recumbent, spatulate sand frond	Figures 3.17,3.18, 3.19, and 3.20	None
Kimberella cf. *K. quadrata* IGM 4998	Kimberellomorph	Figures 3.22, 3.23, 3.24, and 3.25	None
Zirabagtaria ovata n. gen. n. sp. IGM 4995	Solzid kimberellomorph	Figures 3.26, 3.27, 3.46, and 3.47	None
Palankiras palmeri n. gen. n. sp. IGM 4997	Praecambridiid bilateralomorph	Figures 3.28,3.29, 3.30, and 3.31	None
Vendamonia truncata n. gen. n. sp. IGM 4995-4996	Praecambridiid bilateralomorph	Figures 3.33, 3.34, 3.35, and 3.36	Tomopteroid
Korifogrammia clementensis n. gen. n. sp. IGM 4998	Aculiferan (?stem)	Figures 3.22, 3.38, 3.39, 3.40, 3.41, 3.42, and 3.43	None
Clementechiton sonorensis IGM 7461	Aculiferan (?stem), possibly a chiton	Figure 1.2	Chiton
Eophyton IGM 5003	Toolmark/ichnofossil	Figure 3.4	*Monomorphichnus*
Lozenge and tear-drop structures IGM 4999	Dubiofossils; toolmarks or ichnofossil (submat burrows?)	McMenamin (1996, his Fig. 2D);	*Lockeia* sp.
?*Vermiforma* IGM 363	Ichnofossil	McMenamin (1998, his Figs. 3.2 and 3.3)	Probable metazoan traces, pseudofossils, *Vermiforma*

(continued)

Table 3.1 (continued)

Taxon/IGM number	Interpretation	Illustrations	Previous name(s)
Palaeophycus tubularis IGM 5000	Ichnofossil	McMenamin (1996, his Fig. 2E)	None

Fig. 3.12 *Aspidella* sp. Sole of bed. IGM 7455. Width of specimen 5.2 cm (Photograph by Mark McMenamin)

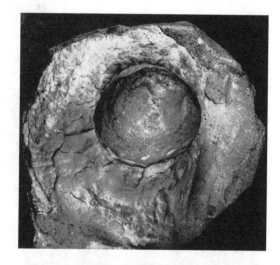

Fig. 3.13 *Sekwia* sp. Sole of bed. Asymmetry of the fossil is probably caused by current tug on the Ediacaran stalk and frond that was anchored by this holdfast structure. IGM 7454. Width of fossil 1 cm (Photograph by Mark McMenamin)

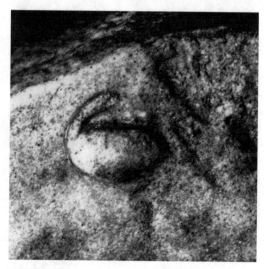

Certain comparisons with previously described forms can nevertheless be made. Erniettomorph I compares closely to an erniettomorph specimen with serrate edge recovered from southern Nevada (McMenamin 1998, Fig. 2.16 therein). Erniettomorph II (Fig. 3.14) shows smaller modules or sand tubes intercalated between larger, longer modules, all with wavy/undulose/bulging tube edges, very similar to the "Erniobeta"

Fig. 3.14 Erniettomorph
II with swollen tubular
module ends. Sole of bed.
IGM 7456. Greatest
dimension of fossil 3.4 cm
(Photograph by Mark
McMenamin)

Fig. 3.15 *Pteridinium* cf.
P. simplex with broad
medial zone (left edge; cf.
Glaessner and Wade 1966,
p. 617) and four isomers
visible. The red matrix
visible on the left edge of
the specimen is the terra
rosa shown (as the opposite
side of this specimen) in
Fig. 2.13. IGM 7460, field
specimen 8 of 3/16/95.
Scale bar in mm
(Photograph by Mark
McMenamin)

Fig. 3.16 *Pteridinium* cf.
P. simplex. Sketch of
specimen in previous
image. Note partial
impression of second
specimen at lower left.
IGM 7460. Greatest
dimension of specimen
2.5 cm

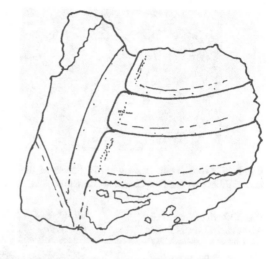

Fig. 3.17 *Beothukis* cf. *B. mistakensis.* Photograph of nearly complete specimen. IGM 4993.
Scale bar in cm (Photograph by Mark McMenamin)

variant of erniettomorphs (Pflug 1972). Plate 39, Fig. 4 and Plate 39, Fig. 8 (upper half
of specimen) of Pflug (1972) clearly portray this distinctive bulging and intercalation
of the tube modules. Pflug's (1972) first illustration mentioned in the previous
sentence is reproduced as fig. 13,4a in Glaessner (1979). It is possible that erniettomorph I and erniettomorph II represent members of the two subfamilies defined by
Pflug (1972), the Erniettinae and the Erniobetinae, respectively. Future discovery of
more complete material may help resolve this question.

    In addition to *?Vermiforma*, trace fossils in the Clemente biota are limited to the
horizontal burrow *Palaeophycus tubularis*. The latter ichnofossil (field sample 7 of
3/16/95) constituted part of the original report on the Clemente Biota (McMenamin
1996), and analysis of the peristaltic character of the burrow confirmed the ichnofossil identity of the specimen (McMenamin 2006). Additional specimens of this
ichnofossil show two horizontal *Palaeophycus tubularis* burrows crossing at

Fig. 3.18 *Beothukis* cf. *B. mistakensis.* Sketch of frond in previous image. This is the largest known Ediacaran from the Clemente Formation biota. Greatest dimension of fossil 10.1 cm

Fig. 3.19 *Beothukis* cf. *B. mistakensis.* Detail of sand-filled modules. IGM 4993. Scale bar in cm (Photograph by Mark McMenamin)

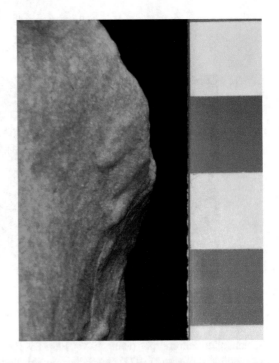

approximately right angles (field samples 6 of 3/16/95 [IGM 5004] and possibly 7′ of 3/16/95 [IGM 5002]). Thus both peristaltic and non-peristaltic animal burrows (the latter presumed to have been formed by nematodes; Parry et al. 2017) are known from Ediacaran strata.

Here I report seven additional creatures from the Clemente biota, including *Pteridinium* cf. *P. simplex* (Figs. 3.15 and 3.16), a recumbent sand frond *Beothukis* cf. *B. mistakensis* (Figs. 3.17, 3.18, 3.19, and 3.20), the kimberellomorph *Kimberella* cf. *K. quadrata* (Figs. 3.22, 3.23, 3.24, and 3.25), the solzid kimberellomorph *Zirabagtaria ovata* n. gen. n. sp. (Figs. 3.26 and 3.27), the praecambridiid *Palankiras palmeri* n. gen. n. sp. (Figs. 3.28, 3.29, 3.30, and 3.31), *Vendamonia truncata* n. gen.

Fig. 3.20 *Beothukis* cf. *B. mistakensis*. Partial frond with section of medial groove preserved. Field sample 6b of 3/16/95; IGM 4994 (Photograph by Mark McMenamin)

Fig. 3.21 *Beothukis mistakensis* from the Avalon assemblage of Mistaken Point, Newfoundland, fiberglass cast by A. Seilacher. Scale bar in cm (Photograph by Mark McMenamin)

n. sp. (Figs. 3.33, 3.34, 3.35, and 3.36), and *Korifogrammia clementensis* n. gen. n. sp. (Figs. 3.38, 3.39, 3.40, 3.41, 3.42, and 3.43).

Pteridinium cf. *P. simplex* is shown in Figs. 3.15 and 3.16. The fossil shows four modules on the right side of a frond, plus the medial region of the frond. The left side of the frond is not preserved. The specimen is somewhat unusual for the genus, as instead of the usual zig-zag medial suture between the modules of the left and right sides of the frond, there is a smooth area with a low keel running parallel to the proximal edge of the right side of frond partitions. Glaessner and Wade (1966), in their description of Australian specimens of *Pteridinium* cf. *P. simplex*, note that a "broad (5 mm.) median zone was found in one specimen...instead of the medial narrow zigzag line." One of their illustrations shows a straight proximal edge to the right side modules, and an upward module curvature (Glaessner and Wade 1966,

Fig. 3.22 *Kimberella* cf.
K. quadrata (two
specimens, center left) and
*Korifogrammia
clementensis* n. gen. n. sp.
(holotype below center
right; anterior of organism
pointed towards base of
photograph) on the same
bedding plane (bedding
sole surface). IGM 4998.
Scale bar in cm
(Photograph by Mark
McMenamin)

their Plate 101, Fig. 3, upper specimen), that is virtually identical to what is seen on
the Sonoran *Pteridinium* specimen (Figs. 3.15 and 3.16). Preservation of this specimen appears to be by early diagenetic cementation with silica (Tarhan et al. 2016).

A recumbent sand frond with irregular primary frond divisions (*Beothukis* cf. *B.
mistakensis*) is shown in Figs. 3.17, 3.18, 3.19, and 3.20. The Sonoran examples
consist of two specimens, one largely complete (Figs. 3.17, 3.18, and 3.19; IGM
4993) and a portion of a left side of a frond (Fig. 3.20; IGM 4994). The larger
specimen is the largest known Ediacaran from the Clemente Formation biota, and
exceeds 10 cm in length.

These fossils are relatively smooth and featureless compared to classic *Beothukis
mistakensis* from the E surface at Mistaken Point, Newfoundland (Fig. 3.21; Brasier
and Antcliffe 2009). However, this is also true for a specimen assigned to *Beothukis
mistakensis* from South Australia (South Australia Museum SAM P41109;
Laflamme et al. 2013). These differences are taken to represent various modes of
preservation (Narbonne 2005), with relatively featureless specimens associated
with more sand-rich depositional environments (Laflamme et al. 2013). This seems
reasonable, as gravitational settling of sand in a sand-filled frond would tend to
stretch the cuticle and smooth out the detailed folds and crenulations characteristic
of the Newfoundland *Beothukis* (these are preserved in much finer-grained sediment, with surface relief possibly enhanced by perimortem contraction?; Brasier
and Antcliffe 2009). Rangeomorph detail is nevertheless still visible on the Sonoran
specimens of *Beothukis* cf. *B. mistakensis*; the medial line can be seen (Figs. 3.17,
3.18, and 3.19) as can as series of bluntly pointed modular partitions (Fig. 3.20).
Also, these specimens show the primary characteristic of *Beothukis*, namely,
irregular primary frond divisions that are often oriented at fairly large angles to one
another rather than being subparallel as in *Rangea*.

Interestingly, the recumbent habit of *Beothukis* cf. *B. mistakensis* would accord
with light capture and a photosymbiotic lifestyle as has also been proposed for
Pteridinium specimens from Namibia (McMenamin 1998). Meyer et al. (2014b)

Fig. 3.23 *Kimberella* cf.
K. quadrata. Sketch of
specimen in previous
image. Length of organism
1.5 cm

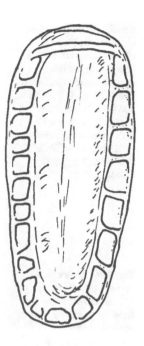

Fig. 3.24 *Kimberella* cf.
K. quadrata. SEM
photomicrograph of edge
crenulations. IGM 4998.
Scale bar 1 mm
(Photograph by Mark
McMenamin)

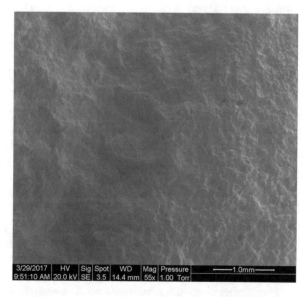

agreed with McMenamin (1998) that *Pteridinium*, with its "very flexible integument," probably had a "semi-endobenthic or epibenthic lifestyle" (Meyer et al. 2014b). Jarosite (an Fe-bearing sulfate mineral) crystals encrust the surface of *Rangea* fronds associated with *Pteridinium*, suggesting that both genera "lived within an oxygenated estuarine or fluvial setting and were transported during storms

Fig. 3.25 *Kimberella* cf.
K. quadrata, undulose
edge of specimen as seen
in petrographic thin
section. Sandstone, Unit 4
of the Clemente Formation,
Sonora, México. Scale bar
in mm. Specimen with
Palaeophycus tubularis;
field sample 7 of 3/16/95
(Photograph by Mark
McMenamin)

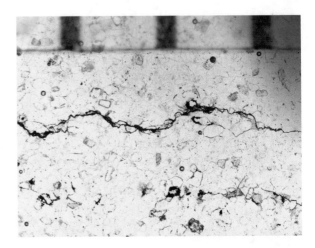

Fig. 3.26 *Zirabagtaria
ovata* n. gen. n. sp.
Concave hyporelief
showing relatively smooth
cephalic region and
post-cephalic region
covered with papillae or
weakly-mineralized
sclerites. Many of the
papillae show transverse
elongation, and are
organized in longitudinal
columns, especially in the
central (axial lobe?) part of
the organism. Scale bar in
mm. IGM 4995; field
sample 6 of 3/16/95
(Photograph by Mark
McMenamin)

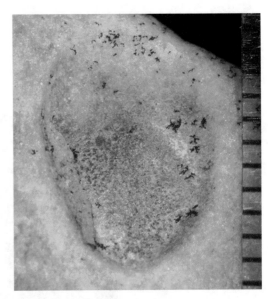

to anoxic, ferruginous environments where they were exquisitely preserved" and
experienced "rapid encrustation of pyrite on the surface of the organisms as they
decomposed and were consumed by sulfate-reducing bacteria within the sandy, near
shore sediments" (Hall et al. 2013).

The kimberellomorph *Kimberella* cf. *K. quadrata* is shown in Figs. 3.22, 3.23,
3.24, and 3.25. It shows typical preservation style for *Kimberella*, namely, an
impression that occurs at a shale/clay to fine sandy bed transition (Ivantsov 2009).
In the Sonoran specimens (n = 2) we see the underside of the organisms as a convex
hyporelief. Edge crenulations, characteristic for *Kimberella*, can be seen in SEM
photomicrograph (Fig. 3.24) and in thin section (Fig. 3.25), preserved as an undulose

Fig. 3.27 *Zirabagtaria ovata* n. gen. n. sp. Reconstruction of organism in previous image. Greatest dimension of specimen 10.5 mm

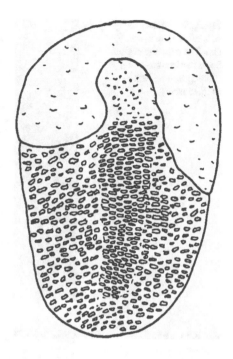

Fig. 3.28 *Palankiras palmeri* n. gen. n. sp. Photograph of entire specimen. Scale bar in mm (Photograph by Mark McMenamin)

iron oxide and/or clay rich layer. The crenulations as preserved here are quite similar to those illustrated by Ivantsov (2007, his Plate 2, Fig. 1) for the small Vendian transversely articulated fossil *Lossinia lissetskii*, but the Sonoran specimen lacks a well-defined head region and is therefore assigned here to *Kimberella* cf. *K. quadrata*.

Fig. 3.29 *Palankiras palmeri* n. gen. n. sp. Detail of cephalic region. Scale bar in mm (Photograph by Mark McMenamin)

Fig. 3.30 *Palankiras palmeri* n. gen. n. sp. SEM image of anterior lobe or 'glabella.' The image is rotated clockwise with respect to the long axis of the organism. Scale bar 2 mm (Photograph by Mark McMenamin)

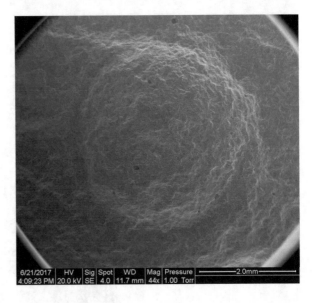

Zirabagtaria ovata n. gen. n. sp. (IGM 4995) is shown in Figs. 3.26 and 3.27. It is a solzid kimberellomorph with an 'oval' body shape (*sensu* Schwabe 2010). Its genus name refers to the riveted mail and plate armor of Mughal India, the *zirah bagtar* or *zirah bakhtar*, and its species name refers to the ovoid outline of the fossil. Its preservation is similar to that of *Korifogrammia clementensis* n. gen. n. sp., namely preservation as a concave hyporelief. The specimen is 10.5 mm long and 7 mm wide. The anterior 3.5 mm consists of a largely smooth head region. The rest of the body lacks a dorsal keel but has a very faint, slightly curved axial lobe (?) and

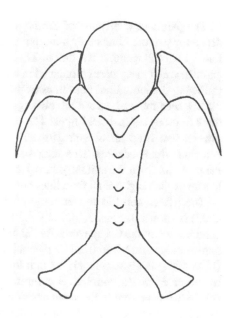

Fig. 3.31 *Palankiras palmeri* n. gen. n. sp. Sketch of reconstructed organism. Greatest dimension of specimen 18 mm

is covered by numerous dorsal epidermal papillae or weakly mineralized sclerites that are in many cases elongated in a direction perpendicular to the main body axis. The papillae resemble tiny rectangular tubercles. Many of the papillae show transverse elongation, and are organized in longitudinal columns, especially in the central (axial lobe?) part of the organism. There are approximately ten longitudinal columns of mostly transversely-elongated papillae in the axial lobe region, and roughly 8–10 irregular columns on either side of the axial lobe in the 'pleural' fields. This gives a rough maximum of up to 30 longitudinal rows of sclerites. The papillae/scerites were preserved by a thin parting of clay (now shale) that settled on the dorsal surface of the *Zirabagtaria* before burial in sand. The presumably aragonitic sclerites (if they were mineralized at all) were subsequently dissolved out and replaced by the same siliceous cement that lithified the sandstone. Ivanstov (2017) notes that *Kimberella quadrata* bore a dorsal protective cover "armored with fine sclerites, apparently mineral, but rapidly dissolving after burial." This comment also applies to *Zirabagtaria*.

The array of papillae in *Zirabagtaria ovata* n. gen. n. sp. is somewhat similar to the tubercles described by Ivantsov (2007, his Pl. 2, Figs. 1–3) of *Lossinia lissetskii*, although unlike *Lossinia*, tubercles on the cephalic region of *Zirabagtaria ovata* n. gen. n. sp. are sparse and the 'scleritome' is mostly restricted to the post-cranial region. The *Zirabagtaria* papillae are also somewhat similar to those of *Solza margarita*, however, those in *Zirabagtaria* are arrayed in longitudinal columns whereas those in *Solza* have a more radial arrangement (Ivantsov and Leonov 2008). Note that "*Praecambridium*" *sigillum* on p. 283 in Fedonkin et al. (2007) may also be a solzid kimberellomorph. Interestingly, a lopsided cephalic region, reminiscent of and perhaps homologous (McMenamin 2003b) to the effaced cephalon of an agraulid trilobite (McMenamin 2004), is apparent on both *Lossinia lissetskii* Ivantsov (2007, his Pl. 2, Fig. 3) and *Zirabagtaria ovata* n. gen. n. sp. (Figs. 3.26 and 3.27).

The papillate scleritome of *Zirabagtaria ovata* n. gen. n. sp. shows clear evidence for strong influence by a morphogenetic field in accordance with the Second Law of Morphogenetic Evolution. This evidence is twofold. First, the individual papillae are in many cases elongated in a transverse direction, and thus appear to be following the latitudinal field lines of the toroidal metazoan arrangement. Second, the sclerites themselves are largely organized into longitudinal columns, and thus track the longitudinal field lines. The papillae at the posterior tip of *Zirabagtaria ovata* n. gen. n. sp. are of approximately the same size as the rest of the sclerites that constitute the scleritome. This suggests that the *Zirabagtaria* anus is not directly at the posterior tip of the animal, otherwise the posterior most sclerites would diminish in size as the longitudinal field lines converge at the posterior pole of the torus.

The praecambridiid bilateralomorph *Palankiras palmeri* n. gen. n. sp. (IGM 4997) is shown in Figs. 3.28, 3.29, 3.30, and 3.31. Its genus name is formed by a combination of the Greek words for "ancient" and "anchor," and its trivial or species name honors American trilobite paleontologist A. R. "Pete" Palmer. The holotype (IGM 4997) is approximately 18 mm long and 11 mm wide. The specimen is preserved as a convex epirelief. A prominent and well preserved anterior axial lobe (Fig. 3.30) is present in the same position as a glabella would be on a trilobite. Also prominent is a genal spine that arches downward almost to the midline of the animal (Fig. 3.29). The incomplete nature of the fossil suggests the possibility that it represents a shed molt, which if so would make it the oldest known such structure, and the oldest direct evidence for ecdysis, in the fossil record.

An illustration of *Praecambridium* in Glaessner and Wade (1971, their Figure 1C) shows certain similarities to the Sonoran specimen shown in Figs. 3.28, 3.29, 3.30, and 3.31, particularly as regards the first 'metamere' or isomere partition. In both *Praecambridium sigillum* and *Palankiras palmeri* n. gen. n. sp. the first partition is drawn out into an arching spine or spine-like structure. These features in *Palankiras palmeri* n. gen. n. sp. confirm inferences regarding the ecdysozoan aspect of *Spriggina*, particularly as regards the presence of trilobitoid features on this distinctive Ediacaran fossil (McMenamin 2003b):

> Convincingly specific homologies between Ediacarans and members of recognized animal phyla have remained elusive, thus casting a shadow of doubt over the animal affinity interpretation of Ediacaran phylogeny. *Spriggina floundersi*, for example, has been described as a tomopterid annelid, an arthropod, and a frondose vendobiont. Reanalysis of *Spriggina* demonstrates the presence of genal spines (comparable to those of fallotaspidoid and paradoxidid trilobites), a cephalic region homologous to the effaced cephalons of agraulid and skehanid trilobites, and a dual cephalic region…that compares to the parts of a trilobite cephalon anterior and posterior of the facial suture. *Spriggina* is thus a trilobitoid ecdysozoan…This result is among the first confident phylogenetic linkages between an Ediacaran and a Cambrian animal, and thus helps to demonstrate that Paleozoic animals could indeed be descended from Ediacarans. If *Spriggina* is a direct ancestor of trilobites, then a case can be made that *Spriggina* itself (or a direct descendant) served as the predator taxon that initiated the Cambrian ecotone transformation. Such heterotrophy, however, does not preclude the possibility of photosymbiosis or other types of mixotrophy in Ediacarans with high surface area such as *Marywadea* and *Dickinsonia*.

Fig. 3.32 *Deiphon forbesi,*
a cheirurid trilobite from
Silurian strata of St. Iwan,
Bohemia. Public domain
image from Zittel (1913).
Length of trilobite
approximately 2.5 cm

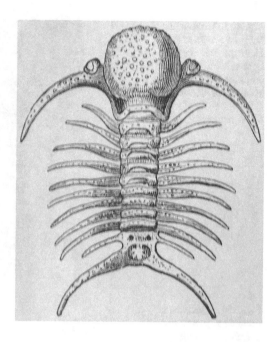

Three key morphological features indicate that these same arguments linking *Spriggina* to trilobites may be applied to *Palankiras palmeri* n. gen. n. sp. These are: presence of a genal spine, presence of a prominent axial lobe (glabella), and rudimentary facial sutures. The genal spine is clearly visible in Figs. 3.28 and 3.29. A linear depression at the proximal-posterior edge of the genal spine (Fig. 3.31) is interpreted here as a rudimentary facial suture. This interpretation strongly supports the concept that the new praecambridiid and *Spriggina* were both ecdysozoans able to molt. The prominent ovoid-spherical anterior lobe in the new praecambridiid, glabella-like in terms of its placement, is very reminiscent of the anterior lobe in *Praecambridium sigillum* as reconstructed by Glaessner and Wade (1971, their fig. 3).

The overall body form of *Palankiras palmeri* n. gen. n. sp. (Fig. 3.31) is very intriguing. Several trilobite orders have produced strange looking forms with what has been called a 'classic fish skeleton' appearance: large spherical glabella ("eye of fish"), long curved genal spines ("skull of fish"), separated pleural segments ("fish ribs") and two prominent spines of the pygidium ("fish tail"). These trilobites include the Silurian cheirurid *Deiphon forbesi* (Fig. 3.32) and the Devonian giant lichid trilobite (>60 cm long) *Terataspis grandis*. No one has yet found a complete specimen of *Terataspis grandis* but the search continues. The similar and odd body form of *Deiphon* and *Terataspis*, which by the way are not closely related trilobites, strongly suggests that their joint similarity to *Palankiras palmeri* n. gen. n. sp. is a result of two separate, convergent cases of evolutionary atavism. *Palankiras palmeri* n. gen. n. sp. shares the entire "fish skeleton" outline, lacking only the 'ribs' (separated pleural segments). Curiously, the "fish skeleton" trilobite form is unknown

Fig. 3.33 *Vendamonia truncata* n. gen. n. sp. Holotype. Bedding sole hyporelief. Scale bar in cm. IGM 4996; field sample 4 of 3/16/95 (Photograph by Mark McMenamin)

Fig. 3.34 *Vendamonia truncata* n. gen. n. sp. Sketch of holotype. Abbreviations: **cb**, cirri base; **bu**, buccal cavity; **ba**, bifurcate distal appendage. Greatest dimension of specimen 21 mm

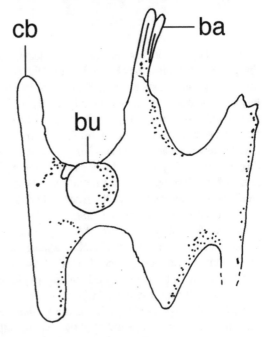

from Cambrian strata. The evolutionarily-convergent atavisms thus appear to be post-Cambrian phenomena of trilobite evolution.

Vendamonia truncata n. gen. n. sp. (Figs. 3.33, 3.34, 3.35, and 3.36) represents a second type of praecambridiid bilateralomorph. I first reported *Vendamonia truncata* n. gen. n. sp. in 2006 (McMenamin 2006). The genus name is a combination of the Russian Vendian System and the Greek word (αμονι) for "anvil." The species

Fig. 3.35 *Vendamonia truncata* n. gen. n. sp. Juvenile specimen (Photograph by Mark McMenamin)

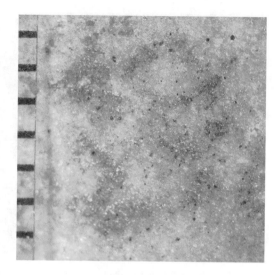

Fig. 3.36 *Vendamonia truncata* n. gen. n. sp. Sketch of specimen in previous image. Circular structure at the top of the image is interpreted here as the buccal cavity. Greatest dimension of specimen 5 mm

name refers to the oddly truncated posterior of the creature, a feature somewhat reminiscent of the truncated foliage of the desert plant *Haworthia truncata*.

The specimen is a soft-bodied impression in fine quartz sand on a bedding plane top surface. It is 23 mm in width and its length is 21 mm. The specimen is preserved as a convex epirelief. It consists of a bar-shaped structure (the parapodial component of the great cirri or cirri base; Meyer 1926), a domal anterior axial lobe (buccal cavity), and two metamers (neuropodial or parapodial pairs) posterior to the anterior axial lobe (Figs. 3.33 and 3.34). A bifurcate distal appendage extends from the left parapodium of the first postcephalic pair.

A second specimen of *Vendamonia truncata* n. gen. n. sp., an apparent juvenile, was discovered in June 2017 (Figs. 3.35 and 3.36). This fossil is 4.3 mm in width

Fig. 3.37 Modern *Tomopteris* (Photograph by Uwe Kils. Image used here per Creative Commons BY-SA 3.0 license)

and 5 mm in length. Initially this creature was identified as the anterior portion of a *Tomopteris*-like worm, with the posterior of the animal excised to form a straight posterior margin. However, discovery of the second specimen indicates that anvil-shaped body and flat posterior margin (bottom of the anvil) in fact accurately represent the entire outline of the body of the creature.

There could nevertheless still be a relationship between *Vendamonia truncata* n. gen. n. sp. and the bioluminescent polychaete worm *Tomopteris helgolandicus*. In living members of the Tomopteridae, the bioluminescence is associated with certain parapodial glands (Phillips Dales 1971), raising the intriguing possibility that if *Vendamonia truncata* n. gen. n. sp. is indeed a tomopteroid, it may imply the presence of Proterozoic animalian bioluminescence.

Living tomopterids are nectonic marine predators. They are usually classified as a unique group of polychaete worms, characterized by an elongate, flattened metameric body, paddle-shaped parapodia, and long anterior cirri (Fig. 3.37). Glaessner's (1958) suggestion of a phylogenetic link between the Ediacaran bilateralomorph *Spriggina* and modern tomopterids was challenged by subsequent researchers (Briggs and Clarkson 1987). *Vendamonia truncata* n. gen. n. sp. is remarkable both for its great antiquity, its size, and for the support that it may provide to Glaessner's (1959) original suggestions concerning the link between *Spriggina* and tomopterids.

Shortly after I found the holotype of *Vendamonia truncata* n. gen. n. sp. in the field, Dave Evans nicknamed it "the anvil" on account of its being shaped like an old-fashioned iron anvil. Note the appearance of a very similar anvil pattern in the anterior portion of *Praecambridium sigillum* as shown in Glaessner and Wade (1971), their Figure 1C. The anterior most first two "segments" in this image appear larger than the other isomers and separate from the posterior part of the fossil (Glaessner and Wade 1971); this part ends up looking like an anvil that has been so

Fig. 3.38 *Korifogrammia clementensis* n. gen. n. sp. Holotype. Anterior end of organism towards top of image. Note column of small valves running vertically in the image. Scale in mm. IGM 4998; field sample 7 of 3/16/95 (Photograph by Mark McMenamin)

Fig. 3.39 *Korifogrammia clementensis* n. gen. n. sp. Reconstruction of holotype. Maximum dimension of organism 1 cm

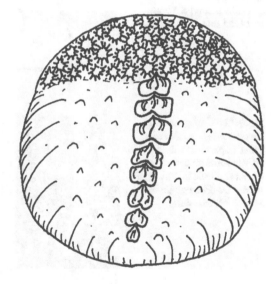

heavily pounded that its horns are bent. I propose that the tomopteroid had a similar morphology, with a smaller and fragile posterior segment series that separated from the more robust first two in the Sonoran fossil.

Korifogrammia clementensis n. gen. n. sp. is shown in Figs. 3.22, 3.38, 3.39, 3.40, 3.41, 3.42, and 3.43. It has a broadly oval body shape (Schwabe 2010), and is interpreted here as a chiton or chiton-like aculiferan. The genus name is derived from the Greek word (κοριφογραμμι) for "crest" or "ridge," and the species name is in reference to the Clemente Formation.

Fig. 3.40 *Korifogrammia clementensis* n. gen. n. sp. Trapezoidal valve IV on the holotype. Width of valve 1 mm (Photograph by Mark McMenamin)

Fig. 3.41 *Korifogrammia clementensis* n. gen. n. sp. Sketch of trapezoidal valve IV on the holotype. Width of valve 1 mm

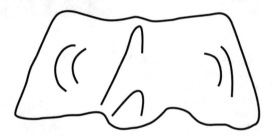

Fig. 3.42 *Korifogrammia clementensis* n. gen. n. sp. Three trapezoidal valves IV-VI visible on partial specimen. Scale in mm. IGM 4998 (Photograph by Mark McMenamin)

Korifogrammia clementensis n. gen. n. sp. occurs on the same slab of rock as *Kimberella* cf. *K. quadrata* (IGM 4998; field sample 7 of 3/16/95), but is preserved as a concave impression on the sole of the bed. The organism is shaped like a swollen circular disc 9 mm long and 9 mm wide. A dorsal medial keel divided into eight carinated trapezoidal valves (*sensu* Schwabe 2010) occurs in a sagittal position.

Fig. 3.43 *Korifogrammia clementensis* n. gen. n. sp. Partial specimen. Sketch of trapezoidal valves IV-VI in previous image. Width of widest approximately 750 microns

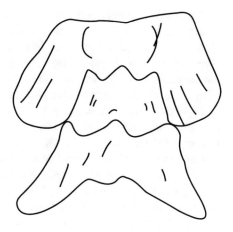

A lunate head region is present at the anterior end of the creature. A fragmentary specimen 750 microns long and 600 microns wide consists of three trapezoidal valves (valves IV–VI). Valve morphology is more distinct on the fragmentary specimen (Figs. 3.42 and 3.43).

Close inspection of the dorsal median keel shows that it is divided into eight small carinated mucro-anterior plates with beaked anterior margins (Figs. 3.38 and 3.39). These form a series of eight keeled impressions that constitute the dorsal medial ridge or crest. On the holotype, a slightly curved sagittal groove divides each plate roughly in half. Faint, paired lateral riblets are preserved on the right sides of the anterior most three plates. The lunate head region is covered with numerous stellate tubercles. These are best preserved on the left side of the specimen, but apparently covered the entire cephalic region (Fig. 3.39). The edge of the abdomen is marked by faint curving partitions that, adjacent to the edge, appear to break up into fine marginal spines. Very faint triangular impressions, probably tubercles, are seen in the pleural field between the valves and the curved sculpturing at the margin.

The stellate tubercles of the head region of *Korifogrammia clementensis* n. gen. n. sp. are reminiscent of similar structures seen in *Onega* and *Lossinia* from the White Sea Vendian biota (Ivantsov 2007). The spacing and arrangement of the stellate tubercles in *Korifogrammia clementensis* n. gen. n. sp. is similar to that of a "retracted" specimen of *Kimberella quadrata* from the White Sea biota (Ivantsov 2009). However, the tubercles in the Russian specimen cover the body of the animal and are not merely restricted to the head region. Thus *Korifogrammia clementensis* n. gen. n. sp., as it bears eight aculiferan (chiton and related forms) plates, appears to represent an intermediate form between mollusc-like Ediacarans such as *Kimberella* and the more arthropod- or annelid-like "Proarticulata" forms, and thus shows an intriguing mosaic of aculiferan and proarticulate features. *Korifogrammia clementensis* n. gen. n. sp. may potentially be ancestral to both crown mollusks and crown arthropods.

Alternatively, Vinther (2015) may be incorrect in claiming that crown Aculifera do not appear until the Ordovician. Basing his assessment on molecular clock data, Vinther (2015) "rejects an older, hidden history of crown chitons" and has them diverging at 450 million years ago. It may be time (yet again) to check the setting on the molecular clock. In any case, the concept of a monophyletic clade Aculifera has been confirmed as a well-resolved branch of molluscan phylogeny (Sigwart 2017).

Now that we have a more comprehensive list (including two aculiferans, *Clementechiton* and *Korifogrammia*) of Clemente biota taxa (this publication more than doubles the number of body fossil taxa from the Clemente Formation), it is possible to make proper assessment of the implications of the find. The preservational style of *Clementechiton sonorensis*, *Palankiras palmeri* n. gen. n. sp., *Zirabagtaria ovata* n. gen. n. sp., *Kimberella* cf. *K. quadrata*, *Korifogrammia clementensis* n. gen. n. sp. and *Vendamonia truncata* n. gen. n. sp. are quite similar; shallow impressions or low relief casts, in several cases preserving a surprising amount of fine detail, in a very fine to fine sandstone with siliceous cement.

It is now clear that the Clemente Biota is part of the White Sea assemblage (~555–550 Ma), primarily due to its age (now placed at 555 Ma thanks to biostratigraphy and the Shuram excursion). This assemblage assignment is in accord with the Clemente biota's high biotic complexity and diversity. The assemblage includes kimberellomorphs, bilateralomorphs, plus undoubted animals as represented by aculiferans and the trackway of a peristaltic burrower. It is important to note, however that the Clemente biota also includes a representative of the preceding Avalon assemblage (the rangeomorph *Beothukis* cf. *B. mistakensis*) plus representatives more characteristic for the subsequent Nama assemblage (erniettomorphs such as *Pteridinium* cf. *P. simplex*; Fig. 2.14). The Clemente biota, in paleobiogeographic terms, may be somewhat endemic as there is not a single demonstrable species level connection to any other region, with the possible exception of *Pteridinium* cf. *P. simplex* (Fig. 3.15), a fossil which closely resembles its Australian counterpart (Glaessner and Wade 1966).

Taphonomic considerations indicate that the animalian interpretation of many Ediacarans is becoming increasingly plausible. In their study of Ediacaran-style decay experiments with sea hares and anemones, Gibson et al. (2017) concluded:

> Our results indicate that labile tissues tend to decay faster than the rate of precipitation of FeS or aluminosilicate minerals, and that these dark 'halos' possibly represent diffusion of organics. Implications of these results are that Ediacaran tissues may have been less labile than previously understood, exhibiting a preservational bias against specific cnidarian and molluscan characters—affinities previously determined for some Ediacarans. Our study provides evidence for the likelihood of both diploblastic and triploblastic organisms during this time.

Ediacaran bilateralomorphs and undoubted animals with shells lived together in the Clemente biota in the same sea floor environment. We can look at this interesting fact in two different ways, and the dual perspective takes us right to the heart of the Ediacaran conundrum as it was initially posed by Pflug (1972; and associated papers), articulated by Sokolov (1965), agonized over by Glaessner and Wade (1971), and finally developed by Seilacher (1989) and others who were, fundamentally,

Fig. 3.44 *Pteridinium simplex*, Nama Group, Namibia. Cast showing twist in the body, probably due to soft sediment deformation. Scale in cm (Artificial cast provided by Hans Pflug; photograph by Mark McMenamin)

perplexed by the odd morphologies of organisms such as *Ernietta* as seen in the Nama assemblage. We might consider the bilateralomorph Ediacarans to be no close relatives to Eumetazoa, but rather representative of a massive case of convergent evolution. Alternatively, we can see a continuum, with bilateralomorph Ediacarans such as *Spriggina* related to, and even giving rise to, Eumetazoans such as trilobites. It is not yet possible to decide unequivocally between these two alternate interpretations of the data. Even those who favor the ancestral animal interpretation are forced to admit the unsettling, bizarre body plan of the rangeomorphs and erniettomorphs.

We can solve this problem by applying the Second and Third Laws of Morphogenetic Evolution. The strangeness of the rangeomorphs, the oldest of the Ediacarans, and many of the other Ediacarans as well such as the erniettomorphs, has to do with the extreme complexity of their morphogenetic field as manifest in their outer cuticle. Forms such as deformed *Pteridinium* (Figs. 3.44 and 3.45) and the strangely elongate *Andiva ivantsovi* (Ivantsov 2007; Ivantsov and Leonov 2008) show that Ediacaran cuticle had unusual properties of extension (perhaps due to an infolded fractal morphogenetic field ultrastructure?) and could be weirdly stretched. Grazhdankin (2004), in discussing fossils "preserved within the convoluted sandstone as casts of randomly oriented bodies engulfed by the sediment," notes that "preservation in the absence of microbial biofilms—such as seen in winnowed examples of *Inaria* or in three-dimensional casts—is by no means inferior to that seen in association with the microbial textures." Grazhdankin's (2004) observations strengthen the concept of the strangeness of the Ediacaran cuticle at the expense of other taphonomic models such as the death mask hypothesis (Gehling 2004). Either the death mask was not a factor in Ediacaran preservation (Meyer et al. 2014a: "any microbes growing on *P. simplex* vanes within mass flow deposits were unlikely to have formed thick mats as envisioned in the death mask model") or strange films or

Fig. 3.45 *Pteridinium simplex*, Nama Group, Namibia. Cast showing extreme elongation to the Ediacaran cuticle, probably due to soft sediment deformation. Scale in cm (Artificial cast provided by Hans Pflug; photograph by Mark McMenamin)

membrane-like structures (part of the *Pteridinium* or fragments of biofilm?; Elliot et al. 2011) were a factor in *Pteridinium* preservation. Continuing research will hopefully clarify this issue.

The cuticle question has posed severe difficulties for Ediacaran interpretation. A specimen identified as *Rangea schneiderhoehni* from Namibia (McMenamin 1998, his Fig. 4.9) was subsequently reinterpreted as a problematic fossil "showing some similarity to *Rangea*" (Grazhdankin and Seilacher 2005, their Fig. 8a). I disagree with the latter interpretation. These types of fossils are all *Rangea*, and represent expressions of the extreme complexity of the morphogenetic field in this early organism. This is an expected outcome of the Second and Third Laws of Morphogenetic Evolution. In the case of rangeomorphs, fronds are literally spilling out of the creature's complex cuticle, joined at their tips in some cases as a representation of convergence of some of the complex field lines. Consider the illustration selected to portray *Rangea* in the *Treatise on Invertebrate Paleontology* (fig. 13,4a in Glaessner (1979)). Here we see a *Rangea* frond with each primary division or branch of the frond resembling a mature frond. Furthermore, the stalk of the frond is also a frond in its own right, and shows its own integrity as a subfrond (thus resembling one of the branches itself) even as the large primary branches of the master frond proceed from its edges. Where one might expect a smooth axial stalk as in *Charniodiscus*, we have in *Rangea* a central frond stalk that is itself undergoing fractal partitioning. The rangemorph *Beothukis* takes this phenomenon into two dimensions (i.e., surface of frond as opposed to single stalk), with the frond surface showing an irregular array of subfronds in both distended (sand filled; Figs. 3.17, 3.18, and 3.19) or contracted (impression in fine clastic sediment; Fig. 3.21) modes of preservation. Note especially the intersecting field lines as seen in Fig. 3.21.

The only way to understand this is by means of the concept of a primitive (and hence highly complex in accordance with the Third Law of Morphogenetic Evolution) morphogenetic field in the vendobiont/vendozoan Ediacarans. Pflug's

(1972) underappreciated and misunderstood Ediacaran research results (supposedly burdened by an "elaborate and complex terminology"; Glaessner 1979; see also Tojo et al. 2007) actually quite correctly emphasized the peculiar body plan characteristics of the rangeomorphs and the erniettomorphs. Pflug's text-figure 2 in his paper (Pflug 1972) on erniettomorphs is a case in point. In this line art illustration, Pflug (1972) shows nascent frond after nascent frond bundled together in a fractal-like space-fill that represents a highly complex cuticular morphogenetic field. This should be considered as a complexly infolded membrane that 'seeks' to pack a three-dimensional morphogenetic field into the thin, curved yet roughly planar surface of the cuticle itself. The 'subsidiary quilts' and double-layered frond structure of *Rangea* (Grazhdankin and Seilacher 2005) are outpocketings or outgrowths of this complex morphogenetic field. Of course, such a complexity-packed cuticle might be expected to stretch and deform in unexpected ways (Figs. 3.44 and 3.45). One might also expect unusual morphologies to appear without prior warning.

Before his untimely death in an automobile accident, Martin D. Brasier (1947–2014) produced an interesting analysis of the holotype of *Charniodiscus concentricus* Ford 1958 (Brasier and Antcliffe 2009). Brasier had ready access to the holotype, so he made it a focal point of study. Brasier and Antcliffe (2009) inferred, correctly in my opinion, that the frond of *Charniodiscus concentricus* was "composed of several fronds oriented at angles to one another and compressed into the same plane" (Liu et al. 2016). Brasier and Antcliffe (2009) identified six different fronds with irregular spacing along the axis in the *Charniodiscus concentricus* holotype. Remarkably, in a volume dedicated to Brasier's memory, Liu et al. (2016) revised Brasier and Antcliffe's (2009) interpretation with a different rendering of which particular parts of the fossil represented separate subfronds. Liu et al. (2016) determined that the fossil consisted of a numbered total of five subfronds. It seems very likely that there are in fact more than 5–6 subfronds, as the colored images in both Brasier and Antcliffe (2009, their Fig. 12) and Liu et al. (2016, their Fig. 3) leave sections of the frond cluster uncolored and unidentified. This confusion is a direct result of the fact that we see in the frond of *Charniodiscus concentricus* the product of a complexly infolded membrane that unfurls at irregular intervals along the main stem or axis to generate the complex pattern observed in the fossil. The morphology of this holotype provides a powerful endorsement of the concept that Ediacarans possessed a complexly-infolded cuticular morphogenetic field.

The contentious glide symmetry of the Ediacaran midline also results from a densely packed morphogenetic field. The 'isomers' of *Paravendia janae* are crowded and intercalated to give the glide symmetry, reminiscent of the crowded, intercalated tubular modules in erniettomorph II (Fig. 3.14; Erniobetinae). Erniettomorph I (Erniettinae), by comparison, has nicely parallel tubular modules and neatly pointed tube terminations. Erniettomorph I may thus have a more advanced, and hence simpler, morphogenetic field (Third Law of Morphogenetic Evolution) in comparison to erniettomorph II, in the same way that, say, a trilobite has a simpler and more advanced morphogenetic field in comparison to the strange *Paravendia janae* with its overlapping segments.

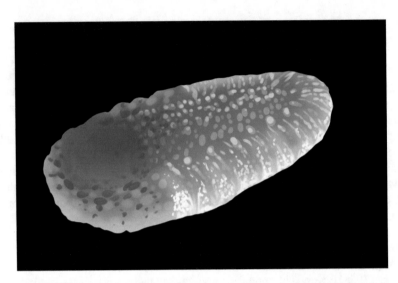

Fig. 3.46 *Zirabagtaria ovata* n. gen. n. sp. Reconstruction of organism (Artwork by Kay A. Hughes. Greatest dimension of specimen 10.5 mm)

Fig. 3.47 *Zirabagtaria ovata* n. gen. n. sp. Reconstruction of organism in its sandy, seafloor habitat (Artwork by Kay A. Hughes. Greatest dimension of specimen 10.5 mm)

It is therefore more than likely that an ancestor-descendant relationship exists between the peculiar Ediacarans and the Eumetazoa. It represents a case of reduction of complexity in their respective morphogenetic fields in accordance with Second and Third Laws of Morphogenetic Evolution. The morphogenetic field in *Zirabagtaria ovata* n. gen. n. sp. is still quite complex (Figs. 3.46 and 3.47), as seen in the columns of individually elongate papillae/sclerites on its dorsal surface, but its morphogenetic field is not nearly as complex as in a rangeomorph. The apparent complexity of the Cambrian (and earlier) scleritome actually represents a decrease in complexity, analogous to Williston's Law, the Law of Reduction of Parts. Williston (1914) declared it "a law in evolution that the parts in an organism tend toward reduction in number, with the fewer parts greatly specialized in function." The deceivingly simple body plan of the Ediacarans actually conceals tremendous complexity in the Ediacaran morphogenetic field. We may call this the Ediacaran Cuticle Paradox.

The development of the scleritome was an outcome of this simplification of the Ediacaran morphogenetic field. The development of the radula from the anterior field of the scleritome led to kimberellomorphs and aculiferans grazing on the marine biomat sea floor, using radular teeth. These were so effective at tearing into the biomats, those textured organic surfaces, that these tough flexible surfaces were effectively destroyed (in a geochemically critical acceleration of heterotrophy). The ensuing pulse of submat biomatter injected a lot of light carbon into sea water, altered the geochemistry of the planet and triggered the Shuram excursion.

References

Alpert SP (1976) Trilobite and star-like trace fossils from the White-Inyo Mountains, California. J Paleo 50(2):226–239

Billings E (1872) On some fossils from the primordial rocks of Newfoundland. Can Nat Geol 6:465–479

Brasier MD, Antcliffe JB (2009) Evolutionary relationships within the Avalonian Ediacara biota: new insights from laser analysis. J Geol Soc Lond 166(2):363–384

Briggs DEG, Clarkson ENK (1987) The first tomopterid, a polychaete from the Carboniferous of Scotland. Lethaia 20(3):257–262

Bykova N et al (2017) A geochemical study of the Ediacaran discoidal fossil *Aspidella* preserved in limestones: implications for its taphonomy and paleoecology. Geobiology 15:572–587

Crimes TP (1987) Trace fossils and correlation of late Precambrian and early [*sic*] Cambrian strata. Geol Mag 124:97–119

Cui H, Kaufman AJ, Xiao S, Zhou C, Liu X-M (2017) Was the Ediacaran Shuram Excursion a globally synchronized early diagenetic event? Insights from methane-derived authigenic carbonates in the uppermost Doushantuo Formation, South China. Chem Geol 450:59–80

Darroch SAF et al (2017) Inference of facultative mobility in the enigmatic Ediacaran organism *Parvancorina*. Biol Lett 13:20170033

Dzik J, Martyshyn A (2015) Taphonomy of the Ediacaran *Podolimirus* and associated dipleurozoans from the Vendian of Ukraine. Precambrian Res 269:139–146

Elliot DA et al (2011) New evidence on the taphonomic context of the Ediacaran *Pteridinium*. Acta Palaeontol Pol 56(3):641–650

Evans SD et al (2017) Highly regulated growth and development of the Ediacaran macrofossil *Dickinsonia costata*. PLoS One 12(5):e0176874

Fedonkin MA (2003) The origin of Metazoa in the light of the Proterozoic fossil record. Paleontol Res 7(1):9–41

Fedonkin MA et al (2007) The rise of animals: evolution and diversification of the Kingdom Animalia. Johns Hopkins University Press, Baltimore

Gehling JG (2004) Microbial mats in terminal Proterozoic siliciclastics: Ediacaran death masks. PALAIOS 14(1):40–57

Gehling JG et al (2000) The first named Ediacaran body fossil, *Aspidella terranovica*. Palaeontology 43:427–456

Gibson BM et al (2017) Ediacaran-style decay experiments with anemones and sea hares. Geol Soc Am Abstr Progr 49(6). https://doi.org/10.1130/abs/2017AM-304261

Glaessner MF (1958) New fossils from the base of the Cambrian in South Australia. Trans R Soc S Aust 81:185–188

Glaessner MF (1959) The oldest fossil faunas in South Australia. Geol Rundsch 47(2):522–531

Glaessner MF (1979) Precambrian. In: Berggren WA et al (eds) Treatise on invertebrate paleontology, part A. Geological Society of America and the University of Kansas, Boulder/Lawrence, pp 79–118

Glaessner MF, Wade M (1966) The late Precambrian fossils from Ediacara, South Australia. Palaeontology 9(4):599–628

Glaessner MF, Wade M (1971) *Praecambridium*-a primitive arthropod. Lethaia 4:71–77

Godfrey-Smith P (2016) Other minds. Farrar, Straus and Giroux, New York

Grazhdankin DV (2004) Patterns of distribution in the Ediacaran biotas: facies versus biogeography and evolution. Paleobiology 30(2):203–221

Grazhdankin D, Seilacher A (2005) A re-examination of the Nama-type Vendian organism *Rangea schneiderhoehni*. Geol Mag 142(5):571–582

Hall M et al (2013) Stratigraphy, palaeontology and geochemistry of the late Neoproterozoic Aar Member, southwest Namibia: reflecting environmental controls on Ediacara fossil preservation during the terminal Proterozoic in African Gondwana. Precambrian Res 238:214–232

Hoekzema RS et al (2017) Quantitative study of developmental biology confirms *Dickinsonia* as a metazoan. Proc R Soc B 284(1862). https://doi.org/10.1098/rspb.2017.1348

Hoyal Cuthill JF, Conway Morris S (2017) Nutrient-dependent growth underpinned the Ediacaran transition to large body size. Nat Ecol Evol 1:1201–1204

Ivantsov AY (2007) Small Vendian transversely articulated fossils. Paleontol J 41(2):113–122

Ivantsov AY (2009) A new reconstruction of *Kimberella*, a problematic Vendian metazoan. Paleontol J 43(6):601–611

Ivantsov AY (2017) The most probable Eumetazoa among late Precambrian macrofossils. Invertebr Zool 14(2):127–133

Ivantsov AY, Leonov MV (2008) Otpechatki vendskix zhivotnykh-unikal'nye paleontologicheskie ob'eky Archangel'skoi oblasti. Arkhangel'sk, Russia

Jensen S, Mens K (2001) Trace fossils *Didymaulichnus* cf. *tirasensis* and *Monomorphichnus* isp. from the Estonian Lower Cambrian, with a discussion on the Early Cambrian ichnocoenoses of Baltica. Proc Estonian Acad Sci Geol 50(2):75–85

Laflamme M et al (2013) The end of the Ediacara biota: extinction, biotic replacement or Cheshire Cat? Gondwana Res 23:558–573

Lang FG et al (2005) Paleoproterozoic Mojave Province in northwestern Mexico? Isotopic and U-Pb zircon geochronologic studies of Precambrian and Cambrian crystalline and sedimentary rocks, Caborca, Sonora. Geol Soc Am Spec Pap 393:183–198

Liu AG et al (2016) Martin Brasier's contribution to the palaeobiology of the Ediacaran-Cambrian transition. In: Brasier AT et al (eds) Earth system evolution and early life: a celebration of the work of Martin Brasier. Geological Society, London, Special Publications 448: https://doi.org/10.1144/SP448.9

MacGabhann BA (2007) Discoidal fossils of the Ediacaran biota: a review of current understanding. In: Vickers-Rich P, Komarower P (eds) The rise and fall of the Ediacaran biota. Geological Society of London Special Publications 286, London, pp 297–313

McCall GJH (2006) The Vendian (Ediacaran) in the geological record: enigmas in geology's prelude to the Cambrian explosion. Earth-Sci Rev 77:1–229

McMenamin MAS (1984) Paleontology and stratigraphy of lower Cambrian and upper Proterozoic sediments, Caborca region, northwestern Sonora, Mexico. Ph.D. Dissertation, University of California at Santa Barbara. University Microfilms International, Ann Arbor

McMenamin MAS (1993) Osmotrophy in fossil protoctists and early animals. Invertebr Reprod Dev 22(1–3):301–304

McMenamin MAS (1996) Ediacaran biota from Sonora, Mexico. Proc Nat Acad Sci 93:4990–4993

McMenamin MAS (1998) The garden of Ediacara: discovering the first complex life. Columbia University Press, New York

McMenamin MAS (ed) (2001) Paleontology Sonora: Lipalian and Cambrian. Meanma Press, South Hadley

McMenamin MAS (2003a) Origin and early evolution of predators: the ecotone model and early evidence for macropredation. In: Kelley PH et al (eds) Predator-prey interactions in the fossil record. Kluwer, New York, pp 159–169

McMenamin MAS (2003b) *Spriggina* is a trilobitoid ecdysozoan. Geol Soc Am Abstr Progr 35(6):105

McMenamin MAS (2004) The ptychoparioid trilobite *Skehanos* gen. nov. from the Middle Cambrian of Avalonian Massachusetts and the Carolina Slate Belt, USA. Northeast Geol Environ Sci 24(4):276–281

McMenamin MAS (2006) New data on the earliest animals and Ediacarans from Sonora, Mexico. Geol Soc Am Abstr Progr 38(7):303

McMenamin MAS (2011) Fossil chitons and *Monomorphichnus* from the Ediacaran Clemente Formation, Sonora, Mexico. Geol Soc Am Abstr Progr 43(5):87

McMenamin MAS (2016) Dynamic paleontology: using quantification and other tools to decipher the history of life. Springer, Cham

McMenamin MAS, Schulte McMenamin DL (1990) The emergence of animals: the Cambrian breakthrough. Columbia University Press, New York

McMenamin MAS et al (1983) Precambrian-Cambrian transition problem in western North America: Part II. Early Cambrian skeletonized fauna and associated fossils from Sonora, Mexico. Geology 11:227–230

Meyer A (1926) Die Segmentalorgane von *Tomopteris catharina* (Gosse) nebst Bermerkungenueber das Nervensystem, die rosetten-förmigen Organe und die Cölombewimperung. Zeitschrift für Wisenschaftliche Zoologie 127:297–402

Meyer M et al (2014a) Taphonomy of the Ediacaran fossil *Pteridinium simplex* preserved three-dimensionally in mass flow deposits, Nama Group, Namibia. J Paleontol 88(2):240–252

Meyer M et al (2014b) Three-dimensional microCT analysis of the Ediacara fossil *Pteridinium simplex* sheds new light on its ecology and phylogenetic affinity. Precambrian Res 249:79–87

Narbonne GM (2005) The Ediacara biota: Neoproterozoic origin of animals and their ecosystems. Ann Rev Earth Planet Sci 33:421–442

Parry LA et al (2017) Ichnological evidence for meiofaunal bilaterians from the terminal Ediacaran and earliest Cambrian of Brazil. Nat Ecol Evol. https://doi.org/10.1038/s41559-017-0301-9

Paterson JR et al (2017) Rheotaxis in the Ediacaran epibenthic organism *Parvancorina* from South Australia. Sci Rep. https://doi.org/10.1038/srep45539

Pettijohn FJ et al (1972) Sand and sandstone. Springer, New York

Pflug HD (1972) Zur Fauna der Nama-Schichten in Südwest-Afrika. III. Erniettomorpha, Bau und Systematik. Palaeontographica Abt A 139:134–170

Phillips Dales R (1971) Bioluminescence in pelagic polychaetes. J Fish Res Board Can 28(10):1487–1489

Savazzi E (2015) The early Cambrian Eophyton toolmark and its producer. Paleontol Res 19(1):61–75

Schiffbauer JD et al (2016) The latest Ediacaran wormworld fauna: setting the ecological stage for the Cambrian explosion. GSA Today 26(11):4–11

Schwabe E (2010) Illustrated summary of chiton terminology (Mollusca, Polyplacophora). Spixiana 33(2):171–194

Seilacher A (1989) Vendozoa: organismic constructions in the Proterozoic biosphere. Lethaia 22:229–239

Seilacher A (1994) Early multicellular life: late Proterozoic fossils and the Cambrian explosion. In: Bengtson S (ed) Early life on earth. Columbia University Press, New York, pp 389–400

Seilacher A et al (2003) Ediacaran biota: the dawn of animal life in the shadow of giant protists. Paleontol Res 7(1):43–54

Sigwart JD (2017) Zoology: molluscs all beneath the sun, one shell, two shells, more or none. Curr Biol 27:R702–R719

Sokolov BS (1965) Paleontologiya dokembriya i organicheski mir k nachalu fanerozoya. Bsesoyuznyi simposiu po paleontologii dokembriya. Novosibirsk, pp 3–7

Sour-Tovar F et al (2007) Ediacaran and Cambrian index fossils from Sonora, Mexico. Palaeontology 50(1):169–175

Stewart JH et al (1984) Upper Proterozoic and Cambrian Rocks in the Caborca Region, Sonora, Mexico—physical stratigraphy, biostratigraphy, paleocurrent studies and regional relations. US Geol Surv Prof Pap 1309:1–36

Tarhan LG et al (2016) Exceptional preservation of soft-bodied Ediacara biota promoted by silica-rich oceans. Geology 44:951–954

Tojo B et al (2007) Theoretical morphology of quilt structures in Ediacaran fossils. In: Vickers-Rich P, Komarower P (eds) The rise and fall of the Ediacaran biota. Geological Society of London Special Publications 286, London, pp 399–404

Vinther J (2015) The origins of molluscs. Palaeontology 58(1):19–34

Williston SW (1914) Water reptiles of the past and present. Chicago University Press, Chicago

Zhuravlev AY et al (2012) New finds of skeletal fossils in the terminal Neoproterozoic of the Siberian Platform and Spain. Acta Palaeontol Pol 57(1):205–224

Zittel KA (1913) Text-book of paleontology. Macmillian, London

Chapter 4
Crystal Creature

> *That's how fossil hunting is: It takes over, like a hunger, and nothing else matters but what you find. And even when you find it, you still start looking again the next minute, because there might be something even better waiting.*
>
> Tracy Chevalier

Abstract An agglutinated animal (or possibly protist) of the Late Proterozoic Clemente Formation biota cemented an array of tourmaline crystals (trigonal prisms; schorl/dravite composition) to its dorsal surface, presumably as ballast, in the earliest known case of an agglutinated animal. Quantitative confirmation of the spatial association (clustering) of the tiny crystals is demonstrated here by means of Kappa (K) value analysis. This discovery of this Proterozoic "crystal creature" reveals both the earliest known case of agglutination, the earliest known case of monomineralogic agglutination, and the earliest known bioaccumulation of tabular crystals of the same mineral. This case compares to the preferential selection of ilmenite to form an agglutinated exoskeleton in the Cambrian agmatan *Volborthella*. It also compares to the preferential selection of muscovite flakes to form the agglutinated Cambrian worm tube *Onuphionella*. Agglutinated dorsal skeletons of the Proterozoic and Early Cambrian utilized selected mineral types, including white mica, ilmenite, anatase, and tourmaline. All of these minerals potentially afford protection from UV-B radiation, and may have been deployed on the dorsal surfaces of these early animals to serve as sunscreen.

Keywords Clemente Formation · Proterozoic · *Volborthella* · *Salterella* · *Campitius* · *Onuphionella* · Tourmaline · Anatase · Ilmenite · Kappa value

In an anticipation of the Second and Third Laws of Morphogenetic Evolution, Signor and McMenamin (1994) wrote that the overall appearance of a scleritome "is often remarkably complex, but the morphogenetic program necessary to generate the multi-iterated units in the scleritome is fairly simple. The gradual replacement

© Springer International Publishing AG 2018
M.A.S. McMenamin, *Deep Time Analysis*, Springer Geology,
https://doi.org/10.1007/978-3-319-74256-4_4

of complex scleritomes by more simple exoskeletons parallels other apparent evolutionary trends toward simplification of skeletal structures," i.e., Williston's Law pertaining to the evolutionary reduction of parts. The kimberellomorph *Zirabagtaria ovata* n. gen. n. sp. discussed in the previous chapter (Figs. 3.2, 3.27, 3.46, and 3.47) is a remarkable case of an early scleritome or "papillotome" with many individual bits. The dorsal surface of this animal may perhaps be more properly regarded as a cuticula, a tough dorsal tissue that in the caudofoveatan solenogastres (aplacophorans or 'shell-less mollusks') may also bear a scleritome of aragonitic sclerites or spicules. The day ten larva in the solenogastre *Wirenia argentea* has well developed epidermal sclerites (Todt and Wanninger 2010).

Zirabagtaria ovata n. gen. n. sp. is a small fossil but its morphology is comparable in some ways to that of the unified scleritome of the Early Cambrian *Halkieria evangelista* from Greenland (Conway Morris and Peel 1995). *Zirabagtaria's* 'scleritome' of tubercular nodes (seen also in some taphomorphs of *Kimberella*; Vinther 2015) is of particular interest because of its great age.

Surely the most useful part of a mollusk's scleritome (from the animal's perspective!) is its radula, the rasping feeding organ that consists of a geometrical array of miniaturized sclerites forming the functional equivalent of a sanding belt. The radula was given its name by Alexander von Middendorf (von Middendorf 1847). For decades the oldest reported fossil radulae belonged to cephalopods (the two main examples were the Carboniferous goniatite ammonite *Eoasianites* (Closs 1967) and an Ordovician orthocone nautiloid (Gabbott 1999)), but no one believed that these reports described the very earliest radulae (Firby and Durham 1974; Butterfield 2008). Ever since, a search has been underway to locate the oldest radula in the fossil record.

Firby and Durham (1974) claimed to have discovered an Early Cambrian cephalopod radula in the Montenegro and Andrews Mountain Members of the Campito Formation. The fossil consisted of a strew field of conical denticles. Mineralogical and elemental analysis showed these denticles to be composed of ilmenite ($FeTiO_3$). This odd composition rendered the fossils unusually titanium-rich, hence Firby and Durham (1974) named the fossils *Campitius titanius*. This unusual putative case of titanium biomineralization gains mention in biochemistry literature as a potentially important example (Buettner and Valentine 2012):

> A lively ensuing controversy centered around whether the fossils were not actually teeth but instead were individuals of *Volborthella* (members of an extinct wormlike group), or perhaps were horns on a worm or mollusk like animal. The point that the grains were ilmenite attracted little notice and has not been followed up.

Volborthella is a strange creature that was placed in its own phylum, Phylum Agmata, by Ellis Yochelson (1977). Yochelson's (1977) proposal has been controversial, with not all paleontologists willing to add Agmata to the list of Early Cambrian phyla. However, the Phylum Agmata concept has merited renewed consideration (Signor and McMenamin 1994; Peel 2016) largely due to the strangeness of the included taxa (*Volborthella* [= *Campitius*], *Salterella* and *Ellisell*

Fig. 4.1 *Salterella* from Sonora. Field sample 801q, Proveedora Hills section, Sonora, Mexico (Sample is from Cooper et al.'s (1952) original collection, provided by Christina Lochman). Scale in mm

yochelsoni) that were at one time classified as tiny chambered cephalopod shells. Acceptance of Agmata as a valid (extinct) phylum adds to the perceived magnitude of the Cambrian Explosion.

The cephalopod shell and cephalopod radula hypotheses for the agmatans has been rejected in favor of an interpretation of these curious fossils as cone-shaped sclerite analogs that are agglutinated, in other words, composed of inorganic sediment particles held together by biological cement (Hagadorn and Waggoner 2002). The individual cones are combined to form an agglutinate scleritome, surely one of the most curious bioconstructions in the fossil record. Figure 4.1 shows disarticulated *Salterella* specimens forming a coquina in the Lower Cambrian Puerto Blanco Formation in a specimen collected by Mount Holyoke College paleontologist Christina Lochman.

Agglutinated skeletons are those formed by gluing together bits of sediment or other particles from the surrounding environment. Modern examples include the Recent marine annelid worm *Pectinaria*. *Pectinaria* builds a tube of oriented sand grains. Caddisfly larvae (Order Trichoptera) build larval cases out of coarse sand and other materials. French artist Hubert Duprat has induced captive caddisfly larvae to form strikingly beautiful golden cases formed of small gold nuggets.

Agmatans apparently had a preference for dense minerals as they formed their agglutinated sclerites. The enhanced density of these minerals perhaps served as ballast and helped the creatures stay anchored to the sea floor (Peel 2016). This solves the ilmenite problem with regard to *Campitius titanius*. *Campitius* was preferentially selecting dense ilmenite sediment grains to build its agglutinated sclerites. As far as we know, no animal, living or extinct, is able to secrete ilmenite as a biomineral. In a bizarre case of Early Cambrian agglutinization, the agmatan *Salterella* used both authigenic anatase (TiO_2) grains and anatase-coated small shelly fossils (worm sclerites of *Hadimopanella apicata*) to form its agglutinated cone (Peel 2016).

Authentic Cambrian radulae were reported by Butterfield (2008) who recovered tooth rows, of organic composition, from the Lower Cambrian Mahto Formation of Alberta, Canada. These oldest radulae, however, are less informative about the origins of the radula than they might otherwise be, because they "already share derived characters associated with aplacophorans and gastropods" (Smith 2012). Smith (2012) concludes that unfortunately, "the organ's origin is ill-constrained," but he infers that the parallel grooves scratched by *Kimberella* (i.e., the ichnofossils *Kimberichnus teruzzii* and *Radulichnus*; see Gehling et al. 2014) indicate that *Kimberella* had a simple stereoglossan-type radula.

Clusters of tiny, elongate black crystals occur on several of the Sonoran Ediacaran fossil samples described in the previous chapter. Several of the Clemente biota samples are coated with a shiny desert varnish (e.g., IGM 4997), and some of them are dotted with tiny black dendritic (manganese oxide; probable microbial influence) deposits that represent a patchy, dark variant of desert varnish as seen on the holotype specimen of *Zirabagtaria ovata* n. gen. n. sp. (Fig. 3.26). These black dendritic lichen-like deposits hide the tiny crystals, because they are approximately the same color. Being tiny and elongate, as are some of the manganese deposits, the crystals are well camouflaged. I only discovered them by accident on June 8, 2017, while preparing the holotype of *Clementechiton sonorensis* for a session on the scanning electron microscope. As I was examining the holotype (IGM 7461) under a binocular microscope, I managed to focus the image on the wrong spot, some distance from the chiton valve fossil. A cluster of tiny elongate crystals came into view. After further scrutiny of the tiny crystals, I was soon to grasp the implications of the find.

A rapid search of the rest of the Clemente biota fossil suite showed the presence of these crystals on IGM 7461 (with *Clementechiton sonorensis*), IGM 4995 (with *Zirabagtaria ovata* n. gen. n. sp.) and IGM 4998 (with both *Kimberella* cf. *K. quadrata* and *Korifogrammia clementensis* n. gen. n. sp.). The crystals are embedded in the sandy rock matrix and encased in the siliceous cement that binds the sand grains together, and thus form part of the fabric of the sedimentary rock. Rather than being more-or-less randomly scattered, however, the crystals are loosely clustered into groups.

A total of sixty associated crystals occur on samples IGM 4995 (n = 3), IGM 4998 (n = 34) and IGM 7461 (n = 23). Each crystal is individually numbered as follows: IGM 4995[1], IGM 4995[2], IGM 4995[3], etc. Five sketch maps plot the location of each crystal (Figs. 4.2, 4.3, 4.4, 4.5, and 4.6). Most crystals under high magnification show parallel, terrace-like ledges along their edges. The crystals show evidence for both pre-lithification and post-lithification wear. Tiny crystal fragments, either fragments of broken crystals or very small individual crystals that are part of the associations, do occur in the crystal clusters but are not listed here.

Two crystals are directly juxtaposed (IGM 7461[17]-[18]; Fig. 4.7). The crystals are acicular with pointed (or at least tapering) ends. The upper surfaces of the crystals are abraded, likely due to natural erosion of the surface of the rock that contains them, to form reticulate erosional surfaces. The intact crystal faces, downslope from the eroded tops, show the characteristic parallel, terrace-like ledges. The crystal to the right is more broad (IGM 7461[18]) and the one to the left is more narrow

Fig. 4.2 Sketch map of tourmaline crystals on the type specimen of *Clementechiton sonorensis* (IGM 7461). Bedding sole surface. Numbers in parentheses represent numbers of individual tourmaline crystals. Colinear short lines at the edges of the rock outline indicate the position of the greatest dimension line. Scale bar 1 cm

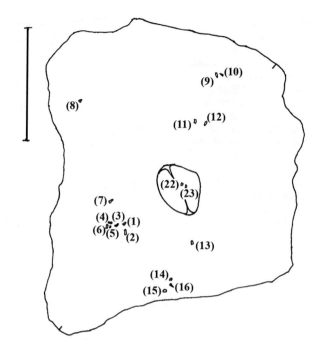

Fig. 4.3 Sketch map of tourmaline crystals on the type specimen of *Clementechiton sonorensis* (IGM 7461). Top bed surface. Large letters and numbers represent the original field sample label. Numbers in parentheses represent numbers of individual tourmaline crystals. Colinear short lines at the edges of the rock outline indicate the position of the greatest dimension line. Scale bar 1 cm

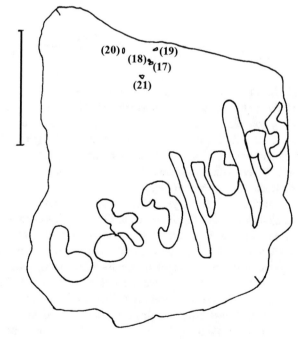

Fig. 4.4 Sketch map of
tourmaline crystals on the
type specimen of
Zirabagtaria ovata n. gen.
n. sp. (IGM 4995).
Bedding sole surface.
Numbers in parentheses
represent numbers of
individual tourmaline
crystals. As there are only
three crystals on this
surface, no position is
shown for a greatest
dimension line (five is the
minimum required for
K analysis). Scale bar 1 cm

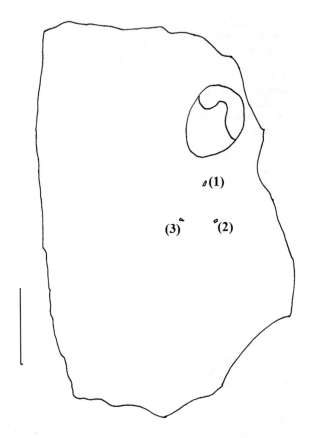

(IGM 7461[17]). Both crystals are between 100–200 microns in length, barely visible to the naked eye. The narrow crystal ([17]) appears to be broken at both ends. Both crystals appear to have a triangular cross section.

A second crystal is shown Fig. 4.8 (IGM 7461[19]). The crystal shows a distinct acicular habit, and again a triangular cross section. Two of the crystal faces are visible in the SEM photomicrograph, with a linear edge representing the juncture between the two crystal faces.

Another acicular crystal with two broken ends and a rounded-triangular cross section (IGM 4995[2]) is shown in Fig. 4.9. This crystal tapers and thins towards its broken tip, but it nevertheless shows the terrace-like edges. The impression of a satellite crystal occurs on the upper surface of the crystal near the tip. The small crystal impression is oriented subparallel to the main axis of the crystal. This crystal habit indicates that these two crystals once formed part of a crystal fan in a metamorphic or igneous source rock. The satellite crystal impression also decisively proves that these are not authigenic minerals, grown in place, because if they were the satellite crystals might still be present, or in any case, there should be a nearby cluster of satellite crystals forming the crystal fan.

These Clemente Formation crystals all appear to have the same mineralogy. It is very likely that they represent some form of tourmaline, as it is the only common

Fig. 4.5 Sketch map of tourmaline crystals on the type specimen of *Korifogrammia clementensis* n. gen. n. sp., bedding sole surface (IGM 4998). Numbers in parentheses represent numbers of individual tourmaline crystals. Colinear short lines at the edges of the rock outline indicate the position of the greatest dimension line. Scale bar 1 cm

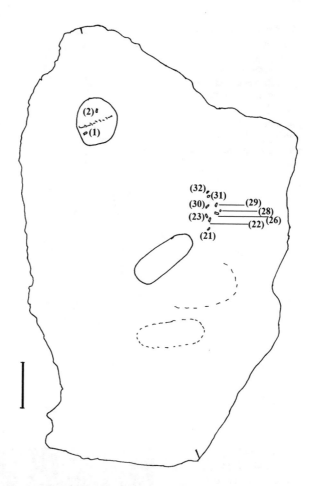

acicular mineral that forms three-sided needles. Scattered rare tourmaline crystals (presumed acicular habit) apparently occur in the sandstone matrix (Fig. 3.7). Tourmaline is distinguished by its trigonal prisms, and can often be confidently identified because it is the only commonly-encountered mineral that forms crystals of this distinctive shape. The three prisms may develop a rounded look due to parallel striations running along the three crystal faces. This is precisely what we see in the Clemente Formation crystals. The crystal shown in Fig. 4.10 (IGM 7461[7]) clearly shows this rounded triangular habit. The upper left edge of the crystal shown in Fig. 4.10 shows transport abrasion that occurred before the crystal was lithified as part of the sandstone, further demonstrating that this is a detrital tourmaline crystal.

Figure 4.11 shows the results of SEM-EDS analysis of crystal IGM 4995[2]. This crystal is also illustrated in Fig. 4.9. This EDS analysis provides evidence for the composition of the tourmaline crystal. Tourmaline is a boron silicate mineral, but boron's atomic number (5; lighter than Carbon at 6) is too low for it to register reliably on the EDS spectrum. The EDS analysis, however, shows a significant

Fig. 4.6 Sketch map of tourmaline crystals on the holotype specimen of *Korifogrammia clementensis* n. gen. n. sp., bed top surface (IGM 4998). Numbers in parentheses represent numbers of individual tourmaline crystals. Scale bar 1 cm

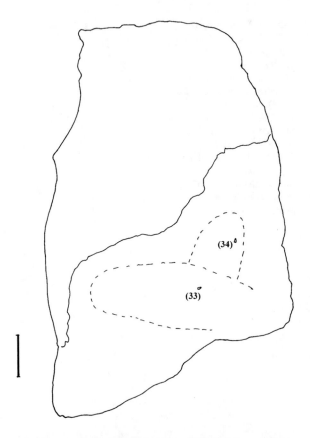

Fig. 4.7 Acicular crystals IGM 7461[17]-[18]. The upper surfaces of the crystals have been naturally abraded, forming a reticulate surface. Scale bar 100 microns

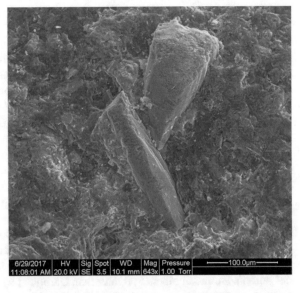

Fig. 4.8 Acicular crystal IGM 7461[19]. Note the straight aspect and the sharp keel or edge between crystal faces. Scale bar 100 microns

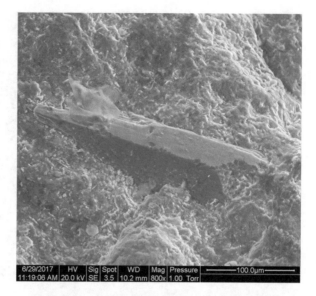

Fig. 4.9 Acicular crystal IGM 4995[2]. Note taper of broken tip and the impression of a satellite crystal on the upper surface of the crystal near the tip that is oriented subparallel to the main axis of the crystal. This provides evidence that the crystal cluster was once part of a crystal fan. Scale bar 100 microns

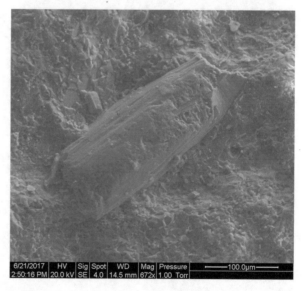

(if low) peaks for sodium and magnesium. The most common Na-rich tourmaline is schorl; thus these appear to be crystals of detrital schorl, although they may also be detrital specimens of the less common tourmaline dravite. In any case, the minerals are crystals of either the schorl (Na and Fe) or the dravite (Na and Mg, respectively) species of tourmaline. The potassium peak (K) on the spectrum indicates an element not known to occur in tourmaline, but may represent clays or other contaminants on the surface of the crystal.

The formula for dravite, first determined by Gustav Tschermak in 1884, is roughly $NaMg_3(Al,Mg)_6B_3Si_6O_{27}$ (Ertl 2007). This composition is a reasonably

Fig. 4.10 Acicular crystal IGM 7461[7] showing rounded triangular crystal habit. The upper left edge of the prismatic crystal shows transport abrasion that occurred before the crystal was lithified as part of the sandstone, thus demonstrating that this is a detrital grain of tourmaline. Scale bar 100 microns

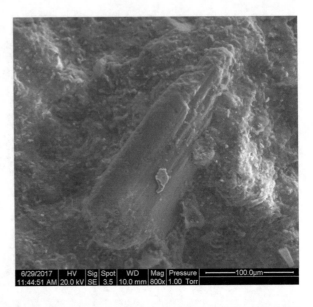

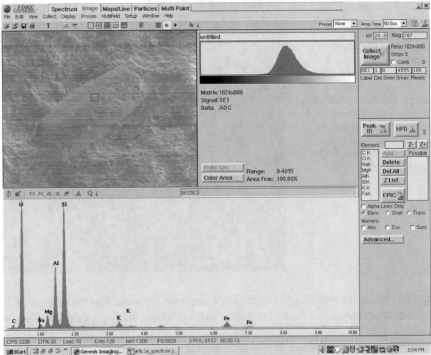

Fig. 4.11 Results of SEM-EDS analysis of crystal IGM 4995[2]. The sodium, iron and magnesium peaks indicate that the crystal is a specimen of either the schorl (Na and Fe) or the dravite (Na and Mg) species of tourmaline

Fig. 4.12 Acicular crystal IGM 7461[1]. "Spear-head" shape. Scale bar 100 microns

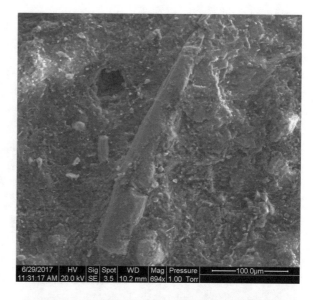

Fig. 4.13 Element mapping SEM-EDS analysis of "Spear-head" shape crystal IGM 7461[1] indicating enhanced Fe content of crystal (FeK-magenta). Same crystal as seen in previous illustration

close match to the EDS spectrum as shown in Fig. 4.11. The Na and Mg peaks are lower than one might expect, possibly because the rock is coated in a thin layer of siliceous desert varnish. The potassium and iron peaks may be components of the desert varnish as well. EDS element mapping on several of the Sonoran crystals shows localization of sodium and magnesium (Figs. 4.12 and 4.13) content on the crystals (and relative depletion in the surrounding matrix), supporting a schorl or dravite composition for the Sonoran crystals.

Figure 4.14 shows a tourmaline crystal (IGM 7461[2]) with impressions of up to three satellite crystals at its tip (left side of tip) and the development of botryoidal tourmaline at the tip to the right. Figure 4.15 shows a close up of the botryoidal development in this tourmaline crystal. Interestingly, fine striations run across the

Fig. 4.14 Acicular
tourmaline crystal IGM
7461[2] showing
impressions of two or three
satellite crystals at the tip
(left) and development of
botryoidal crystal habit at
the tip (right). Scale bar 50
microns

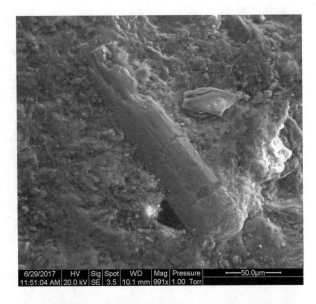

Fig. 4.15 High
magnification
photomicrograph of the tip
of the crystal (IGM
7461[2]) in the previous
image, showing detail of
satellite crystal impressions
at the tip (left) and
botryoidal habit with
striations perpendicular to
the long axis of the crystal.
Scale bar 20 microns

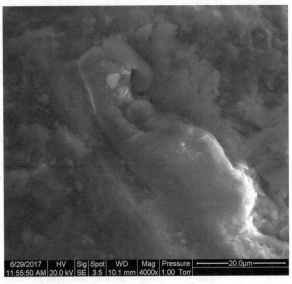

botryoidal fabric perpendicular to the long axis of the crystal; the usual case for
these crystals is for striae to run perpendicular to the long axis of the crystal
(Fig. 4.16). A very flat crystal face, with the crystal deeply embedded in fine
sandstone matrix and the flat surface partly obscured by desert varnish, is shown in
Fig. 4.17.

The evidence for the original occurrence of these crystals in a crystal fan (i.e., the
impression of a satellite crystal in Fig. 4.9) demonstrates that these are detrital crystals of tourmaline (schorl or dravite), and not some type of authigenic occurrence

Fig. 4.16 High magnification detail of acicular crystal IGM 7461[7]. Note broken base of crystal and surface striations running parallel to the long axis of the crystal. Scale bar 20 microns

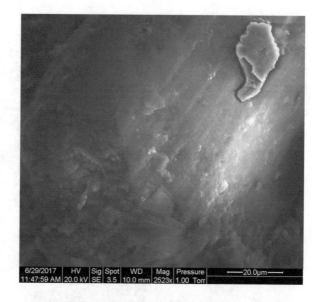

Fig. 4.17 Tourmaline crystal IGM 7461[3], deeply embedded in sandstone matrix and coated in Sonoran desert varnish. Scale bar 50 microns

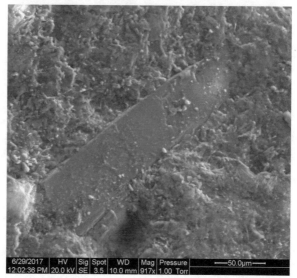

that formed in place. Authigenic overgrowths can occur on Precambrian detrital tourmaline crystals from quartzite occurring north of Jamua, India (Chatterjee et al. 1975). The overgrowths on the Indian crystals are colorless to pale blue (Na-dravite composition), and are in optical continuity with the nucleus (detrital) crystal, growing from its C-axis end (Chatterjee et al. 1975).

Unlike the acicular Sonoran crystals, the Jamua crystals are "roughly idiomorphic, with prisms terminating in sub-rounded to rounded asymmetrical pyramids" (Chatterjee et al. 1975). The Clemente Formation crystals may be derived from

Fig. 4.18 Striated jet black schorl (tourmaline) from the Municipio de Baviácora, Municipio de Santa Cruz, Sonora, México (Photograph by Rob Lavinsky, used here by Creative Commons license CCBY-SA3.0). Greatest dimension of crystal cluster 4.3 cm

weathering of an aplite (aphanitic granite) as these are known to host acicular tourmaline crystals in radiating clusters that resemble flowers. Pectinate (comb-like) occurrences of acicular tourmalines in granite are another possible source (Beurlen et al. 2011). Yet another possible source rock for the Clemente Formation crystals is a tourmaline hornfels, a black metamorphic rock (known to form at the edges of tourmaline granites) that is characterized by needle-like crystals of schorl.

A contact metamorphic aureole of tourmaline hornfels up to 600 m wide surrounds the Brice Creek and Campion Creek granodiorite stocks of the Bohemia mining district of Lane County, Oregon (Taylor 1974). Sonora, México is known for its impressive striated jet black schorl crystals (Municipio de Baviácora, Municipio de Santa Cruz; Fig. 4.18). These schorls are apparently derived from localized breccias on top of a regional batholith, but the geology of the occurrences awaits further study. Unique radiating crystal clusters are known from the Santa Cruz locality. Interestingly for our purposes, radiating clusters of schorl crystals resembling icicles (stalactitic schorl or *velvet tourmaline*) are best known from Santa Cruz, Sonora. According to the 2017 auction catalog of Mineralauctions.com describing a specimen 6.1 cm in greatest dimension (mineral names are capitalized in the catalog text):

> One of the most interesting Tourmaline discoveries has to be those made at Santa Cruz, Sonora, Mexico. This material is rarely seen on the market, but immediately recognizable. These 'velvet' Tourmalines have the most distinctive exterior and thin prismatic form which grew into eye-catching 'sprays.' The piece features several dozen crystals of jet-black Tourmaline with a secondary overgrowth creating the 'velvety' appearance which measure up to 2.6 cm, aesthetically sitting on an off-white color Quartz matrix. The find was made in the 1990s…

These icicle-like crystals could be easily broken off by natural mechanical processes and incorporated into sediments as acicular fragments or clasts. This is apparently what occurred (if not from the modern velvet tourmaline deposits, then from a Proterozoic equivalent) during deposition of the Proterozoic Clemente Formation. There was likely to have been an impressive deposit of velvet tourmaline nearby to provide sufficient material to make it worthwhile for the crystal creatures to search for and acquire the crystals of this particular type.

I propose here that these crystals have been selected by size, shape and density by some sort of biological process, and I propose further that this process was intentional selection by a stem molluscan-type metazoan that was able to form an agglutinated dorsal coat of black tourmaline crystals. These dark crystals would be quite effective at blocking excess solar radiation. Before proceeding further with these hypotheses, however, it is worthwhile to obtain a quantitative measure of the degree of clustering or aggregation of the tiny tourmaline crystals as they occur on the bedding plane surfaces.

To do this we will use a calculation of Kappa value (K) to determine the degree of spatial aggregation of the crystals. In any set of scattered elements, those that show a higher K value are more likely to have been aggregated around a single point or, in the case of the hypothesis outlined above, the dorsal surface of a single animal. Samples with low K value, down to the limit of zero, show less tendency for initial clustering or aggregation.

Kappa value is calculated in the following fashion. A graphical convention is required to perform the calculation. The sample under analysis consists of a collection of *objects*, treated as points, arrayed over a surface (such as, for example, the bedding plane surface of a sedimentary rock sample) that will be referred to here as the *field surface*. The perimeter of the field surface may be of any shape, but a roughly rectangular outline will facilitate measurement of the various input values required for the calculation. The first step in the calculation of K is to determine the position and length of the *greatest dimension line*. This is a line that connects two points on the perimeter of the field surface to give the greatest possible distance, in other words, the greatest dimension line.

The greatest dimension line divides the field surface into two *sectors*. Both sectors may have an array of objects, or the objects may be confined to one of the two sectors. Kappa value may be calculated for sectors that have at least five objects. For each object in a particular sector, draw a line from the object to meet the greatest dimension line at a right angle. The length of this line is unimportant for the K calculation; its purpose is to divide the greatest dimension line at the point of intersection. After these *object lines* are drawn, for n objects the greatest dimension line is divided into $n + 1$ *segments*. These segments will ordinarily be of unequal length. K for a given sector may now be calculated using the Kappa equation:

$$K = (\beta + 1)\left[\sqrt{\frac{\sum_{i=1}^{N}(x_i - \bar{x})^2}{n}}\right] \tag{4.1}$$

Where: n = the number of objects in a particular sector (for $15 \geq n \geq 5$); $N = n + 1$, the total number of segments of the greatest dimension line; $\{x_1, x_2...x_N\}$ represent successive segment length measurements; $\bar{x}$ = the mean of all segment length

Fig. 4.19 "Random" dot
pattern for a test
calculation of Kappa value
(*K*). A *greatest dimension
line* splits the pattern into
two *sectors*. Two triangular
field surfaces as shown,
one marked NW and the
other SE. Lines are drawn
from each *object* in the
NW field surface that are
perpendicular to the
greatest dimension line,
dividing it up in to
segments. See text for
further discussion

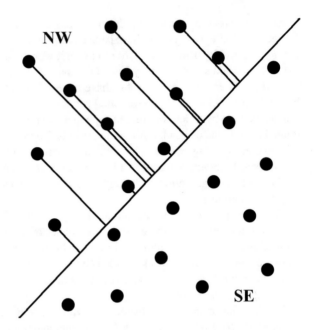

measurements. The β value is determined as follows. Ignoring x_1 and x_N (that is, the first and last segments of the greatest dimension line), β is the total count of the number of times that both x_i and $x_{(i+1)}$ (that is, adjacent segments of the greatest dimension line) are both <10. T denotes the length of the greatest dimension line. The segment length measurements and T measurement may be in arbitrary units (for example, measured from of photographs of specimens at any convenient scale of magnification), as long as the scale used is consistent.

Figure 4.19 shows a "random" dot pattern used here to demonstrate the calculation of Kappa value (*K*). I place the word "random" in parentheses to make a philosophical statement; true or pure randomness may not exist at all and in any case is "a notoriously difficult property to test" (Scarani 2010). Nevertheless, we may say that the objects in Fig. 4.19 are, at least roughly speaking, evenly spaced and without any obvious geometric relationship to one another (although we cannot rule out the possibility that one exists). We therefore will use this pattern as a starting point to show how one generates K values. As the dot pattern is on a square field surface, we will be able to run the simulation twice as the two diagonals of the square field surface serve as two greatest dimension lines of equal length. This will not usually be the case with regard to natural specimens such as rock samples, but even here a single specimen could potentially have two (or more) greatest dimension lines of equal length.

For the first simulation, we will use the greatest dimension line that runs from the upper left corner of the square to the lower right corner of the square. This neatly divides the field surface into two triangular sectors, the northeast (NE) sector and the southwest (SW) sector. We will calculate the K value for the NE sector first. For this sector, n = 12 and there are thirteen segments. The arbitrary unit values for consecutive sector lengths (i.e., $\{x_1, x_2...x_N\}$) are as follows: 57, 34, 4, 33, 3, 18, 16, 5, 23, 19, 18, 37, and 43. As no two adjacent x values are both less than 10, the β value for NE sector is zero. Applying Eq. 4.1 yields a K value of 16.24 for the NE sector.

Using the same greatest dimension line, we will now calculate the K value for the SW sector. Here again, n = 12 and there are thirteen segments. The arbitrary unit values for consecutive sector lengths (i.e., $\{x_1, x_2...x_N\}$) are as follows: 68, 37, 4, 16, 4, 12, 18, 19, 25, 21, 16, 37, 32. No two adjacent x values are both less than 10, so the β value for SW sector is zero. Applying Eq. 4.1 yields a K value of 17.03 for the SW sector. This is fairly close to the K value for the NE sector.

Now let's switch and use the other diagonal of the square as our greatest dimension line, this time the one that runs from the lower left corner of the square to the upper right corner of the square. We will begin by considering the northwest (NW) sector. Once again, n = 12 and there are thirteen segments. The arbitrary unit values for consecutive sector lengths (i.e., $\{x_1, x_2...x_N\}$) are as follows: 70, 29, 30, 15, 6, 2, 17, 19, 16, 2, 36, 3, 67. Notice that $x_5 = 6$ and that $x_6 = 2$. This means that two adjacent segment values are less than 10, and thus the β value is nonzero and $\beta = 1$. Calculating the K value for the NW sector gives the result $K = 45.17$. Note that this K value is approximately 2.7 times greater than the K values calculated for the NE and SW sectors. This is so because the Kappa equation is very sensitive to clustering of objects, and two adjacent NW sector objects had low respective segment lengths and hence $\beta \neq 0$.

Finally, let's take a look at the K value for the southeast (SE) sector. Again, n = 12 and there are thirteen segments. The arbitrary unit values for consecutive sector lengths (i.e., $\{x_1, x_2...x_N\}$) are as follows: 43, 52, 4, 35, 2, 17, 18, 21, 5, 13, 31, 30, 33. No two adjacent x values are both less than 10, so the β value for SE sector is once again zero. Applying the Kappa equation yields a K value of 15.51 for SE sector.

Kappa values for the four sectors are thus: $K_{NE} = 16.24$; $K_{SW} = 17.03$; $K_{NW} = 45.17$; and $K_{SE} = 15.51$. The average of these four K values is 23.49. We may infer from this to a first approximation that Kappa values of $K < \sim 50$, will be characteristic for collections of objects that do not show obvious clustering or spatial associations. Within a certain range, the actual Kappa value calculated will in part depend on the orientation of the greatest dimension line; nevertheless, K value is fairly easily calculated and robust within certain limits.

Let's now turn to actual data. The Clemente Formation samples provided five different bedding plane surfaces with collections of tiny tourmaline crystals. These were the *Zirabagtaria ovata* n. gen. n. sp. holotype (IGM 4995), both sides of the *Clementechiton sonorensis* holotype (IGM 7461), and both sides of the *Korifogrammia clementensis* n. gen. n. sp. holotype (IGM 4998). The association with several different Ediacaran taxa suggests that the crystals are unlikely to be derived from any one of them (that is, *Zirabagtaria ovata* n. gen. n. sp., *Clementechiton*, or *Korifogrammia clementensis* n. gen. n. sp.) but rather that the tiny tourmalines were derived from the same species that *first*, did not otherwise preserve as a soft-bodied fossil, and *second*, was a reasonably common member of the Clemente Ediacaran biota.

Table 4.1 Kappa value (K) calculations

Field surface	Figure	n	β	K	$K > 50$?
Dot pattern NE	4.19	12	0	16.24	No
Dot pattern SW	4.19	12	0	17.03	No
Dot pattern NW	4.19	12	1	45.17	No
Dot pattern SE	4.19	12	0	15.51	No
IGM 7461 sole NW sector (with *Clementechiton sonorensis*)	4.2	12	5	101.24	Yes
IGM 7461 sole SE sector (with *Clementechiton sonorensis*)	4.2	6	1	56.71	Yes
IGM 7461 top	4.3	5	3	165.44	Yes
IGM 4998 sole (with *Korifogrammia clementensis* n. gen. n. sp.)	4.5	9	7	395.45	Yes

Importantly, detrital tourmalines were only detected on bedding plane surfaces of rock samples that also contained Ediacaran body fossils of very small to small size. It may be the case that the crystal creatures were juveniles of a large adult species that shed their dorsal tourmaline armor once they reached a particular body size or ontogenetic stage. Indeed, the tiny tourmaline crystal associations under analysis here may well represent crystal 'scleritomes' shed in a rite of passage similar to the molting of an exoskeleton or the shedding of deciduous teeth. We might speculate further and infer that there was some sort of allometric growth pattern or metamorphosis associated with casting off the crystals. Alternatively, of course, the crystal clusters may represent all that remains of the corpse of a crystal creature, quite possibly one with a diminutive adult body size.

The number of crystals was large enough ($n \geq 5$) on three of the five crystal-bearing bedding plane surfaces to permit use of the Kappa equation for a total of four sectors and four K values. The *Clementechiton sonorensis* holotype (IGM 7461) has tourmaline crystal accumulations on both sides of the rock that are of sufficient number to allow Kappa calculation. The sole surface, with the *Clementechiton sonorensis* fossil, has $n = 12$ in its NW sector (with sector lengths 55, 1, 1, 1, 1, 3, 4, 28, 30, 3, 22, 2, 22) and $n = 6$ in its SE sector (with sector lengths 76, 1, 19, 21, 2, 4, 50). The NW sector gives $K = 101.24$ and the SE sector gives $K = 56.71$. The bedding top surface of IGM 7461 has only one sector with $n = 5$, and sector lengths of 34, 7, 3, 1, 2, and 106, respectively. This gives a K value of 165.44.

The *Korifogrammia clementensis* n. gen. n. sp. holotype (IGM 4998) specimen has tiny tourmaline crystals on both sides of the rock, but in sufficient numbers for Kappa calculation only on the sole surface, that is, the sole surface with the *Korifogrammia clementensis* n. gen. n. sp. holotype. Here we see $n = 9$, with sector lengths 112, 3, 5, 1, 3, 0.5, 0.5, 3 5, and 127, respectively. The Kappa value calculates to 395.45. Table 4.1 summarizes the K values calculated thus far.

Recall that above we inferred, to a first approximation, that Kappa values of $K < {\sim}50$ will be characteristic for collections of objects that do not show obvious clustering or spatial associations. An inference of clustering appears to be borne out by data from the fossiliferous slabs. All four show K values in excess of 50. The highest value of K calculated so far is for IGM 4998, a value of over 395. This is due to a very close clustering of tourmaline crystals as seen in Fig. 4.5. The lowest value for the actual specimens, IGM 7461 sole surface, southeastern sector, is $K = 56.71$. The Kappa value here, although low, is still above 50 and thus provides some support for the hypothesis that there is crystal clustering in this sector. Crystals IGM 7461[14-16] in particular form a triad that may have been derived from a single animal. In taphonomic terms, the K value for a particular crystal creature will potentially decline over time as currents, burrowers, etc. disturb and disperse the original cluster.

By the criteria provided above for interpretation of K values, all of the bedding plane sectors with sufficient quantities of crystals show evidence for spatial association of tourmaline crystals (that is, $K > 50$). Note, interestingly, that crystals IGM 7461[22] and IGM 7461[23] are superimposed on the holotype valve specimen of the aculiferan *Clementechiton sonorensis* (Fig. 4.2). Similarly, crystals IGM 4998[1] and IGM 4998 [2] are superimposed on the holotype of *Korifogrammia clementensis* n. gen. n. sp. (Fig. 4.5). Note also that the tight cluster of tiny crystals, in the sector with the highest K value, are adjacent to the anterior end of the best-preserved specimen of the kimberellomorph *Kimberella* cf. *K. quadrata*. These associations may be entirely accidental, but could also represent evidence for the aculiferan and the kimberellomorph interacting with crystal creatures, such as feeding on a shed crystal-ornamented molts, or even preying upon the crystal creatures themselves.

There is one other feature worth mentioning on the sketch map image of Fig. 4.2. Going from near the center of the image, and right on top of the fossil valve of *Clementechiton sonorensis*, there are three pairs of tourmaline crystals. The pairs are approximately evenly spaced, and delineate a line that runs north and slightly to the east (if north is considered to be at the top of the map). These pairs, in order from most southerly to most northerly, are [22, 23], [11, 12] and [9, 10] respectively. The pattern here is limited to three occurrences of pairs in a line, but nevertheless suggests several possible interpretations. First, the pairs may be due to the fact that the two crystals of each pair were still attached by a scrap of dorsal cuticle that originally held the crystals in place on the dorsal surface of the animal. In this case we see the distended remains of a disarticulated invertebrate. Alternatively, the pairs may have been dropped by the animal as pairs, with shedding perhaps occurring in an episodic fashion that accounts for the relatively even spacing.

Finally, and more speculatively, we might consider the crystal pairs as pairs of structure shed from something like a radula. The radula of *Kimberella* is thought to have consisted of two radular teeth. An agglutinated radula is an unknown occurrence in either the modern world or the fossil record; however, the concept of an agglutinated radula is not impossible in principle assuming that an animal capable of forming an agglutinated radula was able to obtain the crystals (and the shape of the three-edged acicular tourmaline crystals might be a good choice for a radula mineral!) to its radular belt. With a Mohs hardness scale value of 7–7.5, tourmaline is as hard or harder than quartz. The attachment site, even if continuously reinforced with bio-glue, might be subject to materials fatigue and wear, thus radular 'teeth' might very well fall off in pairs. Pointed radular teeth are well suited for grazing on algal tissue (Padilla 1998). If this interpretation is correct, then of course it has the potential to introduce a certain amount of noise into calculations of K value.

Fig. 4.20 Reconstruction of acicular tourmaline-bearing crystal creature. Length of crystal creature estimated to be 2 cm (Illustration by Caitlin Fisher)

Note that the bases of all the Sonoran crystals are broken in roughly the same way. Most interesting, however, is the fact that all of the noteworthy crystals are approximately the same size, with lengths in the 150–300 micron range. The process I propose to accomplish this biological sorting is collection of these crystals by an early metazoan that, like the later (Early Cambrian) agmatans and *Onuphionella*, selected distinctive particles from its sedimentary environment to form, not in this case an agglutinated ilmenite 'scleritome' nor mica-flake tube (Signor and McMenamin 1988), respectively, but rather a dorsal array of acicular tourmaline crystals. Although presumably small in terms of body size, these "crystal creatures" would have been a striking presence on the Proterozoic sea floor with their array of glossy greenish black acicular tourmaline crystals. As they grazed the Proterozoic sea floor along with the creatures (described in the previous chapter) preserved as mostly soft-body impressions, the crystal creatures would have resembled tiny crawling black crystal fans.

Motility is implied in the crystal creature life style for two reasons. First, crystal creatures would be required to move around to locate crystals of appropriate size, color and composition to add to their dorsal array. Second, the collection of crystals probably provided ballast to keep the animals securely grounded to the sea floor as they moved about on the sediment-water interface. The dorsal crystal array would also have provided protection from predators, although this may have been a secondary consideration at the time considering that macrophagous predators appear to have been rare or absent on the Proterozoic sea floor. Figure 4.20 shows a reconstruction of a crystal creature. The creature as drawn here somewhat resembles a modern solenogastre, and is in the size range for a small solenogastre, but solenogastres have biomineralized calcareous spicules (Todt and Wanninger 2010) and are not known to form agglutinated skeletons. Nevertheless, I hypothesize here that the crystal creature represents a Proterozoic solenogastre that selects acicular tourmaline crystals to form its agglutinated dorsal skeleton.

The animal will be left here in open nomenclature, pending discovery of a better-articulated specimen that shows the arrangement of the tourmalines in the crystal 'scleritome.' Nevertheless, this discovery represents both the earliest known case of animal agglutinization, the earliest known animalian case of monomineralogic agglutinization, and the earliest known bioaccumulation of separate, individual crystals of the same mineral.

One possible problem with this interpretation should be considered. If the reconstruction shown in Fig. 4.20 is correct, then why are there not more crystals clustered together in the places where they occur in groups? One possible solution to this question is that these groups of crystals were shed by accident, similar to the occurrences of single, isolated crystals as seen on some of the Clemente Formation Ediacaran fossil-bearing rock samples (e.g., Figs. 3.7 and 4.6), but shed in small groups. Possibly some of the crystals were tangled by sea floor biomat filaments and were pulled free from the dorsal side of the animal. This might be the case if, say, the crystal creature had been passing through a particularly dense patch of biofilm or microbial mat.

The dark color of the tourmaline crystals would have better absorbed solar radiation, and could have provided warmth for the crystal creature as suggested to me by Dianna L. Schulte McMenamin. Alternatively, the dark crystals may have served as a sunshade and UV screen to protect the animal from any harmful effects of solar radiation. One can get sunburned under water, as UV-B can pass through several meters water depth. Caldwell and Natchwey (1975) determined that UV-B "penetration in clear waters off the coast of Puerto Rico was remarkably good: intensities were attenuated to 50 per cent of surface levels at 4 m. Certainly the widely held misconception that UV-B radiation does not penetrate into water must be dispelled." Interestingly, protection against UV-B might also have induced the development of a biomineralized dorsal scleritome (as an alternative crystal screen strategy) in Proterozoic shallow marine animals, at a time prior to the expansion of macrophagous predation at the beginning of the Cambrian. Mica is opaque to UV radiation with a wavelength below 260 nanometers and attenuates the wavelengths of UV-B radiation as well (280–315 nm; Espe 2013). Thus the agglutinated mica tube of *Onuphionella*, like the tourmaline crystals of crystal creature, and possibly also the titanium-bearing (ilmenite and anatase) dorsal agglutinated skeletons of agmatans (Firby and Durham 1974; Peel 2016), may have served as a sunscreen.

References

Beurlen H et al (2011) Geochemical and geological controls on the genesis of gem-quality "Paraíba Tourmaline" in granitic pegmatites from northeastern Brazil. Can Mineral 49(1). https://doi.org/10.3749/canmin.49.1.277

Buettner KM, Valentine AM (2012) Bioinorganic chemistry of titanium. Chem Rev 112:1863–1881

Butterfield NJ (2008) An early Cambrian radula. J Paleontol 82(3):543–554

Caldwell MM, Natchwey DS (1975) Introduction and overview. Chapter 1. CIAP monograph 5. United States Department of Transportation, Washington, DC

Chatterjee SR et al (1975) Authigenic tourmaline in the Precambrian metasediments around Jamua, District Bhagalpur, Bihar, India. Sediment Geol 13(2):153–156

Closs D (1967) Goniatiten mit radula und kieferapparat in der Itararé Formation von Uruguay. Palaeont Zeitschr 41:19–37

Conway Morris S, Peel JS (1995) Articulated halkieriids from the Lower Cambrian of North Greenland and their role in early protostome evolution. Philos Trans R Soc B 347(1321):305–358

Cooper GA et al (1952) Cambrian stratigraphy and paleontology near Caborca, northwestern Sonora, Mexico. Smithson Misc Collect 119(1):1–184

Ertl A (2007) Über die typlokalität und die nomenklatur des minerals dravit. Mitt Österr Miner Ges 153:265–271

Espe W (2013) Materials of high vacuum technology. Volume 2, silicates. Pergamon Press, Elsevier, Oxford

Firby JB, Durham JW (1974) Molluscan radula from earliest Cambrian. J Paleontol 48(6):1109–1119

Gabbott SE (1999) Orthoconic cephalopods and associated fauna from the late Ordovician Soom Shale Lagerstätte, South Africa. Palaeontology 42:123–148

Gehling JG et al (2014) Scratch traces of large Ediacara bilaterian animals. J Paleontol 88(2):284–298

Hagadorn JW, Waggoner B (2002) The early cambrian problematic fossil *volborthella*: new insights from the Basin and range. In: Corsetti F (ed) Proterozoic-Cambrian of the Great Basin and beyond, book 93. The Pacific section of the Society of Economic Paleontologists and Mineralogists (SEPM), Tulsa, pp 135–149

Padilla DK (1998) Inducible phenotypic plasticity of the radula in *Lacuna* (Gastropoda: Littorinidae). Veliger 41(2):201–204

Peel J (2016) Anatase and *Hadimopanella* selection by *Salterella* from the Kap Troedsson Formation (Cambrian Series 2) of North Greenland. GFF 139(1). https://doi.org/10.1080/11 035897.2016.1227365

Scarani V (2010) Information science: guaranteed randomness. Nature 464:988–989

Signor PW, McMenamin MAS (1988) The Early Cambrian worm tube *Onuphionella* from California and Nevada. J Paleontol 62(2):233–240

Signor PW, McMenamin MAS (1994) Lower Cambrian fossil *Volborthella*: the whole truth or just a piece of the beast?—reply. Geology 22(7):666

Smith MR (2012) Mouthparts of the Burgess Shale fossils *Odontogriphus* and *Wiwaxia*: implications for the ancestral molluscan radula. Proc R Soc B. https://doi.org/10.1098/repb.2012.1577

Taylor HP (1974) The application of oxygen and hydrogen isotope studies to problems of hydrothermal alteration and ore deposition. Econ Geol 69:843–883

Todt C, Wanninger A (2010) Of tests, trochs, shells, and spicules: development of the basal mollusk *Wirenia argentea* (Solenogastres) and its bearing on the evolution of trochozoan larval key features. Front Zool 7(6). https://doi.org/10.1186/1742-9994-7-6

Vinther J (2015) The origins of molluscs. Palaeontology 58(1):19–34

von Middendorf AT (1847) Beiträge zu einer malacolozoologica rossica. Chitonen, St. Petersburg

Yochelson EL (1977) Agmata, a proposed extinct phylum of Early Cambrian age. J Paleontol 51(3):437–454

Chapter 5
Trace Fossil Geometry

Abstract The ichnofossils *Treptichnus* and *Multina* provide clues about the development of behavior in burrowing metazoans. A particular behavior type may be condensed and superimposed on the original burrowing pattern of programmed behavior. For example, a fine scale sinusoidal pattern may be superimposed on a large scale sinusoidal pattern. Side to side small scale branching may be superimposed on side to side main burrow elongate loops. Both burrow types are convergent behaviors with the same function, namely, more thorough deposit feeding in deep water sediments and/or low organic content sediments where the animal is required to thoroughly sift the sediment and must avoid reprocessing sediment that has been already fed upon.

Keywords Ichnofossils · Graphoglyptids · *Helminthopsis* · *Cochlichnus* · *Multina* · *Treptichnus* · *Paleodictyon* · *Thalassinoides* · *Vaqueroichnus* · *Zoophycus*

Back in the 1970s, Russian paleobiologist Mikhail A. Fedonkin studied the relationship between trace fossils of the Proterozoic and the behaviors exhibited by the animals that made them (Fedonkin 1978). He noted that increase in behavioral complexity was the result of two factors. The first was the appearance of fundamentally novel types of trackways. These would include (assuming that all of these structures have been properly interpreted as burrows) sinusoidal burrows, spiral burrows, branching burrows, and so on (McMenamin 1998). The second was combination of two or more of these fundamental ichnofossil types to form more complex traces that could help the tracemaker tap into new food sources.

Figure 5.1 shows a sinusoidal burrow belonging to the ichnotaxon *Helminthopsis* from the Proterozoic-Cambrian sequence of Sonora, México. Similar specimens have been reported from the Cambrian Shiyantou Formation, Meishucun, China (Weber et al. 2007). The Sonoran trace fossil could arguably be placed in the ichnogenus *Cochlichnus* or even ichnospecies *Cochlichnus anguineus*, but as these ichnotaxa are more regularly and evenly sinusoidal (Buatois and Mángano 2003) than in the Mexican trace fossil, it will be retained here in *Helminthopsis*. A combination of the spiral trace type and the sinusoidal type to form a spiral-sinusoidal trace could lead to more complete (and hence efficient) deposit feeding of the sea floor (Fig. 5.2). Organisms with large enough brains or sufficiently flexible behavior to engage in

Fig. 5.1 *Helminthopsis* from Cambrian strata of the Caborca Region, Sonora, México. Note two specimens with sinusoidal curves of different amplitude. A bilobed trace intersects the end of one of the *Helminthopsis* specimens. Field sample 10 of 3/23/90, lower Unit 2 of the Puerto Blanco Formation, Cerro Rajón, Sonora, México. Scale in cm

Fig. 5.2 Combination of the spiral trace type and the sinusoidal type to form a spiral-sinusoidal trace could lead to more complete (and hence efficient) deposit feeding of the sea floor. This hypothetical burrow shows a combination of sinusoidal burrowing behavior (imparting the wavy curve) and a spiral burrowing behavior (giving the trace its overall shape) (Modified from McMenamin (1998))

combinations of burrowing behaviors would potentially have considerable selective advantage over forms capable of only a single burrow type.

Treptichnid ichnofossils represent successive, three-dimensional probing through sediment, allowing the tracemaker to process sediment for deposit feeding in three dimensions. Treptichnid trace fossils are known from the base of the Cambrian (or perhaps just before the Cambrian begins) and are subsequently found throughout geologic time up to the present day (Getty et al. 2016). This era-defining trace fossil is associated with the Cambrian explosion, and in fact represents some of the best evidence that the Cambrian explosion was indeed an abrupt evolutionary event (Seilacher 1956; McMenamin and Schulte McMenamin 1990) as the appearance of this new type of trace fossil represents a major advance in animal behavior. The name of the basal Cambrian treptichnid ichnotaxon has varied depending on the preferences of the authors of the various publications that describe its occurrences, but it is now generally referred to as either *Trichophycus pedum* (McMenamin 2016) or *Treptichnus pedum* (Weber et al. 2007). Either way, the trace fossil represents a series of probes through sediment that connect a central burrow or tunnel to iterated side branches.

A similar type of treptichnid, *Treptichnus bifurcus* was reported in Jurassic sedimentary rocks of the Hartford Basin, Massachusetts in 2005 (McMenamin 2005). In an analysis of these exceptionally preserved fossils, Getty et al. (2016) remarked:

> The Holyoke *Treptichnus* did not exhibit significant vertical relief, and the presence and absence of projections is explained by the positioning of new segments at different points along older ones. The bulbous ends of burrow segments resulted from tracemaker behavior, not sediment compaction. Drawing on the analysis of the Holyoke material, a new reconstruction is proposed that presents continental *Treptichnus* as a shallow mole-tunnel-like burrow produced just below the sediments. This reconstruction is consistent with the morphology of Recent *Treptichnus*-like burrows produced by fly (dipteran) larvae, which are considered the most likely makers of the Holyoke *Treptichnus*.

In *Dynamic Paleontology* (McMenamin 2016) I argued that graphoglyptid trace fossils were derived from Early Cambrian microburrow nests, fossils that were originally interpreted as sponge spicules or irregular calcite-filled spaces (Debrenne et al. 1989). The puzzling life history of the unknown animals that fabricate graphoglyptid burrows today can be explained in part by considering that the modern examples (always unoccupied when studied by researchers so far) of the graphoglyptid *Paleodictyon* are actually abandoned nests or nurseries for juveniles of the tracemaker.

This scenario assumes that the original burrow from which graphoglyptids were derived was agrichnial, a 'mushroom farmer' life style as Dolf Seilacher used to call it. There does seem to be some evidence for this assumption if the very ancient Proterozoic trace fossil *Vaqueroichnus stewarti* is interpreted correctly as a submat miner trace that was cultivating protected patches of microbes (the "rudimentary microbial herding" of McMenamin 2016).

Graphoglyptid-type agrichnial trace fossils may also have developed from strictly deposit-feeding precursors. It is possible, even likely, that graphoglyptid ichnofossils evolved more than once, and if so, this would represent a case of behavioral convergence. Another possible route for the development of graphoglyptids, in this case from an initially deposit-feeding lineage, would be via development from the treptichnids. In other words, there may be at least two routes for development of the graphoglyptid burrow morphology, one via Cambrian microburrow nests (resulting in agrichnial network ichnofossils, nesting site ichnofossils, or combinations of the two) and another via Cambrian treptichnids (leading from deposit feeding trace fossils to agrichnial traces). Multiple lineages, by the way, could participate in comparable transformations. It is well established that similar tracks and trails may be made by unrelated types of organisms.

The Holyoke *Treptichnus* ichnofossils were probably made by dipteran larvae digging shallow horizontal burrows (Getty et al. 2016), and thus they made these

Fig. 5.3 *Treptichnus bifurcus* from the Gaulin site, Portland Formation, lower Jurassic, Holyoke, Massachusetts. Note the bulbous ends of the alternating side branches. Scale bar in cm (Photograph by Mark McMenamin)

burrows themselves rather than having the burrows constructed by their parent(s) as was apparently the case with the Early Cambrian microburrowers. As these dipteran burrows from the Jurassic are closely associated with dinosaur tracks in the same lacustrine shales, and can be locally very abundant (Fig. 5.3), they may represent the aquatic larval burrows of a blood-sucking fly that fed upon dinosaurs. In support of this hypothesis is the observation that a culicomorphan fly with a long proboscis (Culicomorpha indet.), known from the Late Triassic (Late Carnian) Newark Supergroup of Virginia, USA, represents the oldest known fossil of an insect with "a structure specialized apparently for blood feeding" (Blagoderov et al. 2007). The Holyoke strata are also considered to be (Portland Formation) part of the Triassic-Jurassic Newark Supergroup (Blagoderov et al. 2007; Olsen et al. 2003).

If this blood-feeding hypothesis is correct, exactly what type of dipteran was biting the Jurassic dinosaurs is unknown. However, biting midges belonging to the modern genera *Culicoides* and *Leptoconops* are known to produce larvae that inhabit aquatic and semi-aquatic sites such as water saturated soils, habitats somewhat comparable to the freshwater aquatic habitats that led to deposition and preservation of Jurassic Newark Supergroup specimens of *Treptichnus*.

An unnamed trace fossil (Fig. 5.4) resembling *?Thalassinoides* also occurs in the Newark Supergroup of Holyoke, Massachusetts, at a regionally famous dinosaur track park on the west bank of the Connecticut River off Route 5. The trace fossil shown in Fig. 5.4 is larger than, and was made by a different tracemaker than, the one that made the *Treptichnus* ichnofossils in the same strata (shales of the Portland Formation). There is nevertheless a comparable feature between the two types of traces, namely, a side branch proceeding from a main burrow or tunnel.

Fig. 5.4 ?*Thalassinoides* from the dinosaur track park on the west bank of the Connecticut River off Route 5, Holyoke, Massachusetts, at the eight acre Dinosaur Footprints site. An unidentified vertical burrow penetrates the plane of bedding at the upper right. Scale bar in cm (Photograph by Mark McMenamin)

Fig. 5.5 *Multina* isp. from Late Triassic Luning Formation of Nevada. Ichnofossil occurs on an upper bedding plane surface. The burrow interior is lined with iron oxide deposits (Zapata et al. 2017). Scale bar in cm (Photograph by Mark McMenamin)

The ichnofossil *Multina* isp. has recently been reported (Figs. 5.5, 5.6, and 5.7) from Late Triassic strata of Nevada (Zapata et al. 2017). *Multina* isp. is a non-graphoglyptid ichnofossil (although a possible precursor to true agrichnial graphoglyptids) consisting of a meandering burrow network that can mimic the regular hexagonal-geometric patterns of graphoglyptids such as *Paleodictyon*. The *Multina* isp. specimens from Nevada were thought to be deposit feeding traces that nevertheless developed iron oxide-stabilized walls (Zapata et al. 2017; McMenamin 2016). A new analysis of the *Multina* from Nevada confirms the deposit-feeding nature of this trace fossil.

Fig. 5.6 *Multina* isp.,
sketch map of best
preserved specimen in
Zapata et al. (2017).
Arrows indicate the
tracemaker's preferred
pathway through the
sediment. Numbers 1–4
indicate the main trunk
burrow loops as shown in
order of formation. Scale
bar 1 cm (Photograph by
Mark McMenamin)

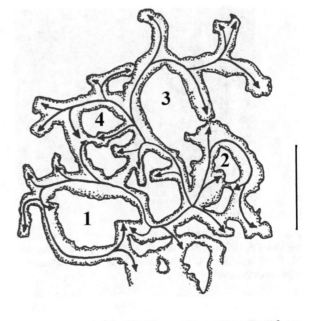

Fig. 5.7 *Multina* isp.,
simplified sketch of the
ichnofossil geometry, with
side branches indicated as
shown. Numbers 1–4
indicate the main trunk
burrow loops as shown in
order of formation. Dotted
lines indicate possible
burrow connections

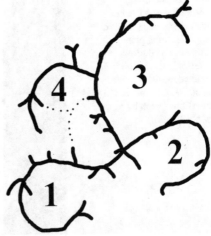

Figure 5.6 shows a sketch map of the best-preserved Nevadan specimen of
Multina isp. Due to the angular relationships between the trunk burrow and the side
branch burrows, it is possible to trace the progress of the tracemaker through the
sediment. The animal's path is shown with arrows in Fig. 5.6. Figure 5.7 shows a
simplified sketch of the ichnofossil geometry, with side branches indicated as shown.

Figure 5.8 shows an idealized geometry of *Multina* isp. This diagram makes
evident the apparent relationship between the main burrow trunk and the side
branches. On any section of main trunk burrow, the side branches split off in an
alternate fashion first on one side of the trunk and then the other, similar to the

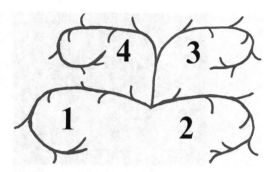

Fig. 5.8 Sketch showing idealized geometry of *Multina* isp. Predicted hypothetical trace type consisting of alternating main loop branches with superimposed small scale sinusoidal waves. Numbers 1–4 indicate the main trunk burrow loops as shown in order of formation

formation of side branches with bulbous ends as seen in *Treptichnus bifurcus* (Fig. 5.3). The main trunk burrow itself consists of four elongate loops. The loops form, one to the left and then one to the right, and then a second pair of trunk burrow loops forms beyond the first, larger pair.

In the sketch map in Fig. 5.6, I have numbered the main trunk burrow loops as shown in order of formation, 1–4. In the first course of main loops, the loop first formed is inferred to be the larger one on the left, and the second (number 2) is the one on the right. In the second course of main loops, the first (and larger) one formed is on the right, and the second-formed (number 4) is on the left. Loop number 4 seems to be somewhat squeezed in between loops 1 and 3.

We see here then a geometrical pattern with fractal characteristics that provides insight into the behavior of the *Multina* isp. tracemaker. The animal's burrow program dictates that main loops are formed in an alternate fashion, left-right first, right-left next, left-right next, and so on. Side branches are also programmed to alternate from one side to the other.

Note the similarity in the idealized geometry of *Multina* isp. (Fig. 5.8) to that of the hypothetical sinusoidal trace shown in Fig. 5.2. In the latter case we also see a combination of behavioral routines, development of sinusoidal geometry superimposed on a spiral pattern. This compares with the initial alternating elongate loops, with superimposed side branches, of *Multina* isp.

These two types of traces give us an important clue about the development of behavior in burrowing metazoans. A new behavior modality may be superimposed on the original burrow program. The fine scale sinusoidal pattern may be superimposed on a larger (and presumably original; Fig. 5.2) spiral pattern. Side to side small scale branching may be superimposed on side to side main burrow elongate loops. Both burrow types are convergent behaviors with the same function, namely, more thorough deposit feeding in deep water and/or low organic content sediments (Seilacher 1967, 1974) where the animal needs to thoroughly sift the sediment and must avoid reprocessing sediment that has been recently processed. It is entirely possible that there exists, or once existed, say, alternating main loop

Fig. 5.9 Scale of the
freshwater osteoglossid
fish *Phareodus encaustus*
(Cope 1871) from the
Eocene Green River
Formation. Trunk scale
showing development of
hexagonal ridges. Scale in
cm (Photograph by Mark
McMenamin)

branches with superimposed small scale sinusoidal waves, although such has not yet been reported from either the modern world or the fossil record. Note, interestingly, that in both this predicted trace fossil and the actual *Multina* ichnofossil (Figs. 5.6 and 5.7), there exists a spiral aspect to the construction of the entire trace fossil as the loops proceed from Loop 1 to Loop 4 in a counterclockwise direction. This suggests that the sweeping spirals of the deep-water trace fossil *Zoophycus* may have originated from *Multina* or a *Multina*-like antecedent.

There is nothing particularly unusual about the hexagonal pattern of the *Paleodictyon* graphoglyptid trace fossil; it certainly does not represent an outpost of a tiny alien civilization (McMenamin 2016). Comparable hexagonal arrangements can occur on body fossils as well, as in the scale of the Eocene Green River Formation osteoglossid fish *Phareodus encaustus* (Cope 1871) (Fig. 5.9; see Grande 1980; Li et al. 1997).

We thus may discern a developing of more complex traces by superimposition of fundamental simple burrowing patterns (alternate branching, sinusoidal, etc.). Each of these fundamental types, when they reach a particular degree of space filling in two dimensions at the sediment-water interface, can potentially lead to the development of a graphoglyptid-type agrichnial lifestyle.

References

Blagoderov V et al (2007) How time flies for flies: diverse diptera from the Triassic of Virginia and early radiation of the order. Am Mus Novit 3572:1–39

Buatois LA, Mángano MG (2003) Early colonization of the deep sea: ichnological evidence of deep-marine benthic ecology from the early Cambrian of northwest Argentina. PALAIOS 18:572–581

Debrenne F et al (1989) Lower Cambrian bioconstructions in northwestern Mexico (Sonora). Depositional setting, paleoecology and systematics of archaeocyaths. Geobios 22(2):137–195

Fedonkin MA (1978) Drevneishie iskopaemy sledi i puti evoluutsi povendeniya grynotoedov. Paleontol Zh 2:106–111

Getty PR et al (2016) A new reconstruction of continental *Treptichnus* based on exceptionally preserved material from the Jurassic of Massachusetts. J Paleontol 90(2):269–278

Grande L (1980) Paleontology of the Green River Formation, with a review of the fish fauna. Geol Surv Wyoming Bull 63:1–333

Li G-Q et al (1997) The species of †*Phareodus* (Teleostei: Osteoglossidae) from the Eocene of North America and their phylogenetic relationships. J Vert Paleont 17(3):487–450

McMenamin MAS (2005) Microbial influence and environmental convergence in marine (Proterozoic) and lacustrine (Jurassic) depositional settings. Geol Soc America Abstr Prog 37(1):7

McMenamin MAS (1998) The garden of Ediacara: discovering the first complex life. Columbia University Press, New York

McMenamin MAS (2016) Dynamic paleontology: using quantification and other tools to decipher the history of life. Springer, Cham

McMenamin MAS, Schulte McMenamin DL (1990) The emergence of animals: the Cambrian breakthrough. Columbia University Press, New York

Olsen P et al (2003) Causes and consequences of the Triassic-Jurassic mass extinction as seen from the Hartford Basin. In: Brady JB, Cheney JT (eds) Guidebook for field trips in the five college region. Five College Departments of Geology and Geography, Amherst/Northampton/Massachusetts, p B5-1-B5-41

Seilacher A (1956) Der Beginn des Kambriums als biologische Wende. Neues Jahrb Geol Palaontol Abh 108:155–180

Seilacher A (1967) Bathymetry of trace fossils. Mar Geol 5:413–428

Seilacher A (1974) Flysch trace fossils: evolution of behavioral diversity in the deep-sea. Neues Jahrbuch für Geologie und Palaontologie, Monatshefte 1974:233–245

Weber B et al (2007) Precambrian-Cambrian trace fossils from the Yangtze platform (South China) and the early evolution of bilaterian lifestyles. Pal Pal Pal 254:328–349

Zapata LP et al (2017) *Multina* isp. from the late Triassic Luning Formation, Nevada. Ichnos 24(1):64–71

Chapter 6
Albion's Ænigma

Abstract The enigmatic Middle Devonian chordate *Palaeospondylus gunni* from the the Achannaras Slate Quarry, Caithness, Scotland, has recently been reinterpreted as a member of the Cyclostomata (the group that includes living jawless fishes). If this interpretation is correct, then *Palaeospondylus* either retains a vertebral column as a vestigial organ, or represents a remarkable case of skeletal homoplasy convergent on bony fishes. The 'hard cartilage' of *Palaeospondylus* may be a unique biomaterial restricted to the Devonian lifespan of this enigmatic genus.

Keywords *Palaeospondylus* · *Dipterus* · Cyclostomata · Sclerotome · Hagfish · Cephalochordates

British marine biologist Walter Garstang (1868–1949) is best known for his speculative hypothesis (Holland 2011) that vertebrates are descendants of sea squirts (tunicates). Walter Garstang, co-publishing with his daughter Sylvia, presented a thorough explication of the hypothesis in 1926 (Garstang and Garstang 1926). In their scenario, vertebrates are descended from a swimming larva (of a sessile adult ascidian) that undergoes paedomorphosis, thereby retaining its juvenile knack for swimming. In evolutionary terms, the lineage sheds its adult phase (Gould 1977; Gee 2007). The idea was championed so forcefully by ascidian expert Berrill and vertebrate paleontologist A. S. Romer that it became known as the Garstang-Berrill-Romer (GBR) hypothesis (Lacalli 2005).

Alas, the GBR hypothesis has not fared well in light of subsequent research. New studies suggest that the common ancestor of the three chordate subphyla (tunicates, amphioxus and vertebrates) was itself a "tadpole-like, free living creature" (Holland 2011), thus eliminating the need for a heterochronic event of paedomorphosis to generate the first vertebrate. According to Holland (2011), "the evidence against Garstang's schemes for chordate evolution is by now compelling."

Annona et al. (2015) published a table listing all the invertebrate animals "that swim by undulating the entire body or its posterior region." The list includes ctenophores, trematode worms, chaetognaths, nematodes, nematomorphs, mayfly larvae, annelid worms including leeches, hemichordates (one species only), and the chordates including cephalochordates (amphioxus, fish of course, some aquatic reptiles) and tunicate larvae (Meinertzhagen and Okamura 2001).

The ctenophores or comb jellies are an interesting addition to the list. Nearly all species are predators, but they are also preyed upon and this explains the undulatory swimming in one species, namely, the escape response of the ribbon-shaped Venus Girdle (*Cestum veneris*). *Cestum veneris* escapes its predators by swimming like a fish. Ctenophores have a fossil record extending back to the Cambrian (but curiously not before, thus adding to the magnitude of the Cambrian explosion).

Tunicates play a key role in the development of evolutionary theory, for it is they that led to the defeat of the classic embranchments classification scheme for Kingdom Animalia. Under the four embranchment theory, promoted by great early natural scientists such as Cuvier, Agassiz and Von Baer, animals could be placed into one of four groups. The groups were distinct and separate and neither evolved from one another nor necessarily shared a common ancestor. The four embranchments included: mollusks, articulates (arthropods and annelids), radiates (cnidarians, ctenophores and echinoderms) and the vertebrates. Embryologist von Baer conceded that microevolutionary changes could occur within the four embranchments, but insisted that no evolutionary linkages existed between the four branches (Annona et al. 2015).

In efforts to establish evolutionary linkages between the vertebrates and the invertebrates, the embranchments concept was put to the test. Alexander Kowalevsky (ironically, von Baer's student) was eventually able to show the presence of a notochord in ascidian tunicates (Kowalevsky 1866). This provided strong support for the concept that tunicates and vertebrates were relatives. Darwin (1871) endorsed the concept, and in a fatal blow to the embranchment concept, wrote that we would be "justified in believing that at an extremely remote period a group of animals existed, resembling in many respects the larvae of our present Ascidians, which diverged into two great branches—the one retrograding in development and producing the present class of Ascidians, the other rising to the crown and summit of the animal kingdom by giving birth to the Vertebrata." Thus was forged an evolutionary linkage between the vertebrates and the invertebrates. Interestingly, a variant of the embranchment concept continues to operate in the paleontological and biological sciences in the guise of the Cambrian Explosion question. The profound differences between the various animal phyla appearing at the base of the Cambrian, a 'basal polytomy' as it has been called, along with insufficient time for a lengthy chain of intermediate forms to bridge the gaps, remains as puzzling now as it was in C. Darwin's day. Even if the Aculifera clade can be traced back into the Proterozoic, the divergences between molluscan clades "occurred at some point relatively early in the Cambrian" (Sigwart 2017).

In their discussion of the difficulty of discerning the evolutionary origins of the notochord, Annona et al. (2015) note that one significant advance has been made, namely, that earlier concepts that amphioxus is more closely related to vertebrates than are tunicates has been overturned. The correct relationship is as follows: amphioxus is the sister group to the clade consisting of the vertebrate-plus-tunicate clade (Desulc et al. 2006). This is interesting, as it suggests that we should reexamine the relationships between tunicates and vertebrates. A key question is as follows: What is the potential for convergence between these two related subphyla?

Palaeospondylus gunni is an enigmatic Middle Devonian chordate fossil that was first discovered in 1890 in the Achannaras Slate Quarry, Caithness, Scotland.

As its scientific name indicates ("Gunn's Ancient Vertebra"), it has long been presumed to be some kind of vertebrate fossil, arguably a fossil fish. Exactly what kind of vertebrate it represents, however, has been far from clear. The fossil has been claimed to be a jawless fish (agnathan), a shark, a placoderm, a holocephalian (chimaeriod), a lungfish, a frog tadpole (notoneurula), and a herring (Thomson 1992; Thomson et al. 2003; Hirasawa et al. 2016). Thomson et al. (2003) call *Palaeospondylus* perhaps "the most enduring of puzzles in palaeontology." I would agree that *Palaeospondylus* represents some kind of chordate, but precisely what kind of chordate is another question well worth careful examination.

Palaeospondylus fossils consist of coalified cartilagenous tissues; no bone material is present. This fact led Thomson et al. (2003) to conclude that *Palaeospondylus* was a larva, namely, the larva of a lungfish. Thomson et al. (2003) further inferred that *Palaeospondylus* was the larva of the ancient lungfish *Dipterus valenciennesi*, the most abundant fish fossil at the Achannaras Quarry. Curiously, however, *Palaeospondylus* reaches up to 6 cm in length, and never shows any sign of impending metamorphosis into a lungfish, nor any features that would link it conclusively to the lungfish *Dipterus*. Joss and Johanson (2007) were unconvinced by the larval lungfish hypothesis, and identified a number of flaws in the concept:

> Morphological features used to identify *P. gunni* as a larval lungfish include enlarged cranial ribs, rudimentary limb girdles, and absence of teeth. However, this combination of features does not characterize the extant lungfish *Neoceratodus forsteri*, even at very young stages, nor early stages of Devonian and younger fossil lungfish. Absence of teeth is problematic because early ontogenetic stages of fossil and living lungfish possess full dentitions including marginal teeth.

Palaeospondylus also lacks any scales. Thomson et al. (2003) argued that this lack of squamation enhanced the preservation of *Palaeospondylus*, for if it had a covering of tough scales as in the bichir *Polypterus*, its cartilaginous skeleton would have rotted away. However, available evidence suggests that *Palaeospondylus* indeed lacks a *scleritome* (of ectodermal origin, if not necessarily *sclerotome* of mesodermal origin). *Palaeospondylus* lacks teeth, ganoid scales, etc., and this strongly limits fruitful comparison with the fishes.

The Mount Holyoke College specimen of *Palaeospondylus* was acquired in an unusual fashion. I was asked to examine the rock and fossil collection of a deceased Mount Holyoke alumna, and to acquire from this collection whatever might be of use to the College's academic program. I was taken up to a dusty attic space with many boxes filled with very dusty rocks and fossils. Not much appeared to be very useful, but one rock stood out from the rest as potentially important. It was completely covered in dust and required several attempts to clean off the upper surface of the rock specimen. It came with an ancient label on crumbling brown paper, written in what appeared to be quill pen ink, that reads "*Paleospondylus gunni*/Old Red Sandstone/Achanarrais,/Sutherlandshire/Scotland".

The Mount Holyoke College *Palaeospondylus* is 2 cm in length (Figs. 6.1, 6.2, 6.3, 6.4, 6.5, and 6.6) and has a cranial region approximately 7 mm in length (Fig. 6.2). The vertebral structure is intact and fairly well preserved (Fig. 6.3). The chondrocranium is preserved in ventral view (Figs. 6.2 and 6.4), but is apparently missing the

Fig. 6.1 *Palaeospondylus*, Mount Holyoke specimen. Entire specimen, ventral view. Scale in mm

Fig. 6.2 *Palaeospondylus*, detail of head/cranial region. Scale in mm

tau-shaped (τ) tauidion structure and its caudal rostral plates (see Hirasawa et al. 2016) are squeezed together and shifted to the side. The fossil does preserve, however, a portion of the left ventral gammation structure (Fig. 6.5) in ventral view (Hirasawa et al. 2016). A slightly curved ridge running across the ventral side of the gammation

Fig. 6.3 *Palaeospondylus,* detail of vertebral region. Scale in mm

Fig. 6.4 *Palaeospondylus,* detail of chondrocranium, ventral view. Width of chondrocranium 2.4 mm

probably represents an anchoring zone for muscle attachment. The animal is distorted into an L-shaped bend, with the head region and anterior caudal vertebrae shown in ventral aspect, and the posterior part of the tail folded over to be visible in right lateral view. The tail has a flattened aspect that no doubt increased its efficiency for tail swimming. Figure 6.6 shows a reconstruction of *Palaeospondylus.*

Palaeospondylus has a very odd anteriormost structure that resembles the 'cow catcher' on an antique locomotive. Thomson et al. (2003) interpreted this structure as a "large attachment organ," supported on its sides by edge thickenings they referred to as rostralia. Thomson et al. (2003) noted that larvae of the gar *Lepisosteus* have comparable sucker organs, but again, gars are characterized by their heavy ganoid scales, and there is no sign of any such squamation in *Palaeospondylus.* As an alternative to the hypothesis of this anterior structure in *Palaeospondylus* serving

Fig. 6.5 *Palaeospondylus*, scanning electron microscope image of the left gammation in ventral view. A slightly curved ridge running across the ventral side of the gammation probably represents an anchoring zone for muscle attachment. Scale bar 300 microns

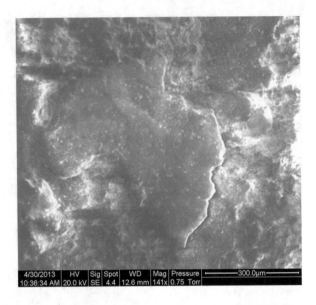

Fig. 6.6 Reconstruction of *Palaeospondylus*. Length of animal approximately 2 cm. Artwork by Smokeybjb, used here by CC BY-SA3.0

as a large attachment organ, resembling say the attachment organ on the anterior end of a tunicate larva, the rostralia in *Palaeospondylus* may perhaps better be interpreted as a sensory organ, comparable to the rostral organ in a coelacanth.

In their 2016 paper "*Palaeospondylus* as a primitive hagfish," Tatsuya Hirasawa, Yasuhiro Oisi and Shigeru Kuratani propose an interesting linkage (Hirasawa et al.

2016) between *Palaeospondylus* and the early ontogenetic stages of a modern jawless fish (cyclostome), the hagfish *Eptatreus*. Hirasawa et al. (2016) argue that "the *Palaeospondylus* cranium cannot be derived from the crown gnathostome [jawed fishes] pattern, which involves a dorsoventrally bifurcated mandibular arch, separate paired nostrils, and adenohypophysis." Instead, Hirasawa et al. (2016) identify features in *Palaeospondylus* that they link to putative homologous structures in members of the hagfish clade of the cyclostome chordates. Hirasawa et al. (2016) see the nasal capsule, the dental/lingual plates, and the subnasal rostral cranium in *Palaeospondylus* as being derived from the anterior nasal process, post-hypophyseal process and mandibular arch of cyclostomes such as the hagfish. I tentatively accept the conclusions of Hirasawa et al. (2016). If they are correct, represents a major advance in our understanding of an important and long-standing paleontological conundrum.

A cyclostome affinity for *Palaeospondylus* represents a truly remarkable case of convergent evolution. Hirasawa et al. (2016) are sensitive to the fact that *Palaeospondylus* possessing a vertebral column is very unusual for a cyclostome, but they support their inference by noting that vestigial vertebrae have been reported from the modern hagfish sclerotome (Ota et al. 2011). It seems possible that the modern hagfish has secondarily lost a mineralized or otherwise hardened vertebral encasement of its notochord. This seems a well-reasoned conclusion, considering that both hagfish and lampreys have cartilaginous fin rays (both dorsal and ventral) and that the origin of vertebrae (and thus vertebrates) begins with patches of cartilage adjacent to the notochord.

A very interesting comparison then may be made between cyclostomes and coelacanth fishes. As discussed at length in the next chapter, a key characteristic of coelacanths is that they lack vertebrae, and rely on their oil-turgid notochord in its primordial function as a stiffening rod. Adult humans still bear remnants of a notochord; these serve as the gel-like nucleus pulposus of our intervetebral discs (or intervertebral fibrocartilages; McCann et al. 2011). Cyclostomes and coelacanths, are both chordates to be sure, and both are cephalochordates, and they are also considered fish due to the cartilaginous tail rays and other piscine features, but in a certain sense they lack backbones. Similar remarks may be made about acipenseriform fishes such as paddlefishes and sturgeons, and they (such as *Condorlepis groeberi* from the Jurassic of Patagonia (López-Arbarello et al. 2013) and *Crossopholis magnicaudatus* from the Eocene Green River Formation of the Fossil Lake area, Wyoming (Fig. 6.7)) are considered a primitive group of bony fish. Nevertheless, we still classify cyclostomes and coelacanths in the subphylum Vertebrata. In the next chapter, we will continue this theme by examining a fascinating, newly discovered vestigial feature of the coelacanth sclerotome (or generally speaking, its bony skeleton).

Palaeospondylus has another connection to coelacanths, and that is the odd rostralia at its anterior end that I propose here served a function very similar to the rostral organ in coelacanths. This advanced prey-seeking sensory array was without close precedent until living coelacanths were captured and studied (Forey 1998). Also, note that *Palaeospondylus* appears to have cartilaginous rays extending poste-

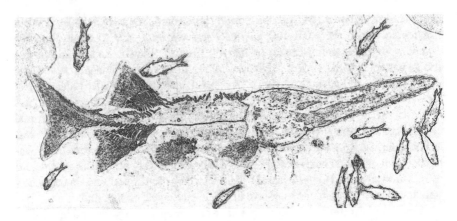

Fig. 6.7 *Crossopholis magnicaudatus* from the Eocene Green River Formation of the Fossil Lake area, Wyoming. The smaller fish are specimens of *Knightia eocaena*. Note heterocercal tail in *Crossopholis,* and the fine projections from its rostrum. These presumably enhanced the rostrum's performance as a sensory organ used to detect prey. Length of *Crossopholis* 97 cm

riorly from its vertebral column (hemal spines; Fig. 6.3), suggesting that these may have once existed on its dorsal side in its ancestral lineage.

Palaeospondylus appears to have taken a middle evolutionary path between that of cyclostomes (reliant on cartilage) such as lampreys and hagfish and that of bony fish (reliant on bone).

The black, shiny smooth material that constitutes the *Palaeospondylus* sclerotome is referred to in the literature as fossilized "cartilage," but I propose here that it may represent a singular type of biomaterial that is intermediate in some fashion between cartilage and bone. This 'hard cartilage' of *Palaeospondylus* may be a unique material restricted to the Devonian lifespan of this enigmatic genus.

References

Annona et al (2015) Evolution of the notochord. EvoDevo 6:30. https://doi.org/10.1186/s13277-015-0025-3

Darwin C (1871) The descent of man, and selection in relation to sex. John Murray, London

Desulc et al (2006) Tunicates and not cephalochordates are the closest living relatives of vertebrates. Nature 439:965–968

Forey PL (1998) History of the coelacanth fishes. Chapman & Hall, London

Garstang SL, Garstang W (1926) On the development of *Botrylloides* and the ancestry of the vertebrates. Proc Leeds Phil Lit Soc (Sci Sect) 1:81–86

Gee H (2007) This worm is not for turning. Nature 445:33–34

Gould SJ (1977) Ontogeny and phylogeny. Harvard University Press, Cambridge, MA

Hirasawa T et al (2016) *Palaeospondylus* as a primitive hagfish. Zool Let 2:20. https://doi.org/10.1186/s40851-016-0057-0

Holland ND (2011) Walter Garstang: a retrospective. Theory Biosci 130:247–258

Joss J, Johanson Z (2007) Is *Palaeospondylus gunni* a fossil larval lungfish? Insights from *Neoceratodus forsteri* development. J Exp Zool B Mol Dev Ecol 308(2):163–171

Kowalevsky A (1866) Entwickelungsgeschichte der einfachen Ascidien. Mém Acad Imp Sci St-Pétersbourg (Sér VII) 10(15):1–19

Lacalli (2005) Protochordate body plan and the evolutionary role of larvae: old controversies resolved? Can J Zool 83:216–224

López-Arbarello et al (2013) A new genus of coccolepidid fishes (Actinopterygii, Chondrostei) from the continental Jurassic of Patagonia. Palaeontol Electron 16(1) 7A 23 p

McCann M et al (2011) Tracing notochord-derived cells using a Noto-cre mouse: implications for intervertebral disc development. Dis Model Mech 5(73). https://doi.org/10.1242/dmm.008128

Meinertzhagen IA, Okamura Y (2001) The larval ascidian nervous system: the chordate brain from its small beginnings. Trends Neurosci 24(7):401–410

Ota KG et al (2011) Identification of vertebra-like elements and their possible differentiation from sclerotomes in the hagfish. Nat Commun 2:373

Sigwart JD (2017) Zoology: molluscs all beneath the sun, one shell, two shells, more or none. Curr Biol 27:R702–R719

Thomson KS (1992) The puzzle of *Paleospondylus*. Am Sci 80(3):216–219

Thomson KS et al (2003) A larval Devonian lungfish. Nature 426:833–834

Chapter 7
Coelacanth Vestiges

Abstract A vestigial second dorsal fin spine in a Carboniferous rhabdodermatid coelacanth (cf. *Rhabdoderma* sp.) provides new data bearing on the evolutionary origins of coelacanths. The fossil was recovered from a cannel coal deposit in the Allegheny Group of Ohio. A vestigial second dorsal fin spine occurs between the second dorsal fin endochondrial support plate and the second dorsal fin. The fin spine is preserved as a carbonized impression showing longitudinal ridges, comparable to the ridges seen on dorsal fin spine of the early osteichthyan *Guiyu oneiros*. A diagonal band of bends in the tracks of the spine ridges is seen in the medial spines of both *Guiyu oneiros* and cf. *Rhabdoderma* sp., as is compression of the ridges to form a reticulate pattern near the base of the spine. The vestigial spine in cf. *Rhabdoderma* sp. indicates the evolutionary derivation of coelacanths from the *Guiyu-Psarolepis* cluster, and furthermore suggests that the *Guiyu-Psarolepis* cluster is more closely allied to sarcopterygians than to basal actinopterygians. The vestigial coelacanth medial spine reported here represents a second type of osteichthyan medial spine and provides new data concerning the origin of coelacanths and possible affinities of the *Guiyu-Psarolepis* cluster. This conclusion is further supported by the reappearance of the *Guiyu* median dorsal plate as an otico-occipital shield in the bizarre Triassic latimeriid coelacanth *Foreyia maxkuhni*.

Keywords Coelacanth · *Rhabdoderma* · *Guiyu* · *Psarolepis* · *Foreyia* · *Polypterus* · *Cardiosuctor* · *Hadronector* · *Polyosteorhynchus* · *Doliodus*

From a paleobiological perspective, coelacanths are interesting fish, well worthy of the fame (Forey 1988) they achieved when a live specimen was caught off the South African coast in 1938. This was not the first time a fish of "prehistoric" aspect was brought to the surface so to speak. The western European discovery of bichirs

Fig. 7.1 *Polypterus senegalus*, albino specimen. Length of scale bar 2 cm (Photograph by Mark McMenamin)

(*Polypterus bichir*) resulted from Napoléon Bonaparte's Egyptian Campaign (Duthiel 1999). Étienne Geoffroy Saint-Hilaire provided a description of the fish in 1802, remarking that "the bichir cannot be shoehorned into any known group of fish, and must be seen as a separate group." He later remarked (Geoffroy Saint-Hilaire 1809) that "had I only discovered this one species in Egypt, it would have been full compensation for the pains endured on the arduous trek."

An albino specimen of *Polypterus senegalus* is shown in Fig. 7.1. Note the interesting bend in the dorsal finlet spines (*cf.* Gayet et Meunier 1996) in this individual of *Polypterus senegalus*. *Polypterus* is an air breather, and will drown if submerged in water too long and prevented from reaching air. To prevent this, in addition to mouth breathing they are able to breathe through a pair of spiracles at the top of the skull. Polypterids are thought to be descended from Triassic scanilepiform fishes (such as *Beishanichthys*). In a recent analysis scanilepiforms are considered to be stem polypterids (Giles et al. 2017). *Polypterus* has a nearly symmetrical diphycercal tail, somewhat resembling the diphycercal tail of coelacanths. The Late Cretaceous polypterid *Bawitius* is thought to have reached lengths up to 3 m (Grandstaff et al. 2012).

The fact that coelacanths still live in the deep sea stands as one of the most remarkable and delightful facts that we have learned in our collective study of natural history. Their body form is very ancient: "anatomically modern coelacanths are known since the Early Devonian" (Cavin et al. 2017). Juveniles of the West Indian Ocean coelacanth (*Latimeria chalumnae*), their blue bodies flecked with white, have been captured on video in their native habitat. On 6 October 2009, a 20 min video of a 31.5 cm juvenile coelacanth was recorded by a self-propelled submersible remotely operated by Japanese researchers at a depth of 161 m in Manado Bay near the coast of Sulawesi Island, Malaysia. The comparatively oversized diphycercal tail with posterior lobe of this coelacanth pup gives it a decidedly strange yet appealing aspect. This young coelacanth wins my vote for the cutest "living fossil". The researchers at Aquamarine Fukushima deserve thanks for transforming their remotely-operated submersible into a time machine.

The coelacanth *baüplan* is characterized by the following traits: "comma-shaped mandible, few abdominal vertebrae and rays in paired fins, and dense covering of large tubercles on the dermal bones and denticles on the scales" (Cavin et al. 2017).

Fig. 7.2 cf. *Rhabdoderma* sp. Base of the neural arches in a coelacanth skeleton showing forked bases that curve around the turgid notochord on both the right and left upper sides of the organ. Fine lines on the SEM photomicrograph are scale impressions. Sample 1 of 2/4/2017. Length of scale bar 2 mm

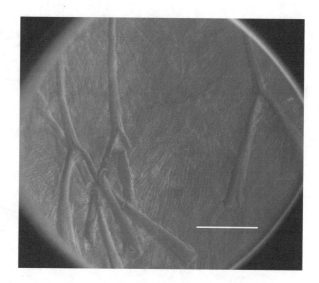

The curious and striking pedunculate fins of coelacanths are, in my opinion, not their most unusual feature. Rather than having a boney spinal column as is the case for teleost fishes and tetrapods, coelacanths retain a large, oil-filled notochord to support their swim musculature. Coelacanths also, of course, develop a bony skeleton. The base of the neural arches in the coelacanth skeleton have forked bases that curve around the turgid notochord on both the right and left upper sides of the organ (Fig. 7.2). In a similar fashion, the hemal arches curve up from below the notochord.

Another important fact about coelacanths is that they possess, ensconced in their braincase's ethmoid area, a unique sense organ that was unknown to science until the discovery of live *Latimeria*. This is the coelacanth rostral organ, evidently used to help the fish locate its prey while hunting as a drift predator (Fricke and Plante 1988). Coelacanths are well known for their strange diphycercal tail with its distinctive supplementary caudal fin lobe that in some specimens (e.g., *Laugia prolata* [Wendruff and Wilson 2013]) almost appears as if a small fish is growing out of the posterior tip of the tail. The tall symmetrical caudal fin of *Rebelletrix divaricerca* Wendruff and Wilson, 2012 is evolutionarily convergent on the split homocercal tail of many modern fishes and is comparable to the tail of the modern Atlantic mackerel (*Scomber scombrus*). In *Rebelletrix*, the supplementary caudal fin lobe is still present, but as fish body size gets larger, the supplementary lobe gets proportionally smaller. The *Rebelletrix* supplementary caudal fin lobe appears to be in the process of becoming evolutionarily lost. Wendruff and Wilson (2012) calculated that "the supplementary lobe exhibits negative allometric growth, becoming proportionally smaller through ontogenetic development." Wendruff and Wilson (2012) reported how this effect has been seen in other coelacanths, including *Rhabdoderma elegans*, where the supplementary lobe is proportionally largest in juveniles (Schultze 1972; Cloutier 2010).

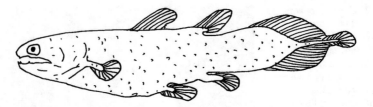

Fig. 7.3 *Cardiosuctor populosum.* Reconstruction of coelacanth from the Lower Carboniferous Bear Gulch Limestone of Montana, USA. Length of fish 25 cm

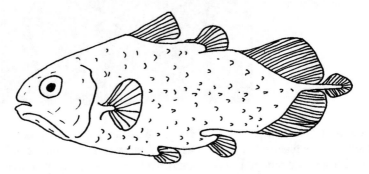

Fig. 7.4 *Hadronector donbairdi.* Reconstruction of coelacanth from the Lower Carboniferous Bear Gulch Limestone of Montana, USA. Length of fish 10 cm

Fig. 7.5 *Polyosteorhynchus simplex.* Reconstruction of coelacanth from the Lower Carboniferous Bear Gulch Limestone of Montana, USA. Length of fish 12.5 cm

Reliance on a notochord rather than an ossified backbone for mechanical support of the swim propulsion musculature did not hinder coelacanths from radiating into wide spectrum of ecological types. Coelacanth fossils with strangely familiar (to the typical American freshwater fisherman) shapes were deposited in the Mississippian Bear Gulch Limestone of Montana. This bay or estuarine deposit hosted three genera of coelacanths with quite distinct body forms (Lund and Lund 1984, 1985; Lund and Poplin 1999). One genus developed a 'rainbow trout' (e.g., *Oncorhynchus mykiss*) morphology (*Cardiosuctor populosum*; Fig. 7.3), a second developed a 'common sunfish or pumpkinseed' (e.g., *Lepomis gibbosus*) morphology (*Hadronector donbairdi*, Fig. 7.4), and a third developed a 'largemouth bass' (e.g., *Micropterus salmoides*) morphology (*Polyosteorhynchus simplex*, Fig. 7.5).

The genus name of the latter is apt, as there are indeed "many bones in the nasal region." Note how the lower lobe of the diphycercal tail in *Hadronector donbairdi* (Fig. 7.4) is slightly advanced in a posterior direction, as is also the case for the albino *Polypterus senegalus* seen in Fig. 7.1. The diphycercal tail in the living coelacanth *Latimeria* is more nearly symmetric, showing only a slight posterior advancement on the posterior lobe of the tail.

This type of evolutionary divergence is not uncommon and is frequently encountered among fishes. One of the best known is the case of Nile and Congo cichlids. Lowe-McConnell (2009) writes that trophic "specialisation is the basis of their radiations in L[ake] Victoria and they have evolved into at least 16 tropic groups, with body form and teeth specialised for life as detritivores, zooplanktivores, insectivores, prawn-eaters, molluscivores or piscivores (of which there are more that 100 species)." Ancient hybridization has been linked to the astonishing cichlid adaptive radiations (Meier et al. 2017).

The Bear Gulch coelacanths represent a form of convergent evolution or iterative evolution that is comparable to that seen in the evolution of fish tails. Louis Agassiz (1832) emphasized what he considered to be a "threefold parallelism in nature, specifically the parallel between the succession of fossil fishes and the chief epochs of creation represented by geological periods. . . [In Agassiz's scheme, fishes] with homocercal tails first appeared in the Jurassic, and Triassic rocks were supposed to be dominated by fishes with heterocercal tails" (McCune 1986; see also Agassiz 1832; Fraas 1861). The evolutionary transition from heterocercal to homocercal tails may be observed in teleost fishes, pycnodonts, and even coelacanths. *Miguashaia*, "the probable sister group to all other coelacanths" (Ahlberg 1992), has a heterocercal tail.

The Devonian fish *Psarolepis* reveals a combination of sarcopterygian and actinopterygian features (Benton 2005). The discovery of a dorsal median spine in *Psarolepis* added an enigmatic character to the mix of its morphological features, as median spines are known in elasmobranchs, acanthodians and the problematic *Doliodus* (Maisey et al. 2017; the "phylogenetic position of *Doliodus* within the chondrichthyan total group is therefore considered ambiguous; it could fall within 'conventionally defined chondrichthyans,' or within 'acanthodians,' or between both of them"). Dorsal median spines, however, had not been reported from other osteichthyans (bony fish; Zhu et al. 2012).

The vestigial coelacanth medial spine reported here, representing a second type of osteichthyan medial spine, provides new data concerning the origin of coelacanths and possible affinities of the *Guiyu-Psarolepis* cluster (Zhu et al. 2012) to sarcopterygians. This report adds to the number of cases of vestigial features known (e.g., snake hindlimbs; Apesteguía and Zaher 2006) from the fossil record.

A vestigial second dorsal fin spine occurs on a newly discovered, unique rhabdodermatid coelacanth (cf. *Rhabdoderma* sp.; Figs. 7.2, 7.6, 7.7, 7.8, 7.9, 7.10, 7.11, 7.12 and 7.13). The cf. *Rhabdoderma* sp. specimen described here consists of a nearly complete post-cranial skeleton (missing only supplementary caudal fin lob; field sample 1 of 2/4/2017; Mount Holyoke College specimen number 8957) from Linton, Ohio. It was collected at the Ohio Diamond Mine, Allegheny Group,

Fig. 7.6 cf. *Rhabdoderma* sp. Photograph of coelacanth specimen from Ohio. The scattered shiny round structures are fossil scales. Sample 1 of 2/4/2017. Scale bar in centimeters (Photograph by Mark McMenamin)

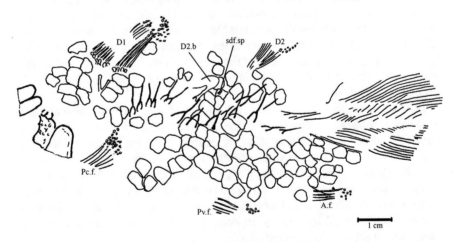

Fig. 7.7 cf. *Rhabdoderma* sp. Sketch of coelacanth specimen from Ohio. **A.f.,** anal fin; **D1,** first dorsal fin; **D2,** second dorsal fin; **D2.b,** second dorsal fin endochondrial support plate; **sdf.sp.** second dorsal fin spine; **Pc.f.,** pectoral fin; **Pv.f.,** pelvic fin. Scale bar = 1 cm

Jefferson County Ohio from cannel coal of Carboniferous (Pennsylvanian) age deposited as sapropelic infill of an oxbow lake (Hook and Baird 1988). A few posterior skull plates are preserved on the specimen (Fig. 7.7). The vestigial dorsal fin spine is not an ingested remain, as it is located in a dorsal position far away from the coelacanth's gut region, unlike the conodont elements recently reported from a Late Devonian coelacanth from the Holy Cross Mountains, Poland (Zaton et al. 2017).

Fig. 7.8 cf. *Rhabdoderma* sp (Photograph of second dorsal fin medial spine. Broadly spaced shiny grooves above spine are ganoine ridges of a posterior scale; finely spaced ridges below the spine are anterior scale ridges) Length of spine 1 mm

Fig. 7.9 cf. *Rhabdoderma* sp. SEM of spine. A scale, which once covered the medial spine due to postmortem overlap, has partially broken away to reveal the spine. Scale bar 400 microns

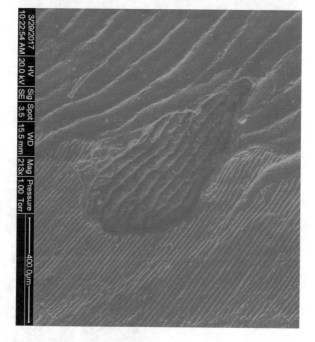

The medial spine occurs as part of an exceptionally preserved, nearly complete coelacanth postcranial skeleton from the Carboniferous of Ohio (Figs. 7.2, 7.6, 7.7, 7.8, 7.9, 7.10, 7.11, 7.12 and 7.13). Bones in the specimen are preserved as external molds, and the medial spine is preserved in the same way. The specimen is unquestionably a coelacanth based on its fin and scale morphology. The scales are of rhabdodermatid type (i.e., bearing ridges that posteriorly converge; Forey 1981), and

Fig. 7.10 Comparison of
the dorsal medial spines in
(**a**), *Guiyu oneiros*, and;
(**b**), cf. *Rhabdoderma* sp.
Sketch of dorsal medial
spine of *Guiyu oneiros* is
based on Figs. 4A and B of
Zhu et al. (2012). Spine
length in A = 3.5 cm; spine
length in B = 1 mm

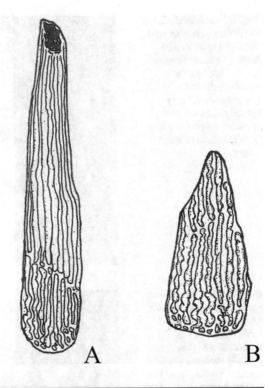

Fig. 7.11   cf.
Rhabdoderma sp.
Rhabdodermatid-type scale
showing microridges that
converge posteriorly to
form a chevron pattern
between the primary
ridges. Scale bar 100
microns

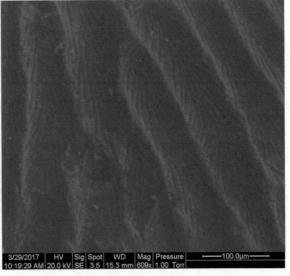

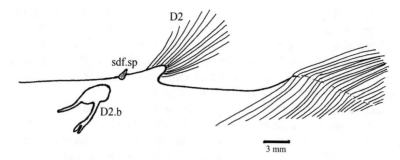

Fig. 7.12 cf. *Rhabdoderma* sp. Reconstruction of the original position of the second dorsal fin medial spine. **D2**, second dorsal fin; **D2.b**, second dorsal fin endochondrial support plate; **sdf.sp.**, second dorsal fin spine; Scale bar = 3 mm

Fig. 7.13 cf. *Rhabdoderma* sp. Lepidotrichia or fin rays of first dorsal fin. Proximal part of fin to top of photomicrograph. Scale bar = 2 mm

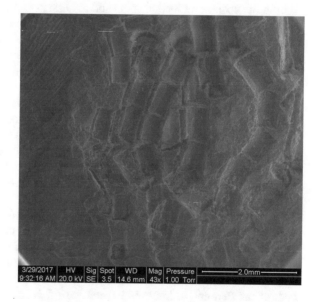

show exceptional preservation by the presence of second order microridges (10 microns wide) between the primary ridges. In a manner similar to the primary ridges, the microridges converge posteriorly to form a chevron pattern between the primary ridges (Fig. 7.11).

The medial spine, surrounded completely by mesenchymal tissue, is oriented with its tip pointing upward and toward the posterior of the fish. The newly-discovered feature is located between the second dorsal fin support and the second dorsal fin. It has shifted to a more ventral location relative to the second dorsal fin, and was apparently dragged down by taphonomic displacement of the tissues of the coelacanth in a ventral direction as indicated both by a downward bend in the neural arches between the first and second dorsal fin, and by the rows of large scales that have been shifted (with the scale rows largely intact) in the same direction (Fig. 7.7). Figure 7.12 shows the reconstructed position of the second dorsal fin medial spine.

The vestigial spine is 1 mm in length, and bears approximately eight preserved longitudinal ridge impressions (Figs. 7.8, 7.9 and 7.10). The ridges show evenly spaced bulges that become more pronounced near the base of the spine. A shared bend or inflection in the ridges occurs some 300 microns from the tip of the spine. This forms a distinctive inflection or disturbance in the track of the spine ridges. The inflection is oriented at 40° to the long axis of the spine.

An alternative hypothesis regarding the spine structure is that it represents some type of pointed dermal denticle as seen in palaeoniscoid fishes as seen in the Carboniferous actinopterygian fish fauna of the Manning Canyon Shale Formation (central Utah; Mickle 2012). These denticles, however, are smooth and tooth-like (Goodrich 1942; Sire et al. 1998). A far better comparison with regard to palaeoniscoid fishes is the three prominent medial spines anterior of the dorsal fin (Mickle 2012) of the palaeoniscoid *Bourbonnella jocelynae*.

Both the coelacanth medial spine and the first dorsal fin spine of *Guiyu oneiros* (Zhu et al. 2012) show compression of the ridges to form a reticulate pattern near the base of the spine (Fig. 7.10). An inflection in the ridges, manifest as diagonal disturbances/dislocations in the ridges and oriented at an approximate 40–60° angle to the long axis of the spine, is also seen in *Guiyu oneiros* (Fig. 7.10) in the lower third of the spine.

There is a real resemblance between the coelacanth dorsal medial spine and fin spines of the early sharklike gnathostome *Doliodus*. According to Maisey et al. (2017), the "fin spines are ornamented by ridges, each bearing a series of closely spaced pectinations. The ridges extend diagonally along the spine length, with new ridges inserted at intervals along the anterior midline and passing to the spine base in a chevron arrangement. The ornamented field terminates just above the base of the fin spine, leaving only a short region of attachment."

The interpretation of the structure in Figs. 7.8, 7.9 and 7.10 as a vestigial median spine (sdf.sp. in Fig. 7.12) is thus strongly supported by details of its ridge morphology. The ridge ornament on the spine is unlike that of parts of the *Rhabdoderma* shoulder girdle, pectoral girdle, or various parts of its skull (Forey 1981). The vestigial spine somewhat resembles the tip of the labial side of one of the cusps of the tooth of the Permian orthocanthid shark *Barbclabornia lendarensis*, but no tooth enameloid is present and the spine ridges are crisply defined, lacking the wear and polish characteristic of *Barbclabornia* teeth (Johnson 2003). The ridge pattern of the vestigial spine does not occur on coelacanth teeth.

Strong support for the vestigial median spine interpretation is provided by a newly described Swiss Triassic coelacanth *Foreyia maxkuhni*, named for the late Peter L. Forey (Cavin et al. 2017). This astonishing fossil fish represents a latimeriid coelacanth with a bizarre bony head shield. The otico-occipital shield projects backwards from the head region, somewhat resembling the bony neck frill of a ceratopsian dinosaur such as *Triceratops*. Cavin et al. (2017) consider *Foreyia maxkuhni* to exemplify "a case of rapid heterochronic evolution likely triggered by minor changes in gene expression." Cavin et al. (2017) suggest the *Pax 1/9* genes as possibly responsible for *Foreyia's* bizarre morphology, but they do note that the search for a "single genetic source is an oversimplification since we know that *Pax* genes work in cooperation with *Hox* genes."

Table 7.1 Rhabdodermatid coelacanths compared

	Rhabdoderma elegans	cf. Rhabdoderma sp.	Hadronector donbairdi
Aanal fin rays (A)	>12–13	10–11	7–9
Pelvic fin rays (V)	14–16	9	8
Vertical scale rows anterior to supplementary caudal fin	~55	~20	~50
Exposed scale length	2 mm	3.5–4 mm	1.5 mm
Vestigial medial spine on D2?	No	Yes	No
Microridge chevrons between primary scale ridges?	No	Yes	No

Cavin et al. (2017) propose heterochrony to explain the bizarre morphology of *Foreyia*, but even in Cavin et al.'s (2017) own Fig. 3D-3E a coelacanth embryo does not much resemble the adult *Foreyia*. It is hard to see how this makes a compelling case for heterochrony. A better interpretation of the bizarre morphology of *Foreyia maxkuhni* is a ring-like inflation of the morphogenetic torus in the neck region of the fish that triggers an atavism in the fish's anterior morphology. The atavism in this case is the reappearance of a median dorsal plate (Zhu et al. 2012) as seen in *Guiyu oneiros*. In the same way that cf. *Rhabdoderma* sp. has a vestigial or atavistic dorsal median spine derived from a *Guiyu*-like ancestor, *Foreyia* has a median dorsal plates (otico-occipital shield; Cavin et al. 2017), also derived from a *Guiyu*-like ancestor.

The coelacanth described here is small (estimated SL = 130 mm; 102 mm preserved) although it appears to be an adult specimen. The most similar species is *R. elegans*, a commons species in Linton coals (Forey 1998). As only fragments of the skull are preserved, the new specimen will be left here in open nomenclature. Nevertheless, significant differences distinguish the new specimen from *R. elegans* as shown in Table 7.1. Of particular interest is the large size of scales in cf. *Rhabdoderma* sp., their microstructure, and the presence of a relict medial spine on the dorsal surface of the fish. An approximately 120 mm long specimen of the rhabdodermatid coelacanth *Hadronector donbairdi* (Lund et al. 1984), by comparison, has roughly 50 vertical scale rows anterior to the supplementary caudal fin in comparison to 55 in *R. elegans* and only 20 in cf. *Rhabdoderma* sp.

The second dorsal fin of cf. *Rhabdoderma* sp. shows both dermal (medial spine) and endoskeletal (second dorsal fin support) associated structures. Whether the coelacanth medial spine reported here represents a species characteristic for cf. *Rhabdoderma* sp., or alternatively represents an isolated atavism in an aberrant individual, is uncertain, although the former possibility seems more likely considering the associated apparent species- to genus-level distinctions shown in Table 7.1. Either way, the discovery of a coelacanth medial spine supports the conclusions of Long (2001), who considered *Psarolepis* to be a basal sarcopterygian. The vestigial spine in cf. *Rhabdoderma* sp. indicates an affinity between coelacanths and the *Guiyu-Psarolepis* cluster, and furthermore suggests that the *Guiyu-Psarolepis* cluster may ultimately prove to be more closely allied to sarcopterygians than to basal actinopterygians (Zhu et al. 2012).

This discovery of the vestigial coelacanth medial spine, buried within the tissues of the fish, recalls the vestigial whale pelvic bones that are deeply buried within its soft tissues and thus no longer of any use for supporting hind limbs. This does not mean, however, that these vestigial bones are without use for the cetacean. Dines et al. (2014) conclude that the highly reduced pelvis of the male whale "seems to serve no other function except to anchor muscles that maneuver the penis . . . . [providing] evidence that sexual selection can affect internal anatomy that controls male genitalia. These important functions may explain why cetacean pelvic bones have not been lost though evolutionary time." Curiously, creationists at the Institute for Creation Research have attempted to use the Dines et al. (2014) study to deny that whale pelvic bones are vestigial by arguing that since these structures do provide a function, then they must not be vestigial organs. This fallacious line of reasoning is easily corrected by considering the case of parvipelvian ichthyosaurs, where the hind limbs and pelvis underwent drastic evolutionary reduction in size, but had not reached the cetacean grade of pelvic reduction as the ichthyosaur hind fins of the Parvipelvia were small but still present (Motani 1999).

The evolutionary linkages between the vertebrates are clearly seen in any case, if we consider the homology between the fin rays or lepidotrichia segments of the coelacanth fin (Fig. 7.13) and the phalanges (singular phalanx) of the human hand. The curious thing about this homology is that the connection is not with the pectoral (or pelvic) fins of the fish, but with its first dorsal fin as shown in Fig. 7.13. This is strange. The homologous features to the phalanges of tetrapods occur, in a coelacanth, in the pectoral and pelvic fins, where we would expect them to be, but also in the first dorsal fin, second dorsal fin, anal fin, and even the rays of the diphycercal tail where they are comparatively longer (Witzmann et al. 2010). Ahlberg (1992) remarked on curious fin similarities in the coelacanth:

> The internal skeletons, musculature and innervation of [the posterior dorsal and anal fins] correspond extremely closely to those of the paired fins. Further, they are supported by complex basal plates that do not resemble those of other lobe-finned fishes but are closely comparable in structure and ossification pattern to the coelacanth pelvis.

Ahlberg (1992) infers that "paired fin structures are being expressed at the posterior dorsal and anal fins" in what should otherwise be (and are assumed to be derived from) non-pedunculate ("lobed") fins. Alberg (1992) infers that the similarities between the paired coelacanth fins and the posterior dorsal and anal fins is due to a "single-step event caused by a simple 'switch' in gene expression . . . [and] may thus have arisen suddenly, rather than through gradual change." The digit phalanx character of the first dorsal fin lepidotrichia (Fig. 7.13), however, imply a correlation between this fin and the paired fins as well.

Clearly, something interesting is going on here. It is as if the dorsal medial spine of coelacanth phylogeny has become embedded in a cortex of outer tissue, a sort of Somatic Biogenic Law if you will, or even a sort of nested pseudoecdysis, with a more ancient body part shifted to the interior. It is as if the fin rays/lepidotrichia of a coelacanth have penetrated the shells of expanded, concentric morphogenetic field surfaces that extend away from the center of the animal, and that these lead to the phalanx segments of the lepidotrichia in the trunk fins. The phalanx homologues in the coelacanth tail fin rays are longer, suggesting distension of the morphological field expansions in this posterior region. The supplementary caudal fin lobe of the coelacanth diphycercal tail represents not merely an elongation of part of the primary surface morphogenetic field, but shows influence of multiple concentrically expanded repeats of the morphogenetic field and that gives the strange midline extension of the tail with its comical little tail tuft at the posterior tip.

Cambrian *Microdictyon* species with concentrically nested trunk plates or sclerites have been considered as failed cases of sclerite ecdysis (Zhang and Aldridge 2007). An alternative (and not necessarily mutually exclusive) interpretation is that these composite sclerites represent a type of expansion of the outer surface of the morphogenetic torus that does not achieve sufficient separation and thus results in what might be called a nested pseudoecdysis. We will return to these considerations to develop the Eighth Law of Morphogenetic Evolution in the next chapter.

References

Agassiz L (1832) Untersuchungen ueber die fossilen Fische der Lias-Formation. Neues Jb Miner Geol Paläont 1832:139–149

Ahlberg PE (1992) Coelacanth fins and evolution. Nature 358:459

Apesteguía S, Zaher HA (2006) Cretaceous terrestrial snake with robust hindlimbs and a sacrum. Nature 440:1037–1040

Benton MJ (2005) Vertebrate palaeontology, 3rd edn. Blackwell, Oxford

Cavin L et al (2017) Heterochronic evolution explains novel body shape in a Triassic coelacanth from Switzerland. Sci Rep 7:13695. https://doi.org/10.1038/s41598-017-13796-0

Cloutier R (2010) The fossil record of fish ontogenies: insights to developmental patterns and processes. Semin Cell Dev Biol 21:400–413

Dines JP et al (2014) Sexual selection targets cetacean pelvic bones. Evolution 68(11):3296–3306

Dutheil DB (1999) The first articulated fossil cladistian: *Serenoichthys kemkemensis*, gen. et sp. nov., from the Cretaceous of Morocco. J Vert Paleo 19:243–246

Forey PL (1981) The coelacanth *Rhabdoderma* in the Carboniferous of the British Isles. Palaeontology 24:203–229

Forey PL (1988) Golden jubilee for the coelacanth *Latimeria chalumnae*. Nature 336(22):727–732

Forey PL (1998) History of the coelacanth fishes. Chapman & Hall, London

Fraas O (1861) Ueber *Semionotus* unde einige Keuper-Conchylien. Jh Ver vaterl Naturk Württ 17:81–101

Fricke H, Plante R (1988) Habitat requirements of the living coelacanth *Latimeria chalumnae* at Grande Comore, Indian Ocean. Naturwissenschaften 75(3):140–151

Gayet M, Meunier FJ (1996) Nouveaux Polypteriformes du gisement coniacien-sénonien d'In Becetem (Niger). C R Acad Sci Paris, sér II a 322:710–707

Geoffroy Saint-Hilaire E (1809) Poissons du Nil, de la Mer Rouge et de la Méditerranée. In: Description de l-Egypte, Vol. 1, Imprimerie impériale, Paris

Giles S et al (2017) Early members of 'living fossil' lineage imply later origin of modern ray-finned fishes. Nature 549:265–268

Goodrich ES (1942) Denticles in fossil Actinopterygii. Quart J Microscop Sci 83:459–464

Grandstaff BS et al (2012) *Bawitius*, gen. nov., a giant polypterid (Osteichthyes, Actinopterygii) from the Upper Cretaceous Bahariya Formation of Egypt. J Vertebr Paleontol 32(1):17–26

Hook RW, Baird D (1988) An overview of the Upper Carboniferous fossil deposit at Linton, Ohio. Ohio J Sci 88:55–60

Johnson GD (2003) Dentitions of *Barbcabornia* from the upper Paleozoic of North America. Mitt Mus Nat Kd Berl Geowiss Reihe 6:125–146

Long JA (2001) On the relationships of *Psarolepis* and the onychodontiform fishes. J Vert Paleo 21:815–820

Lowe-McConnell R (2009) Fisheries and cichlid evolution in the African Great Lakes: progress and problems. Fr Rev 2:131–151

Lund R, Lund WL (1984) New genera and species of coelacanths from the Bear Gulch Limestone (Lower Carboniferous) of Montana (U.S.A.) Geobios 17(2):237–244

Lund R, Lund WL (1985) Coelacanths from the Bear Gulch Limestone (Namurian) of Montana and the evolution of the coelacanthiformes. Carnegie Mus Nat Hist Bull 25:1–74

Lund R, Poplin C (1999) Fish diversity of the Bear Gulch Limestone, Namurian, Lower Carboniferous of Montana, USA. Geobios 32(2):285–295

Maisey JG et al (2017) Pectoral morphology in *Doliodus*: bridging the 'acanthodian'-chondrichthyan divide. Am Mus Novit 3875:1–15

McCune AR (1986) A revision of *Semionotus* (Pisces: Semionotidae) from the Triassic and Jurassic of Europe. Palaeontology 29(2):213–233

Meier JI et al (2017) Ancient hybridization fuels rapid cichlid fish adaptive radiations. Nat Commun 8. https://doi.org/10.1038/ncomms14363

Mickle KE (2012) Unraveling the systematics of palaeoniscoid fishes—lower actinopterygians in need of a complete phylogenetic revision. Ph.D. Dissertation, University of Kansas

Motani R (1999) Phylogeny of the Ichthyopterygia. J Vert Paleont 19(3):472–495

Schultze H-P (1972) Early growth stages in coelacanth fishes. Nature New Biol 236:90–91

Sire J-Y et al (1998) Comparison of teeth and dermal denticles (odontodes) in the teleost *Denticeps clupeoides* (Clupeomorpha). J Morph 237:237–255

Wendruff AJ, Wilson MVH (2012) A fork-tailed coelacanth, *Rebellatrix divaricerca*, gen. et sp. nov. (Actinistia, Rebellatricidae, fam. nov.), from the lower Triassic of western Canada. J Vert Paleont 32(3):499–511

Wendruff AJ, Wilson MVH (2013) New early Triassic coelacanth in the family Laugiidae (Sarcopterygii: Actinistia) from the Sulphur Mountain formation near Wapiti Lake, British Columbia, Canada. Can J Earth Sci 50:904–910

Witzmann F et al (2010) A juvenile early Carboniferous (Viséan) coelacanth from Rösenbeck (Rhenish Mountains, Germany) with derived postcranial features. Fossil Record 13(2):309–316

Zaton et al (2017) The first direct evidence of a late Devonian coelacanth fish feeding on conodont animals. Sci Nat 104(3-4):26. https://doi.org/10.1007/s00114-017-1455-7

Zhang X-G, Aldridge RJ (2007) Development and diversification of trunk plates of the lower Cambrian lobopodians. Palaeontology 50(2):401–415

Zhu M et al (2012) Fossil fishes from China provide first evidence of dermal pelvic girdles in osteichthyans. PLoS One 7:e35103. https://doi.org/10.1371/journal.pone.0035103

Chapter 8
Barasaurus Squamation

Abstract New specimens of the procolophonoid parareptile *Barasaurus* from the Permo-Triassic Sakamena Group of Madagascar show skin preservation in the form of scale patches. Based on its appendicular skeletal anatomy, *Barasaurus* was an aquatic form, the only known aquatic procolophonoid. Its squamation consisting of large (up to 4 mm greatest dimension on an animal approximately 30 cm in length), skink-like ventral scales suggest that this ventral scale configuration was well suited for existence in an aquatic habitat. The *Barasaurus* lifestyle was comparable to that of the crab-eating modern Madagascan skink (*Amphiglossus astrolabi*). Although they had the potential to do so since they survived the Permo-Triassic mass extinction, barasaurian procolophonoids did not diversify into a major group of Mesozoic Marine tetrapods.

Keywords Marine reptiles · Procolophonids · *Barasaurus* · *Amphiglossus* · Sakamena Group · *Mesosaurus* · Zeugopod · *Hovasaurus* · Parareptiles · *Sclerosaurus*

Ever since the Carboniferous tetrapod *Pederpes finneyae* Clack 2002 began to walk around with fully functional, pentadactyl digits, the potential has existed for land tetrapods to say "the heck with dry land" (Tarlach 2017) and return to the sea. Mesozoic aquatic reptiles are a diverse group that includes ichthyosaurs, plesiosaurs (the group includes the short-necked pliosaurs), thallatosaurs, thallatosuchians such

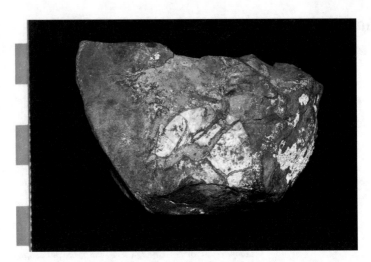

Fig. 8.1 *Heteropolacopros* ichnosp., presumed adult *Shonisaurus* coprolite from the Shaly Limestone Member of the Luning Formation, just east of Berlin-Ichthyosaur State Park, Nevada containing?*Cenoceras* nautiloid. Scale bar in cm (Photograph by Mark McMenamin)

as *Dakosaurus*, nothosaurs, pachypleurosaurs, placodonts, the shovel-jawed marine reptile *Atopodentatus*, the herbivorous marine sphenodontian *Ankylosphenodon* (a tuatara [*Sphenodon*] relative), sea turtles, palaeophilid sea snakes, and mosasaurs. One more group must be added to the list, the procolophonoid parareptile *Barasaurus* (not to be confused with the giant sauropod dinosaur *Barosaurus*) from the Permian of Madagascar. The typical procolophonoid parareptile (such as *Hypsognathus*) is a rather squat creature with a pair of pointed projections that jut out from the back of its skull, giving the animal a resemblance to a desert horned lizard. It is interesting to know that there once existed a successful aquatic procolophonoid.

Some of the marine reptile groups listed above belong to the Superorder Sauropterygia, the "flipper lizards." Ichthyosaurs were the most specialized of the group, leading to a unique body form that Cuvier described as "un museau de Dauphin, des dents de Crocodile, une tête et un sternum de Lézard, des pattes de Cétacé, des vertèbres de Poisson" ("the snout of a dolphin, the teeth of a crocodilian, the head and sternum of a lizard, the flippers of a cetacean, and the vertebrae of a fish"; Piveteau 1926). Indeed, Piveteau (1926) remarks that the entirety of ichthyosaurian evolution tends toward greater and ever greater adaptation to aquatic life. All marine reptiles experience a catastrophic loss of diversity in the end-Cretaceous mass extinction, an event from which the majority of these groups never recovered.

Marine reptiles were important members of Mesozoic marine ecosystems, and new data has recently become available concerning their synecological relationships with other marine creatures (McMenamin 2016) and with one another (Fröbish et al. 2013). For example, nautiloids (cf. *Cenoceras*; Figs. 8.1, 8.2 and 8.3) and

Fig. 8.2 Sketch of
Heteropolacopros coprolite
containing ?*Cenoceras*
nautiloid as seen in
previous figure. Scale bar
1 cm

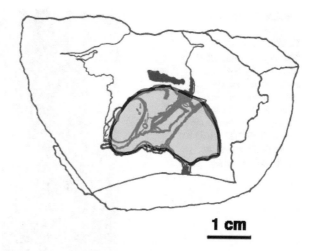

1 cm

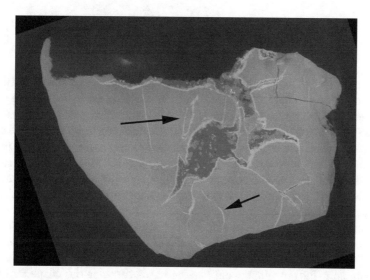

Fig. 8.3 *Heteropolacopros* ichnosp., CT scan showing internal features of ingested ?*Cenoceras* nautiloid as seen in the previous two figures. The upper arrow indicates the position of the nautiloid siphuncle, and the lower arrow shows the position of a nautiloid septum

whiteiidid coelacanths occur in coprolites (*Heteropolacopros* ichnosp.; McMenamin and Hussey 2015; McMenamin et al. 2016) of the Triassic ichthyosaur *Shonisaurus popularis*. Ichthyosaurs acquired a keystone predator role in the Middle Triassic (*Thalattoarchon saurophagis*; Fröbish et al. 2013), and continued as formidable predators well into the Cretaceous (*Platypterygius*; Adams and Fiorillo 2010). Figure 8.4 shows the crown of a tooth of *Platypterygius* sp. from Cretaceous strata of the Van Horn region, Texas, and as per the Cuvier quotation above, it does superficially resemble crocodilian dentition.

Fig. 8.4 Tooth crown of
Platypterygius sp. from
Cretaceous strata of the
Van Horn region, Texas, as
identified by the
longitudinal wrinkles
characteristic of
ichthyosaurian teeth. MAS
McMenamin collection
No. K1/P23/N32. Scale bar
in cm (Photograph by
Mark McMenamin)

An enormous pulse of ichnofossil activity attributed to swimming reptiles has been recorded from the Lower Triassic (Thomson and Droser 2015). In a case of suspected environmental convergence (McMenamin 2005, 2016), Thomson and Droser (2015) argue that, in terms of substrate disturbance, the post-Permian extinction aquatic world was "repressed due to delayed biotic recovery following the end-Permian mass extinction, resulting in extremely low intensities of bioturbation. . . [and a delay in] well-mixed sediment." This presumably led to better chances for preservation of tetrapod swim tracks due to a greater prevalence of fine, undisturbed muddy aquatic substrates.

A normalized analysis shows a huge Early Triassic spike (more than an order of magnitude greater than background levels) in the occurrence of tetrapod swim tracks (Thomson and Droser 2015). Many of these were made by animals probably quite similar to the procolophonoid *Barasaurus* discussed in this chapter. Although there is presumably no connection between the tracemakers, there is an interesting similarity between longitudinal striations on swim tracks from the Circle Cliffs area, southern Utah (Thomson and Droser 2015) and Proterozoic specimens of *Eophyton* (Figs. 3.3 and 3.4); this similarity is no doubt merely due to similarity in the mechanical process of formation.

Thomson and Droser (2015) provide an interesting analysis of their tetrapod swim track data, concluding that "it is not the swimming behavior alone, but the prevalence of unbioturbated substrates resulting from the unique combination of ecological and environmental conditions during the Early Triassic that led to the production and preservation of swim tracks." They thereby place the emphasis on a (temporary) taphonomic bias favoring the production and preservation of the swim tracks. It seems more likely, considering the order-of-magnitude ichnofossil spike and the fact that various tetrapod groups (including some procolophonids) were racing to colonize aquatic habitats, that the main reason for the swim track spike has primarily to do with the activities and abundance of the track makers. *Barasaurus* discussed here is part of this Permo-Triassic swarm of swimming reptiles.

Paleozoic aquatic reptiles are considerably less familiar to the general public than their Mesozoic descendants. The slender aquatic parareptile *Mesosaurus* from the Early Permian (285 Ma) is known to most geology students because, being only 1 m long at full size, it is assumed to have been too small to cross the open ocean and thus serves as a paleobiogeographic marker species for supercontinental reconstruction. *Mesosaurus's* currently disjunct paleobiogeographic distribution (South America and South Africa) was famously used by early advocates of continental drift to reunite the southern continents into the supercontinent Gondwana. Contemporary research accords with this conclusion, and suggests that *Mesosaurus* (although fully aquatic) inhabited hypersaline environments rather than open ocean waters (Piñeiro et al. 2012).

Like *Mesosaurus*, paleogeographic distribution of procolophonoid parareptiles such as *Barasaurus* also played a supporting role in Paleozoic continental reconstructions. The species *Barasaurus besairiei* was described in 1955 by Jean Piveteau (Piveteau 1955). In his description of the genus and species (*Barasaurus* remains a monotypic genus, that is, only one species occurs in the genus), Piveteau (1955) noted:

> Dès le Permien supérieur, la famille de Procolophonidés présentait une vaste répartition géographique. Nous la connaissons maintenant d'Afrique australe, de Madagascar et de Russie (série de Mesen, un peu plus anciennes sans doute que la zone à *Cisticephalus*). Nous avons là un nouvel argument pour rejeter l'individualité d'un continent de Gondwana à la fin des temps paléozoïques. [In the Late Permian, the procolophonid clade showed a vast geographic distribution. The group is currently recorded from South Africa, to Madagascar and to Russia (Mesen series, no doubt a bit older than the *Cisticephalus* zone). We now have a new reason to reject the concept of isolation of Gondwana at the end of Paleozoic time. (translation M McMenamin)]

An unrelated diapsid reptile group, the Late Permian-Early Triassic tangasaurids (Family Tangasauridae), developed aquatic members that have been assigned to the subfamily Tangasaurinae. This group is well represented in strata of southern Madagascar (Ketchum and Barrett 2004). The study of Madagascan Permian tetrapods has an interesting history associated with French colonialism on the island (Piveteau 1926, 1955). Paleontologist M. Boule encouraged French Officer Captain J. Colcanap to organize scientific explorations on the island in search of fossils. Capt. Colcanap's efforts met with success in 1906, and led to the recognition of the famous Karoo Supergroup (Carboniferous to Jurassic of southern Africa; Schlüter 2008) in Madagascar (Battistini and Richard-Vindard 1972), which furthermore helped to demonstrate that Africa and Madagascar were once contiguous as part of Gondwana (and by extension part of Pangea as well).

Two years later Boule (1908) published an initial report on the Permian flora and fauna of Madagascar. In a search for fossils showing exceptional preservation, Capt. Colcanap found the following: the palaeoniscoid fish *Atherstonia*, an amphibian mandible comparable to that of members of temnospondyl family Eryopidae, and a great number of reptile fossils (Piveteau 1926; Boule 1910). Capt. Colcanap met an early death, but he was lauded by French paleontologists who called him their "vaillant et regretté explorateur" ("valiant and lamented explorer"; Piveteau 1926). Perrier de la Bathie (1919) emphasized that, using the vertebrate paleontological evidence, Eduard Suess's (1885) inference was correct regarding the assignment of Madagascar to Gondwana.

Piveteau (1926) considered the Madagascan reptile fossils to be the most important Permian fossils to be recovered from Madagascar, and underscored the fact that they could not be placed into recognized clades. The fossils were recovered from curious small concretions in the Sakamena Group, a Karoo-equivalent sequence of terrestrial and aquatic strata (Ketchum and Barrett 2004). The Lower Triassic part of the Sakamena Group has produced a fully aquatic temnospondyl 'amphibian' (*Wantzosaurus elongatus*). *Wantzosaurus* swam using lateral undulations of its tail and body (Steyer 2002).

Entombment by concretion is a relatively unusual type of preservation for Permian reptiles, and is more usually associated with invertebrate fossils and fossil fish. Jean Piveteau (1926:102), in his superb monograph that was foundational for subsequent work, remarked on the odd character of this mode of preservation:

> These impressions are preserved in fine-grained shaly concretions. It appears that the fossil, particularly the abdominal region which often contains quartzose pebbles, had served as a center of lithification for the concretion. This mode of preservation explains the general absence of the limbs and the head and tail, which were too far from the center to be caught by the concretion. (translation MAS McMenamin)

As we shall see, the Madagascan Permian reptile concretions also provided cases of exceptional preservation.

The family Tangasauridae was named by the Charles Camp (1945a, b), the Berkeley paleontologist perhaps best known for helping to establish the Berlin-Ichthyosaur State Park in Nevada. The genus *Tangasaurus* was established by Haughton (1924), from two poorly preserved skeletons collected in Tanganika (Tanga region; Piveteau 1926). Like *Mesosaurus*, these are relatively small animals. The largest tangasaurine, *Hovasaurus*, reached a maximum length of only 30 cm. But unlike *Mesosaurus*, the tangasaurines are classified as diapsids, that is, the "reptilian" group that includes lizards, snakes, non-avian dinosaurs and birds. Incidentally, *Mesosaurus* has periodically been assigned to the Synapsida, the clade that includes mammals (Huene 1940; Piñeiro et al. 2012). Diapsids are often rare in Permian tetrapod-bearing strata, thus fossils such as *Orovenator mayorum* from Richard's Spur, Oklahoma, although fragmentary, are considered important for understanding early diapsid evolution (Reisz et al. 2011). The lightly built skull of *Orovenator mayorum*, its large orbit, and its teeth lacking in serrations suggest an insectivorous diet (Reisz et al. 2011).

An interesting aspect of tangasaurines such as *Hovasaurus* is that they survive the Permo-Triassic mass extinction event at 251 million years ago, generally recognized as the worst mass extinction of all time. Currie (1981) originally listed

Hovasaurus as a victim of that great extinction, but Ketchum and Barrett (2004) make a special point of illustrating disarticulated caudal ribs and very tall caudal neural spines (clearly an adaptation for swimming) of a fossil from the Middle Sakamena (and hence Lower Triassic) which they identify as *Hovasaurus*. Earliest Triassic fossils (Dienerian regional stage) are known from the middle Sakamena Group of Irarona village, Ambilobe, northwest Madagascar, a site (Takai 1976) that has produced the palaeoniscoid fish *Atherstonia madagascariensis*. The Late Permian (Lower Sakamena) equivalent to this fish (Currie 1981) is *Atherstonia colcanapi*. Coelacanths are also known from the Lower Triassic of northwestern Madagascar (Moy-Thomas 1935; Clement 1999). Coelacanths are represented by *Piveteauia madagascariensis* and other species.

  Barasaurus besairiei also survived the end-Permian extinction event and lived on into the first stage of the Early Triassic, the Induan Stage (Ruta et al. 2011). This makes the Triassic hovasaurs (a tangasaurid group; Currie 1981) and *Barasaurus besairiei* (an owenettid procolophonoid; deBraga 2003) among the few aquatic reptiles to survive the Permo-Triassic mass extinction. This is all the more remarkable considering the small size of these creatures and the fact that subsequent, predatory aquatic reptiles were so thoroughly devastated at the end-Cretaceous mass extinction, which was nevertheless not as severe an extinction as the end-Permian. It has been suggested that the Tangasaurinae were ancestral to the Sauropterygia (Carroll 1981), especially considering that tangasaurs survived into the Mesozoic. Creatures such as *Hovasaurus* therefore have affinity to, and could even be the progenitor of, the great clades of sauropterygian marine reptiles. The procolophonoid *Barasaurus besairiei* was also aquatic and survived the Permo-Triassic mass extinction, but apparently did not give rise to a great lineage of Mesozoic marine reptiles. Marine reptiles are now thought to have radiated rapidly after the Permo-Triassic mass extinction (Jiang et al. 2016), and it is possible that hovasaurs gave rise to the sauropterygian branch.

  Figures 8.5, 8.6, 8.7, 8.8, 8.9, 8.10, 8.11, 8.12, 8.13, 8.14, 8.15, 8.16, 8.17, 8.18, 8.19, 8.20, 8.21, 8.22, 8.23, 8.24, 8.25, 8.26 and 8.27 show images of the pectoral region of *Barasaurus besairiei* from Madagascar, the only known aquatic procolophonoid parareptile. The fossil forms a natural mold in a siltstone concretion that was split in half to reveal the fossil. The estimated total length of animal is roughly 30 cm. Patches of bone remain, particularly at the ends of the long bones. Both dorsal and ventral impressions show the central and anterior parts of the skeleton, the proximal left forelimb bones, and the left manus. The manus has been inverted by twisting, a configuration similar to a person lying on the floor with their chest to the ground and with the back of their left hand pressed against their chest (not a comfortable position!). The left radius and the left ulna cross one another as a result of the twist (Fig. 8.5, 8.6, 8.7, 8.8, 8.9 and 8.10). No skull material is present in the fossil, so the following analysis will be limited to consideration of the appendicular (post-cranial) skeleton. Two different aspects of the massive pectoral girdle complex are visible in the dorsal and ventral impressions on the two halves of the concretion, respectively.

Fig. 8.5 *Barasaurus
besairiei*. Photograph of
dorsal impression slab.
100 g sample; sample 1 of
10/23/2017. Scale bar in
cm (Photograph by Mark
McMenamin)

Fig. 8.6 *Barasaurus
besairiei*. Sketch of dorsal
impression slab. Scale bar
1 cm

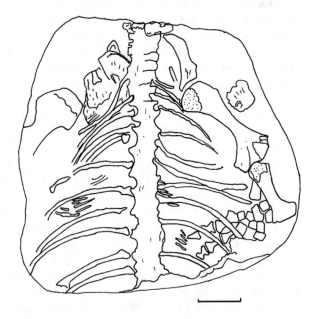

Although precise locality data was not recorded for the specimens described
here, the skeleton is clearly from the Sakamena Group of Madagascar (Tortochaux
1949). These strata crop out primarily in three Madagascan basins (Diego, Majunga,
and Morondava basins; Ketchum and Barrett 2004), but it is not certain from which
of these three basins this fossiliferous concretion is derived. Madagascar and the

Fig. 8.7 *Barasaurus
besairiei*. Photograph of
ventral impression slab.
66 g sample; sample 2 of
10/23/2007. Scale bar in
cm (Photograph by Mark
McMenamin)

Fig. 8.8 *Barasaurus
besairiei*. Sketch of ventral
impression slab. Preserved
scale patches are labelled
S1-S3. Scale bar 1 cm

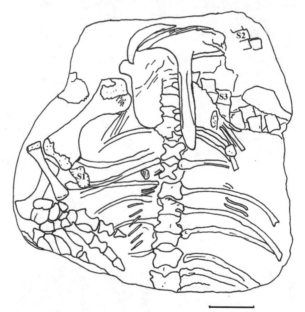

adjacent African mainland (Tanzania) have been called the *Tangasaurid Province* of
Gondwana (Anderson and Cruickshank 1978). It is possible that the parareptile
fossil material described here is from near Ranohira, west of the junction of
Beroraha-Ihosy routes (Currie 1981). Indeed, it seems very likely that the fossils are
from Ranohira, as this site is known (Smith 2000) for its abundant specimens of
Barasaurus besairiei. Smith's (2000) interpretation of the sedimentology of the
Ranohira site indicates that the habitat of *B. besairiei* was the open waters of a
"thermally-stratified deep lake."

Fig. 8.9 *Barasaurus besairiei*. View of left manus, ventral slab, showing crossing of radius and ulna. Scale in cm with mm subdivisions (Photograph by Mark McMenamin)

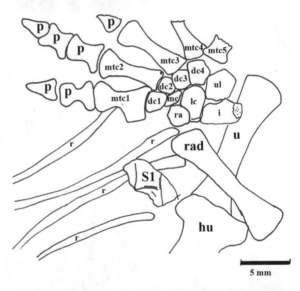

Fig. 8.10 *Barasaurus besairiei*. Sketch of previous image. Abbreviations: **dc1–4**, distal carpal I-V; **i**, intermedium; **hu**, humerus; **lc**, lateral centrale; **mc**, medial centrale; **mtc 1–5**, metacarpals I-V; **p**, phalanx (plural: phalanges); **r**, rib; **ra**, radial; **rad**, radius, **S1**, scale patch; **u**, ulna; **ul**, ulnare. Scale bar 5 mm

Fig. 8.11 *Barasaurus besairiei*. Left humerus, impression of dorsal side with bone material still adhering to the mold. Distal part of bone to upper left. Scale in cm with mm subdivisions (Photograph by Mark McMenamin)

Fig. 8.12 *Barasaurus besairiei*. Left humerus, ventral view, sketch showing scarring of left humerus. The scarring indicates the insertion of the coracobrachialis brevis muscle. Abbreviations: **br**, brachialis inferior; **c**, capitellum; **cb**, coracobrachialis brevis, **E**, entepicondyle. Distal end of bone at bottom of image. Scale bar 3 mm (Photograph by Mark McMenamin)

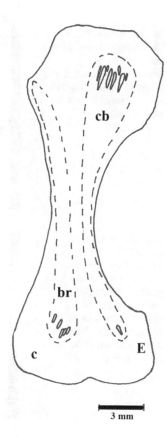

Fig. 8.13 *Barasaurus besairiei*. Ribs, gastralia and section of vertebral column. Scale in mm (Photograph by Mark McMenamin)

Fig. 8.14 *Barasaurus besairiei*. Shoulder girdle region, ventral view, anterior of animal to right. The T-shaped interclavicle characteristic of ankyramorph parareptiles is clearly visible. Scale in mm (Photograph by Mark McMenamin)

Fig. 8.15 *Barasaurus besairiei*. Dorsal impression of vertebral column and shoulder girdle, showing both left and right crescentic clavicle at the base of the image. Width of image 4 cm (Photograph by Mark McMenamin)

Fig. 8.16 *Barasaurus besairiei*. Scanning electron micrograph of cancellous or trabecular bone (spongy bone) showing a patch of spongy bone across a fractured surface preserved at the proximal epiphysis of the right humerus (ventral side) with prominent Haversian canals. Scale bar 2 mm

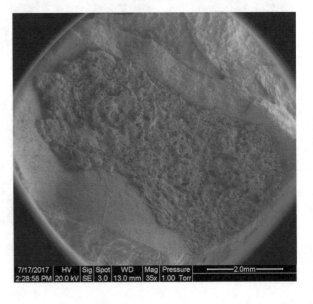

Fig. 8.17 *Barasaurus besairiei*. Enlarged view of spongy bone (cancellous or trabecular bone) in previous image. Scale bar 400 microns

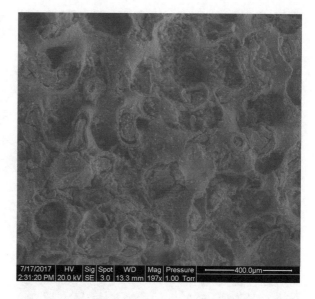

Fig. 8.18 *Barasaurus besairiei*. Trabeculae meshwork (=lamellae), a single trabecula, aligned toward the outer surface of the proximal articulation (and at a right angle to the plane of the microphotograph). Scale bar 20 microns

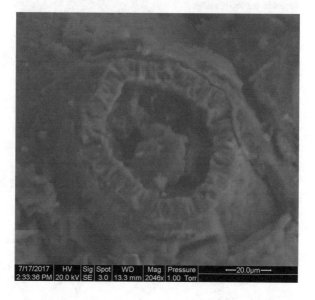

The specimen now consists of two halves of a concretion. The heavier half (100 g; weak fizzing with 3% HCl, thus calcareous siltstone nodule; Figs. 8.5 and 8.6) preserves an impression of the dorsal surface of the animal, while its less massive counterslab (66 g; Figs. 8.7 and 8.8) preserves an impression of the ventral side of the animal. An impression of the manus occurs on the dorsal half, but the impression of a right rib cuts across it as it was above the manus during fossilization. An uninterrupted impression of the same left manus occurs on the bottom piece.

Fig. 8.19 *Barasaurus besairiei*. Belly scale preservation on the ventral counterslab. The large scale patch (patch S3) of scaly skin runs roughly parallel to the ruler; the small patch is near the bottom of the image. Scale in cm with mm subdivisions (Photograph by Mark McMenamin)

Fig. 8.20 *Barasaurus besairiei*. Detailed view of scales in previous image. Scale in mm (Photograph by Mark McMenamin)

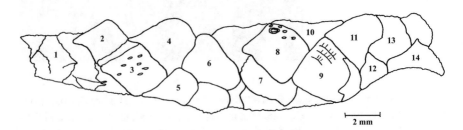

2 mm

Fig. 8.21 *Barasaurus besairiei*. Sketch of ventral scale preservation in patch S3. Anterior to left. Fourteen scales are preserved as indicated by numbering. Scales are shown with their inner surfaces visible. Scales 3 and 8 show the bases of small elongate tubercles. Scale 9 shows faint growth rings and very fine radial ridges. Scale bar 2 mm (Photograph by Mark McMenamin)

Fig. 8.22 *Barasaurus besairiei*. Scanning electron micrograph detail of ventral scales from scale patch S3. Note denticulate edge of scale and fine parallel ridges at the scale margin. The latter is most visible near the triple junction between scales near the center of the image. Scale bar 1 mm

Fig. 8.23 *Barasaurus besairiei*. Scanning electron micrograph detail of ventral scales, showing crenulated edges. Scale bar 2 mm

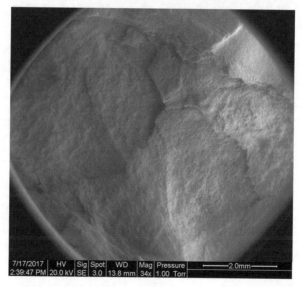

Unfortunately, the skull of this specimen of *B. besairiei* is not preserved (as is typical for most parareptile and reptile specimens from the Sakamena Group due to the concretion mode of preservation, which usually does not capture the skull and the tail). However, its hand morphology suggest that it is closely allied to, if not a member of, the Procolophonidae (Ketchum and Barrett 2004).

Ruta et al. (2011) do not include *Barasaurus* in their cladogram of the Procolophonidae. Rather, *Barasaurus* is considered in these analyses to belong to

Fig. 8.24 *Barasaurus besairiei*. Generalized sketch of a single ventral scale. Some scales in the species are keeled, keel not shown here. Greatest dimension of scale 3.5 mm

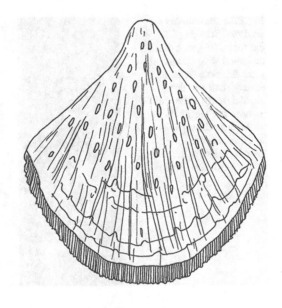

Fig. 8.25 *Barasaurus besairiei*. Reconstruction of ventral, scaled surface of aquatic reptile. Webbed hands and feet are conjectural

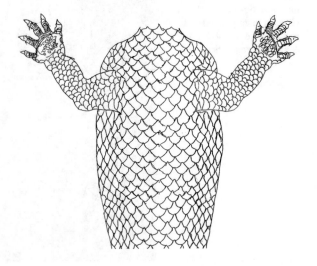

family Owenettidae. Family Owenettidae in this rendering (Cisneros et al. 2004) contains three lineages: *Candelaria* plus "*Owenetta*" *kitchingorum*; *Owenetta rubidgei* and *Saurodektes*; and *Barasaurus* itself (see also Ruta et al. 2011). However, competing analyses (Tsuji et al. 2013) are less definitive regarding placement of *Barasaurus* in the Owenettidae. Other phylogenetic configurations remain possible at this time, and the situation can be improved by adding more characters to the list of traits used for the phylogenetic analyses.

Fig. 8.26 Fossil palaeoniscoid fish (*?Atherstonia* sp.) from the Sakemena Formation, Madagascar, showing squamation and polygonal scales. Scale in cm (Photograph by Mark McMenamin)

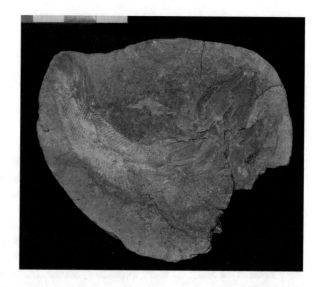

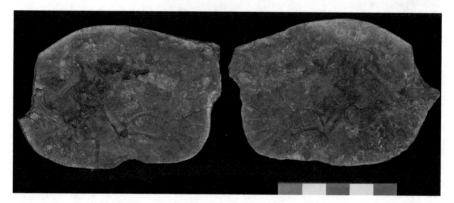

Fig. 8.27 *Barasaurus besairiei*, dorsal (right; sample 3 of 10/23/2017) and ventral (left; sample 4 of 10/23/2017) impressions of the pelvic region of the animal with short sections of tail preserved. Scale in cm (Photograph by Mark McMenamin)

With information from both halves of the concretion (Figs. 8.5, 8.6, 8.7 and 8.8), we can say that the left arm of the animal is nearly complete, and includes the stylopod (left humerus), zeugopod (radius and ulna), most of the carpus and a portion of the manus. The zeugopod is disarticulated, and the radius and ulna overlap. The manus includes digit I with its metacarpal plus two phalanges (mtc1, ph, ph), digit II with its metacarpal plus three phalanges (mtc2, ph, ph, ph), digit III with an intact metacarpal with a small portion of the first phalanx (mtc3, ph), about half of the digit IV metacarpal (mtc4), plus a little less than half of the digit V metacarpal (mtc5). The first two digits show exactly the same pattern as that of *Barasaurus besairiei* described by Meckert (1995; his fig. 24, specimen P5). Preserved carpal elements of the left hand

Taxon	Distal carpals contacting the medial centrale
Hovasaurus boulei	dc1, dc2, dc3, dc4
Acerosodontosaurus piveteaui	dc1, dc2
Barasaurus besairiei	dc1, dc2, dc3

Table 8.1 Contacts between the medial centrale and distal carpals in selected aquatic/semiaquatic Permo-Triassic tetrapods from Madagascar

in the specimen described here include the following: first distal carpal (dc1), second distal carpal (dc2), third distal carpal (dc3), forth distal carpal (dc4), medial centrale (mc), radiale (ra), intermedium (i), lateral centrale (lc) and ulnare (ul). Both the fifth distal carpal (dc5) and the pisiform (p) appear to be lost.

The curious perforating foramen (referred to here as the 'stigmata foramen') between the intermedium and the ulnare as seen in *Hovasaurus*, is reduced to a slight raised projection on the ulnare in *B. besairiei*. The stigmata foramen is well developed in the manuses of the Permian synapsids *Dimetrodon* and *Cotylorhynchus* (Davis 2012).

In *B. besairiei*, the mediale centrale does not contact the forth distal carpal, nor does it prevent articulation between the lateral centrale and the third distal carpal as Currie (1981) demonstrated for *Hovasaurus boulei*. In the enigmatic Permian diapsid reptile from Madagascar *Acerosodontosaurus*, the medial centrale articulates only with the first and second distal carpals (Currie 1980; Ketchum and Barrett 2004). The articulations between the medial centrale and the distal carpals in *Hovasaurus*, *Acerosodontosaurus*, and *B. besairiei* are shown in Table 8.1.

In *Barasaurus besairiei*, the second distal carpal is smaller than the first distal carpal. This is opposite to the situation with *H. boulei*, where the second distal carpal is larger than the first distal carpal and contacts the first and second metacarpals. In *Barasaurus besairiei*, the forth distal carpal does not contact the medial centrale.

Bickelmann et al. (2009) advocated an aquatic lifestyle for *Acerosodontosaurus piveteaui* due to its "relaxed arrangement of carpal elements" and lack of an ossified olecranon on the ulna, a lack they refer "to the phenomenon of skeletal paedomorphosis, which is frequently seen in secondarily aquatic amniotes" as for example in the marine reptile *Helveticosaurus zollingeri* from the Besano Formation, Monte San Giorgio, Switzerland (Rieppel 1989). Scheyer et al. (2017) list the paedomorphic elements in *Helveticosaurus zollingeri* as "reduced size of pectoral and pelvic girdle elements; lack of ossified epiphyses of stylopodial and zeugopodial elements, [and] retention of open neurocentral sutures." This compares to the observations of Piveteau (1926) in his description of *Hovasaurus boulei*: "Le carpe n'est pas ossifié et la main est composée des métacarpiens et de phalanges très réduites, accolées les unes aux autres, comme chez les Ichthyosauriens ou les Siréniens. . . . La main a, en gros, l'aspect d'une palette natatoire comparable, par la forme générale, à la main des Ichthoysauriens, ou mieux de Plésiosauriens; mail il n'y a ni hyperphalangie ni hyperdactylie." ("The carpus is not ossified and the manus consists of metacarpals and phalanges very reduced and adjacent to one another, like those of ichthyosaurs and sirenians. . . The manus has the overall aspect of a swimming appendage comparable, in general form, to the ichthyosaur flipper, or even better that of plesiosaurs, but it does not develop hyperphalangy nor hyperdactyly." [translation M McMenamin]).

Forelimb morphology is of particular interest in the study of Paleozoic tetrapods, as, curiously, phylogenetic trees generated by data collected from the forelimbs and pectoral region are very similar to trees constructed by the whole organism character data set, whereas trees generated from data taken only from the hind limbs and pelvic region show "considerable loss of phylogenetic resolution" (Ruta and Wills 2016). This indicates that forelimbs and the pectoral region are more phylogenetically informative than more posterior parts of the appendicular skeleton. Why this is the case not entirely clear, but it is also true that the skull is one of the most phylogenetically important parts of a tetrapod skeleton. This implies that there is a gradient from nose to tail in the phylogenetic significance of morphological traits, with those traits closer to the head end of the tetrapod being of greater importance for classification, whereas those traits more posterior are of less use in this regard. This general observation has as its corollary the general rule that the more distal parts of the limb skeleton (autopod [wrist plus fingers/ankle plus toes] first, zeugopod [ulna plus radius/tibia plus fibula] second, and stylopod [humerus/femur] third, respectively) are comparatively more phylogenetically informative (Mayor 2000).

With this in mind, let us now consider the forelimbs in *Barasaurus besairiei*. The carpus in *Barasaurus besairiei* (Figs. 8.9 and 8.10) is less relaxed in configuration than is the case in *Acerosodontosaurus piveteaui*, but is nevertheless in a somewhat relaxed configuration considering the loose spacing between the ulnare, the forth distal carpal and the third distal carpal. The phalangeal formula (that is, the number of phalanges for each of the first through fifth digits) for *Hovasaurus boulei* is 2.3.4.5.2, and in *Barasaurus besairiei* it is 2.3.4.5.3 (Meckert 1995). The plesiomorphic condition for the pentadactyl amniote hand is 2.3.4.5.3 (Scheyer et al. 2017), thus *Barasaurus* has the primitive configuration.

The squat triangular terminal phalanges (unguals) of *Barasaurus besairiei* are similar to those of *Eusaurosphargis dalsassoi* as described by Scheyer et al. (2017). Scheyer et al. (2017) characterize these as "wide spade-like terminal phalanges instead of tapering ones," and use this observation to argue for a terrestrial rather than aquatic mode of life for *Eusaurosphargis dalsassoi*. I disagree with this reinterpretation. The squat body, numerous osteoderms, and relatively short limbs of *Eusaurosphargis dalsassoi* do not preclude an aquatic or at least semi-aquatic lifestyle, as attested by modern chelonians (turtles).

Currie (1981) notes that the "increase in symmetry of the metacarpals of *Hovasaurus* could be related to the use of the manus as a paddle, which is apparently the case in nothosaurs and crocodiles (Robinson 1975)." Metacarpal symmetry is considerably less in *Barasaurus besairiei*, with the first metacarpal being proportionally wider than the second through fifth metacarpals. Currie (1981) also notes that the "penultimate phalanx of each digit is as long as or longer than the antepenultimate phalanx (or metacarpal in the case of the first digit)." This is not the case for *Barasaurus besairiei*, for either the first or second digit.

Taxon	Shape of radius
Hovasaurus boulei	Curved
Acerosodontosaurus piveteaui	Twisted
Barasaurus besairiei	Straight

Table 8.2 Radius shape in selected Permo-Triassic aquatic/semiaquatic tetrapods from Madagascar

The radius in *Barasaurus besairiei* is not twisted as it is in *Acerosodontosaurus*. *Barasaurus besairiei* has a straighter radius than either *Hovasaurus* or *Acerosodontosaurus* as seen in Table 8.2.

As is the case for *Hovasaurus* and *Acerosodontosaurus*, the humerus of *Barasaurus besairiei* has a well-developed, rectangular entepicondyle (Figs. 8.11 and 8.12). The relationships between the thickness of the humerus at its midpoint are as follows: *Hovasaurus* > *Acerosodontosaurus* > *Barasaurus besairiei*. Thus, *Hovasaurus* has a more robust humerus than *Barasaurus*, and considering that they lived in the same place (Madagascar) at the same time (Permo-Triassic), it seems possible that the partial corpses preserved in the Malagasy concretions represent the vanquished in battles between *Hovasaurus* and *Barasaurus* at sea. Here may be a case where *Hovasaurus* was the more aggressive animal, driving off the prolcolophonoid competition in shallow waters. The potential for evolutionary radiation in *Barasaurus* was cut short by competitive exclusion by *Hovasaurus* at a critical moment when the oceans were wide open for exploitation by sea-going tetrapods. *Hovasaurus* may have gone on to establish the sauropterygian radiation, whereas *Barasaurus*, somewhat sadly, went extinct without issue. This is unfortunate, as it would have been interesting to study a successful group of procolophonoid Mesozoic marine reptiles and observe how their aquatic adaptive radiation differed from those of the other marine 'reptilian' groups.

Currie (1981) noted in *Hovasaurus* that scarring "on the ventral surface of the proximal head of the humerus posterior to the supracoracoideus marks the insertion of the coracobrachialis brevis." This same scarring is visible at precisely the same place on the ventral side of the left humerus of *Barasaurus besairiei* as seen in Fig. 8.12, at a position close to the proximal articulation. This was presumably for a more effective power stroke as this muscle pulls the humerus towards the torso by means of the glenoid joint. Unlike *Hovasaurus*, however, scarring is also seen in *Barasaurus besairiei* immediately proximal to the capitellum (Fig. 8.12). This may represent a significant distal expansion of the brachialis inferior muscle. A third region of scarring (Fig. 8.12) is interpreted here as the distal end of a very elongate coracobrachialis brevis. Even if the scarring at the distal end of the bone shown in Fig. 8.12 represents unknown muscle insertion sites, we see here evidence for a powerful forearm swimming stroke that involved flexing at the glenoid joint and at the elbow. *Barasaurus besairiei* was a powerful swimmer and would seem to have been a worthy opponent for *Hovasaurus*.

There are fewer ribs anterior to the humerus in *Barasaurus besairiei* than in *Hovasaurus*. The ribs in *Barasaurus besairiei* have longitudinal shallow furrows or grooves. These are weakly developed in *Acerosodontosaurus* and apparently absent

Table 8.3 Variation in
maximum rib length (with
x = the length of the dorsal
centrum) in selected aquatic/
semiaquatic Permo-Triassic
tetrapods from Madagascar

Taxon	Rib length
Hovasaurus boulei	7.5x
Acerosodontosaurus piveteaui	7.8x
Barasaurus besairiei	4.2x

in *Hovasaurus*. These grooves may disappear with phyletic thickening of the ribs as marine reptiles develop increasingly pachyostotic ribs for ballast. The ribs in *Barasaurus besairiei* show only incipient pachyostosis (Fig. 8.13), but are slightly swollen at their ends. *Hovasaurus* has a mix of double headed and holocephalus ribs (Currie 1981), and this seems also to be the case for *Barasaurus besairiei*.

Variations in maximum rib length (Currie 1981), with x = the length of the dorsal centrum, are shown in Table 8.3. Here *Barasaurus besairiei* shows a distinct divergence from *Acerosodontosaurus* and *Hovasaurus*, whose rib length respective to dorsal centrum width is considerable longer. This trait may serve as a distinguishing characteristic for *Barasaurus besairiei*.

Ribs (Figs. 8.5, 8.6, 8.7 and 8.8) in *Barasaurus besairiei* in the mid dorsal region are curved their entire length, as is the case for *Acerosodontosaurus* and *Hovasaurus*. Some of the ribs in *Barasaurus besairiei* bend fairly sharply at their distal ends. Several ribs in *Barasaurus besairiei* have short projections on their anterior edges, in one case giving the rib a pectinate aspect as seen in dorsal view (Figs. 8.6, 8.7, 8.8, 8.9, 8.10, 8.11, 8.12 and 8.13). Gastralia ("belly ribs") are visible at various places on the fossil (Fig. 8.13). Although gastralia are present, they are not extensively developed, suggesting that there was still a degree of flexibility to the torso of the animal.

As in *Hovasaurus*, some of the rib ends in *Barasaurus besairiei* are concave, suggesting that they continued as cartilage (Currie 1981). This suggests that all three animals, *Acerosodontosaurus*, *Hovasaurus* and *Barasaurus besairiei*, were "relatively broad bodied animals" (Currie 1981), as was also the case for *Kenyasaurus*, *Tangasaurus* and *Youngina*.

The anterior thoracic ribs just before the shoulder girdle in *Barasaurus besairiei* are short, evidently to provide a wider range of motion to the forelimbs for a swimming power stroke. Comparable rib shortening near the pelvic girdle is present, evidently to allow greater range of motion in the hind flippers, in plesiosaurs. Three sacral ribs are present, as described below.

The shoulder girdle is preserved on the ventral impression of *Barasaurus besairiei* (Figs. 8.14 and 8.15). A clearly preserved T-shaped or anchor-shaped interclavicle extends nearly to the center of the fossil for a total length of 23 mm. This distinctive interclavicle shape is what gives ankyramorph parareptiles their name (deBraga and Reisz 1996; Reisz and Scott 2002). The anterior end of the interclavicle is very wide (18 mm) and has a slightly polygonal to crescentic anterior edge. The posterior part of the interclavicle is narrow-elliptical with a rounded posterior end. The left clavicle has a crescent shape and is visible in Figs. 8.7 and 8.8. The scapulocoracoids are damaged, but they appear to be roughly rectangular and the glenoid joint matches perfectly to the proximal articulation of the humerus on both the right and left sides (Figs. 8.5, 8.6, 8.7 and 8.8). The sternum plate as developed in *Hovasaurus* appears to be missing or is only very faintly preserved in *Barasaurus besairiei*.

Most or all bones in *Barasaurus besairiei* appear to be finished or mature, but the relatively small size of the example described here suggests that it may have been a juvenile or young adult. Muscle scarring appears on the left humerus (Fig. 8.12). Approximately 15 vertebrae are preserved, ten of these in the thoracic region and five in the shoulder girdle area. Ten ribs are preserved on the right side of the animal, and 11 ribs are visible on its left side.

The thoracic vertebrae (Figs. 8.5, 8.6, 8.7 and 8.8, 8.13, 8.15) in *Barasaurus besairiei* are preserved in articulation and in an upright and vertical orientation, not flopped over on their sides, thus it is difficult to determine the shape in lateral view and especially the height of the neural spines. In dorsal view these vertebrae show transverse processes that are slightly angled forward and even the more anterior vertebrae near the pectoral girdle seem to curve forward forming short horn-like projections. In ventral view, the vertebral centra make rounded, roughly symmetrical X-shaped impressions.

Thirteen presacral vertebrae are present on this specimen of *Barasaurus besairiei*. As is the case for *Hovasaurus*, the "basic structure of a notochordal centrum is hourglass shaped, constricted at the center and expanding anteriorly and posteriorly into round, hollow ends that form the intervertebral articulations" (Currie 1981). The centra appear to be strongly amphicoelous and were most likely notochordal (Bickelmann et al. 2009).

As seen in Figs. 8.5 and 8.6, there is a slight spinal curvature where the spinal column enters the shoulder girdle region. While this may be a postmortem feature, I interpret it here as evidence that the animal had a mild case of congenital scoliosis. Paleozoic examples of this malformation are rare. A more serious case of congenital scoliosis was reported by Szczygielski et al. (2017) in the Permian aquatic parareptile *Stereosternum tumidum* from Paraná, Brazil. In the case of this mesosaur, the deformity (possibly the result of a chromosome deletion) was the result of a congenital hemivertebra, but was evidently sublethal as the animal reached adulthood (Szczygielski et al. 2017). As is the case for the closely related parareptile *Mesosaurus*, *Stereosternum* develops pachyostotic ribs that serve as ballast for its aquatic lifestyle. Permian cases of spinal malformation due to hemivertebra development are also known from temnospondyl amphibians (Witzmann 2014) and in a captorhinomorph reptile (Johnson 1988). The captorhinomorph, as in the case of *Stereosternum*, is an early amniote example (considered to be Early Permian in age; Holterhoff et al. 2013; Szczygielski et al. 2017). I suspect that spinal malformation due to hemivertebra will be eventually found in a fossil of *Mesosaurus*.

Well-preserved cancellous or trabecular bone (spongy bone) is present in *Barasaurus besairiei*. Figure 8.16 shows a patch of spongy bone preserved at the proximal epiphysis of the right humerus (ventral side) with prominent Haversian canals. This trabecular bone (Gomez 2002) appears to be surrounded by a thin shell of compact bone. Figure 8.17 shows a closer view of cancellous bone across the fractured surface. Figure 8.18 shows a high magnification view of part of the trabeculae meshwork (=lamellae), a single trabecula, aligned toward the outer surface of the proximal articulation (and at a right angle to the plane of the microphotograph).

The bone in *Barasaurus besairiei* is flexible and light weight and is comparable to that of the proximal half of the right humerus in a Cretaceous plesiosaur (Kiprijanoff 1883; Liebe and Hurum 2012). It is also comparable to the "spongious, 'osteoporotic-like' bone inner structure" of ichthyosaurs and cetaceans (Houssaye et al. 2014). There is a close resemblance in the alternation of large and small inter-trabecular spaces between the bones of *Barasaurus besairiei* (Figs. 8.16, 8.17 and 8.18) and those of the ichthyosaur *Temnodontosaurus* (Houssaye et al. 2014, their Fig. 6A). Cancellous bone structure is typical for ichthyosaurs.

In a revealing case of exceptional preservation, reptilian cycloid scales (in a con-sistent arrangement) are preserved on the ventral specimen of *Barasaurus besairiei*, adjacent to the ventral impression of the proximal articulation of the right humerus (Figs. 8.19, 8.20, 8.21, 8.22, 8.23, 8.24 and 8.25). Such scale patches are preserved in several known specimens of *Barasaurus besairiei*, but this parareptilian squama-tion was misidentified as fish scales in a Ph.D. thesis undertaken at McGill University, Quebec, Canada (Meckert 1995). This interpretation has apparently gone unchallenged in the literature, although Smith (2000) mentions: "Scale impres-sions on some specimens and abdominal stone clumps confirm that most individu-als were intact when they arrived at their final burial site." Meckert's (1995) thesis contains the following comments: "The nodules frequently contain pebbles, plant debris, and in some cases fish scales which often obscure parts of the skeleton . . . Specimen CM 47514, skeleton in ventral view with skull elements obscured by fish scales." The correct interpretation is that these pesky scales are in fact integral parts of the *Barasaurus* body.

Meckert (1995) made an understandable error, as fossil fish with diamond-shaped scales and squamation somewhat similar to *Barasaurus* do occur in the Sakamena Formation (Fig. 8.26). It is good to finally have this matter resolved, as scale shape may provide a critical character state for resolving the phylogenetic position of *Barasaurus*, and in any case, the more we learn about barasaurian squa-mation the more we can say about the barasaurian morphogenetic field. These odd procolophonoid scales do resemble the ganoid scales of "primitive" fishes, and may in fact have developed thin bony layers similar to those of the bony ganoid scales of fishes such as *Polypterus*. If so, we may have here a parareptilian scale torology (convergent evolution by torologous relationship) with cladistic (i.e., polypteroid) fishes. This is a fascinating possibility that may very well bear on the question regarding whether or not chelonians (turtles and their kin) are closely related to or even descended from procolophonoids as has been occasionally suggested (Laurin and Reisz 1995; but see Lee 2013).

With regard to scale morphology providing a key character state, we must con-sider the Triassic procolophonoid parareptile *Sclerosaurus armatus* from the Buntsandstein of southern Germany and northern Switzerland (Sues and Reisz 2008). Very interestingly, a strip of polygonal dorsal osteoderms (diamond shaped!)

in *Sclerosaurus* has a key resemblance to the *Barasaurus* scales (which may qualify as osteoderms if they are weakly mineralized with bone). If this is the case, then the *Barasaurus* squamation, perhaps in a medial dorsal strip of reinforced scales, transitions from the non-mineralized squamation to the secretome. Both *Sclerosaurus* and *Barasaurus* are procolophonoids; Michael deBraga (2003) considered *Sclerosaurus* "a bona fide member of the family Procolophonidae." A close look at the scale/ osteoderms in *Sclerosaurus* (Sues and Reisz 2008, their Fig. 1 of the holotype of *Sclerosaurus armatus*) shows a series of three dorsal crest scale/osteoderms with doubly pointed anterior and posterior tips. This may represent a doubling or fine division in the dorsal midline morphogenetic field line that runs along the back of *Sclerosaurus*.

Three patches of ventral squamation (S1, S2 and S3) have been preserved in the first *Barasaurus besairiei*-bearing concretion described here. Two patches (S2 and S3) are slightly displaced from their original position on the ventral side of the animal. The smallest patch (S1), close to the distal end of the left humerus on the ventral impression (Figs. 8.8, 8.9 and 8.10) shows a nearly complete scale with portions of others adjacent. This scale is rotated approximately 37° to the left, with the acute point of the scale (i.e., anterior tip of the scale) still pointed in a roughly anterior direction.

The intermediate sized patch (S2) has been rotated approximately 37° clockwise with respect to the long axis of the animal, and the larger patch of scales has been rotated approximately 60° counterclockwise with respect to the long axis. Patch S2 consists of parts of five scales (Fig. 8.8, upper left). Thirteen diamond-shaped scales up to 4 mm in greatest dimension (Figs. 8.8, 8.19, 8.20, 8.21, 8.22 and 8.23) are visible in the larger patch (S3) in a narrow strip of integument that continues beneath the ventral bones of the animal. These scales are not keeled, but they do have a partially granular aspect (see scales 3 and 8 in Fig. 8.21). We will encounter keeled scales in a second *Barasaurus*-bearing concretion, see below.

Growth rings are visible in scale 9 of Fig. 8.21. Evenly spaced short ridges and rounded tubercle-like concretions up to 0.5 mm greatest dimension are present on the scales. A whitish distal edge to the scales may represent plate-like, thin osteoderms as in the pygmy spiny-tailed skink (*Egernia depressa*; Lindgren et al. 2009). Very fine radial ridges or striae (25 to 50 microns wide) occur at the edges and surfaces of some scales (scale 9 of Fig. 8.21), somewhat resembling the ornamentation of fine lines on scales (Davis 2012) of the Permian amphibian *Trimerorhachis*.

Figures 8.24–8.25 show a reconstructed scale and a generalized reconstruction of the of ventral, scaled surface of *Barasaurus besairiei*. The scale is diamond-shaped but the anterior corner of the diamond becomes more acute and is drawn out into a short nipple-like projection. Small round concretions occur on the anterior end of the scale. In the central part of the scale, elongate concretions averaging about 200 microns but up to 0.5 mm in length are present. Nearing the posterior edge of the scale, the concretions begin to merge together to a certain extent, forming a reticulate pattern somewhat resembling the reticulate pattern on the dorsal surface of the scale-osteoderms of *Sclerosaurus*. The edge of the *Barasaurus besairiei* scale has a crenulated aspect (Fig. 8.23) Finally, fine striae ornament occurs in the region near the posterior edge of the scale, not often going right to the posterior edge of the scale, but instead crossing over the anterior edge of the whitish plate-like border at the posterior edge of the scale. The overall shape of each scale is similar to that of the mosasaur *Ectenosaurus clidastoides* with the exception that the *E. clidastoides* scales are keeled (Lindgren et al. 2011).

The patterning formed by the scales is quite similar to the dorsal squamation of the snakes *Boiruna* and *Clelia* from Paraguay and Argentina (Scott et al. 2006), but is even more similar to the ventral squamation of skinks such as those belonging to the genus *Plestiodon*. The main difference is that ventral scales on a typical skink tend to be wider than they are long, and the fossil scales reported here are roughly diamond-shaped/equidimensional, although one of the scales in the smaller patch (Fig. 8.10) has dimensions very similar to those of a ventral scale of a modern skink. Reptilian scales such as these are very likely to have been ectodermally derived and certainly involved EDA/EDR cell-cell signaling interaction and has "similarities with the function of Pax6 in vertebrate and invertebrate eye development where a common genetic pathway is used for the development of functionally similar but morphologically unrelated organs" (Sharpe 2001; see also McMenamin 2016).

Scales and/or scale-osteoderms reported from *Hovasaurus* are round (small scales) to pentagonal and octagonal (larger scales; Piveteau 1926) or oval (ventral surface near sternum; Currie 1981). The oval scales reported for *Hovasaurus* by Currie (1981) are from the same region of ventral integument as the scales reported here for *Barasaurus besairiei*. Ketchum and Barrett (2004) reported on a large suite of Madagascan parareptilian and reptilian material from the Lower Triassic part of the Sakamena Group, but noted that they had not encountered a single case of soft tissue preservation. Currie (1981) remarked that the irregular scales illustrated by Piveteau (1926) from *Hovasaurus* could in fact be osteoderms. The pattern they make somewhat resembles the bone ornamentation seen on a specimen of the Permian amphibian *Acheloma?* (Davis 2013).

The ventral abdomen scales of *Barasaurus besairiei* are superficially similar to the scales of the Cretaceous mosasaur *Ectenosaurus clidastoides*, particularly in the way that the anterior end of the rhomboid scale appears to be drawn out somewhat into a blunt point with concave sides (Lindgren et al. 2011). The ventral scales of *Barasaurus besairiei* lack a central keel and, interestingly, are larger than those of *Ectenosaurus* (2.7 mm), attaining a size nearly as large as those of the russellosaurine mosasaur *Platecarpus* (4.4 mm; Lindgren et al. 2011). The ventral scales of *B. besairiei* have a scalloped-denticulate margin and fine parallel and in some cases branched thin ridges at the edge of the scale (Figs. 8.22, 8.23, and 8.24). The rhomboidal scales are similar enough to mosasaur scales, however, that the comment by Lingren et al. (2010) referring to mosasaurs may well apply to *B. besairiei*: "We hypothesize that the small-sized and morphologically homogenous body scales . . . may have served to stiffen the body and to resist axial compression during rhythmic bending movements, thereby providing a hydrostatic structure that maintained the shape of the animal when it was swimming." Indeed, these remarks would seem even more applicable to a small aquatic animal such as *Barasaurus besairiei* than to the typically much larger mosasaurs.

The large size of the ventral scales is surprisingly similar to those of the skink (superfamily Scincoidea, Family "Scincidae"). These reptiles are members of the Scincomorpha group named in 1923 by Charles Camp. The taxonomic placement of the group is currently debated (Pyron et al. 2013; Hedges 2014). They are a highly successful group and account for one quarter of all lizard species broadly speaking. Almost half of described skink species develop some variant of viviparous or ovoviviparous gestation, which might potentially have exapted (preadapted) them for aquatic existence. In the astonishing case of the skink *Trachylepis ivensi*, the mother reptile develops a placenta that is highly convergent on the eutherian mammalian placenta (Blackburn and Flemming 2011). The largest living skink is the Solomon Islands skink (*Corucia zebrata*) and at 70 cm in length, twice the length of many *Barasaurus besairiei* individuals.

The dorsal squamation of *Corucia* is virtually identical to what we see in *Barasaurus besairiei*. *Corucia zebrata* is one of the very few living reptiles to exhibit social bonding, living in skink communities known as circuli (sing.: circulus). Fossils of adult and juvenile specimens of *Barasaurus besairiei* are known to occur in close juxtaposition (Meckert 1995, specimen CM 47512) but whether or not this constitutes evidence for social bonding is uncertain at present.

Very interestingly, an aquatic skink from Madagascar (*Amphiglossus astrolabi*; see Glaw and Vences 2007) has recently been photographed by biologist Asia Murphy eating crabs (Naish 2016). This skink, with its smooth dark green dorsal coloration and cream yellow ventral surface, seems to bear the countershading and coloration one might expect for a marine reptile of larger size. Another at least semi-aquatic member of the genus in Madagascar, *Amphiglossus reticulatus*, has such a powerful bite force that a collector's hand was numb for more than an hour after receiving a bite from the animal (Glaw and Vences 2007; Naish 2014). The fossil record of skinks is thought to go back to the Early Cretaceous, but the fossils are scrappy and mostly limited to fragmentary jaw material (Estes 1983). It is safe to infer that skinks retain their evolutionary potential to radiate into a major clade of marine reptiles reminiscent of the great marine reptilian lineages of the Mesozoic.

The soft tissue preservation in the pectoral region specimen of *Barasaurus besairiei* is associated with the right anterior region of the animal. This further suggests: first, the animal's method of swimming; second, the striated muscle anatomy of the animal, and third, another hypothesis for cause of death. The strip of enhanced preservation corresponds to the position of the triceps longus medialis. In the Australian estuarine crocodile (*Crocodylus porosus*), this muscle serves as an elbow extensor (Klinkhamer et al. 2017). Interestingly, a taxonomically significant point of divergence in forelimb musculature occurs in the triceps longus medialis of crocodilians. As reported by Klinkhamer et al. (2017), "two tendons of origin for triceps longus medialis have been noted for *Alligator mississippiensis* and other crocodilian taxa, with one tendon originating on the scapula and one from the coracoid. Only the scapular tendon was noted for *C. porosus*." The triceps longus medialis thus seems more liable to evolutionary modification than other, more conserved, parts of the reptilian forelimb musculature.

The triceps longus medialis was very well developed in *Barasaurus besairiei*, presumably with a massive tendon origination on the scapulocoracoid (Currie 1981) suggesting that the parareptile was a forelimb swimmer that propelled itself with a power stroke that extended its elbows by contraction of the triceps longus medialis. In other words, it performed a variation on the breaststroke (perhaps fittingly in this case, also known as the frog stroke).

The cause of death for this and other specimens of Malagasy *Barasaurus besairiei* is currently unknown, although the possibility of aquatic battles with *Hovasaurus* was mentioned above. Smith (2000) hypothesized that the "most plausible cause of death in this [lacustrine] setting is overturn of the thermocline caused by storm-generated flooding." This seems to be an implausible explanation, because the animal was a powerful swimmer and was presumably able to move away from cold, upwelling water and even move on shore if necessary.

Exhaustion might also be the cause of death. In this scenario, the unfortunate *Barasaurus besairiei*, caught in a current of some sort, may have died of exhaustion trying to swim against the stream. Due to overexertion, its triceps longus medialis became heavily depleted in oxygen (oxygen debt) and thus may have suffered lactic acidosis. Oxygen recovery in the muscle perhaps could not occur after the animal's death. Reptiles such as *Crocodylus porosus* are highly susceptible to lactic acidosis, and if they overexert their muscles performing anaerobic energy metabolism (glycolysis), they can become unresponsive or may even die due to pH imbalance in their blood supply (Seymour et al. 1987). This hypothesis has a difficulty similar to the objection noted to the previous hypothesis, namely, this animal was a strong swimmer and would be unlikely to be overwhelmed by currents in its aquatic environment.

I propose here an additional, and potentially important, hypothesis to explain the *Barasaurus* and *Hovasaurus* mortality in the Sakamena Formation. As air-breathing animals, they would have been susceptible to sudden decrease in local air quality. If, for example, the lake surface was to be covered by a blanket of carbon dioxide due to lake overturn (or alternatively, or perhaps in concert with, release of volcanic gases), the animals may have suffocated and then been dragged to the bottom of the lake by the weight of their ingested gastroliths.

Recall Smith's (2000) linkage of the deaths to thermal shock from overturn of the lacustrine thermocline. A somewhat comparable phenomenon is limnic eruption, the motive force behind *exploding lake syndrome*, that leads to the catastrophic release of carbon dioxide from the depths of a lake causing mass mortality of air-breathing organisms nearby. Limnic eruptions are rare phenomena, and today are known to occur in only three African lakes, all associated with volcanic provinces: Lake Nyos and Lake Monoun in Cameroon, and Lake Kivu in the Democratic Republic of Congo. In 1986, a limnic eruption in Lake Nyos killed 1746 people and thousands of head of livestock (Rouwet et al. 2016).

In a triumph of engineering finesse and scientific cooperation, Lake Nyos is now considered safe for the most part due to the installation of artificial degassing measures. Smith (2000) may have been right about the Malagasy Permo-Triassic aquatic parareptile and reptile deaths being due to lake overturn, but the killing mechanism

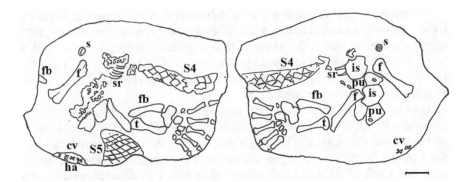

Fig. 8.28 *Barasaurus besairiei*, sketch of specimens in previous image. Abbreviations: **f**, femur; **fb**, fibula; **t**, tibia; **pu**, pubis; **is**, ischium; **cv**, caudal vertebrae; **ha**, hemal arches of caudal vertebrae; **s**, seed; **sr**, sacral ribs; **S4-S5**, scale patches. Scale bar 1 cm

would be suffocation by carbon dioxide rather than thermal shock. Left unexplained by this new hypothesis is the deaths of fish that are also found in Sakamena Formation calcareous concretions, and another mechanism must be sought to explain their deaths, although the thermal shock hypothesis and limnic eruption hypothesis may not be mutually exclusive, for Smith (2000) notes that in "modern large, deep lakes, perhaps the most common cause of mass mortality among teleosts is oxygen deficiency and thermal shock when large volumes of cool anoxic bottom water mix with warm oxygenated surface water ... seismically induced degassing of the anoxic bottom sediments can cause mass mortality in and around the lake."

Soft tissue preservation within concretions is not unusual; indeed, about one third of all fossiliferous concretion sites contain exceptionally preserved soft tissue fossils (McCoy 2015). Rapid "concretion nucleation and growth promotes soft tissue preservation . . . the faster concretions nucleate and grow, the better the preservation" (McCoy 2015). This is so because decay of organisms "in sediment can produce HCO_3^- faster than it diffuses away, creating a local microenvironment of high alkalinity around the decaying organisms that promotes carbonate precipitation" (McCoy 2015). True to this pattern, the Sakamena fossiliferous concretions are calcareous shale that will fizz weakly with dilute hydrochloric acid. The width of the reaction front at the edge of a concretion while it grows is "controlled by the diffusion rate of HCO_3^-, a byproduct of fatty acid (R-COOH) decomposition and rapid $CaCO_3$ precipitation due to pH change at the front . . . the concretions continued to grow until there was no more carbon of organic origin remaining" (Yoshida et al. 2015).

A second specimen of *Barasaurus besairiei* described here consists of both halves of a Sakamena Formation concretion (Figs. 8.27, 8.28, 8.29, 8.30, 8.31, 8.32, 8.33, 8.34, 8.35, 8.36, 8.37, 8.38, 8.39, 8.40, 8.41 and 8.42). This specimen also includes scale impressions, plus a section of the tail as will be described below. Dorsal and ventral impressions are preserved here (Figs. 8.27 and 8.28). Also present in the concretion is a fairly well-preserved seed resembling a tiny coconut

Fig. 8.29 Unidentified fossil seed resembling a tiny coconut from the Permian Sakamena Formation. A keel runs along the midline of the seed. Scale in millimeters (Photograph by Mark McMenamin)

Fig. 8.30 Unidentified fossil seed resembling a tiny coconut from the Permian Sakamena Formation. Seed surface on this side is smooth with longitudinal fibers visible. Scale in millimeters (Photograph by Mark McMenamin)

Fig. 8.31 Ballast gravel ingested by *Barasaurus besairiei*. Scale in millimeters (Photograph by Mark McMenamin)

Fig. 8.32 *Barasaurus besairiei*, right hind limb, dorsal view. Note the bilobed astragalocalcaneum. Scale in millimeters (Photograph by Mark McMenamin)

Fig. 8.33 *Barasaurus besairiei*, detail of right foot in previous image. Note the bilobed astragalocalcaneum with the stigmata foramen at its top center. Scale in millimeters (Photograph by Mark McMenamin)

(Figs. 8.29 and 8.30). A collection of grit and coarse sand on the concretion are evidently clasts ingested by *Barasaurus besairiei* for ballast (Fig. 8.31). This suggests that Smith (2000) was incorrect to state that *Barasaurus* would only ingest larger pebbles, as evidently it would take smaller grit-sized clasts as well.

A portion of the right leg is clearly preserved with the major bones intact, showing the tibia, fibula and details of the tarsus (articulating small foot bones) and pes (metatarsals and phalanges; Figs. 8.32, 8.33, 8.34, 8.35 and 8.36). Preserved here are metatarsals I-V, distal tarsals I-V, and the astragalocalcaneum (formed by fusion of the calcaneum and the astragalus). Only the proximal ends of the proximal-most

Fig. 8.34 *Barasaurus besairiei*, right hind limb, ventral view. Note the bilobed astragalocalcaneum. Scale in millimeters (Photograph by Mark McMenamin)

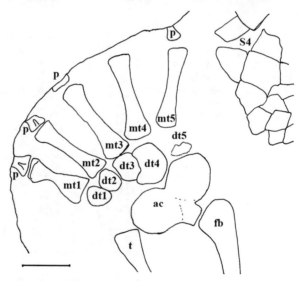

Fig. 8.35 *Barasaurus besairiei*, sketch of specimens in previous image. Abbreviations: **f**, femur; **r**, radius; **u**, ulna; **ac**, astragalocalcaneum; **dt1–5**, distal tarsals I-V; **mt1–5**, metatarsals I-V; **p**, phalanx (plural: phalanges). Scale bar 5 millimeters

Fig. 8.36 *Barasaurus besairiei*, right hind limb, ventral view, detail of tarsus and pes. Note slight scarring on the calcaneum part of the astragalocalcaneum; also, the astragalus part of the astragalocalcaneum appears to have at least two centers of ossification, the one on the right likely being the intermedium. Scale in millimeters (Photograph by Mark McMenamin)

Fig. 8.37 *Barasaurus besairiei*, right hind limb, dorsal view of part of pes showing medial groove in phalanx to left. Scale in millimeters (Photograph by Mark McMenamin)

Fig. 8.38 *Barasaurus besairiei*, right pubis, ventral view. The obdurator foramen is clearly visible as an opening in the pubis. Scale in millimeters (Photograph by Mark McMenamin)

Fig. 8.39 *Barasaurus besairiei*, left pubis, ventral view. The obdurator foramen is clearly visible as an opening in the pubis. Scale in millimeters (Photograph by Mark McMenamin)

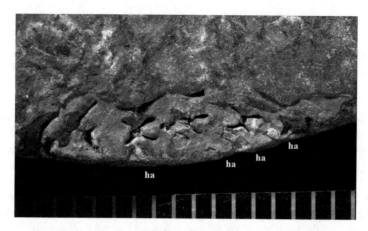

Fig. 8.40 *Barasaurus besairiei*, distal part of tail, caudal vertebrae with hemal arches. Parts of four hemal arches are visible in the photograph. **ha**, hemal arch. Scale in millimeters (Photograph by Mark McMenamin)

Fig. 8.41 *Barasaurus besairiei*, scale strip (S4) on ventral surface showing diamond-shaped scales. Keels are present on several of the scales. Scale in millimeters (Photograph by Mark McMenamin)

Fig. 8.42 *Barasaurus besairiei*, scale patch (S5) on dorsal surface. Scale in millimeters (Photograph by Mark McMenamin)

phalanges are preserved, and the centrale appears to be missing. The astragalocalcaneum is well preserved (Figs. 8.32, 8.33, 8.34, 8.35, and 8.36), and a critical feature,
as it demonstrates both the identity of the fossil as *Barasaurus besairiei*, and also
indicates the procolophonid affinity of the animal as the presence "of a fused astraglocalcaneum indicates that the material belongs to the Procolophonia, a clade that
includes procolophonoids, pareiasaurs, and possibly chelonians [turtles]" (Ketchum
and Barrett 2004). Specimens of *Barasaurus besairiei* illustrated by Ketchum and
Barrett (2004) show fusion of the astragalus and calcaneum to form the astraglocalcaneum, to such a degree that "the junction between the two bones has been completely obliterated proximal to the perforating foramen [stigmata foramen] (with
only a low ridge marking the boundary between the two elements), whereas distal
to the foramen the junction is still visible as a narrow groove (although the two elements are apposed very tightly)." The astragalocalcaneum in *Barasaurus* is a curious and diagnostic feature, and in both the specimen shown here (Figs. 8.32, 8.33,
8.34, 8.35 and 8.36) and in the one illustrated by Ketchum and Barrett (2004), the
fusion between the astragalus and the calcaneum appears to be complete.

Also preserved in the hind limb of *Barasaurus* described here is a small distal
tarsal V (Figs. 8.32, 8.33, 8.34, 8.35 and 8.36) . Ketchum and Barrett (2004) note
that loss of distal tarsal V is a characteristic trait for the procolophonoids and indeed
for the entire parareptilian clade. Of the procolophonoids, only *Barasaurus* retains
distal tarsal V, and this as "a reversal to the primitive amniote condition" (Ketchum
and Barrett 2004). As we can see in (Figs. 8.32, 8.33, 8.34, 8.35 and 8.36), however,
the distal tarsal V in *Barasaurus besairiei* is much reduced and appears to be incom-
pletely ossified, and is represented by a lobate impression on both the dorsal and
ventral halves of the concretion. In addition to the fused astraglocalcaneum, a procolophonoid assignment *Barasaurus besairiei* is "supported by the presence of the
medial flange on the proximal end of metatarsal I" (Ketchum and Barrett 2004;
Meckert 1995).

Currie (1981) described a possible distal tarsal V from *Hovasaurus boulei* in his
Fig. 29h, noting that "there is a gap between the fifth metatarsal and calcaneum in
most [*Hovasaurus boulei*] specimens. The possible fifth distal of MNHN 1908–32-
68/1924–8-14 (Fig. 29[h]) fills part of this space." The presence of distal tarsal V in
this specimen, plus the possible fusion of the calcaneum and the astragalus (the two
are tightly juxtaposed and may form a astragalocalcaneum), indicates that this particular specimen might better be assigned to *Barasaurus besairiei*, not *Hovasaurus
boulei*.

The detailed image in Fig. 8.36 shows the large fused astragalocalcaneum as is
characteristic for *Barasaurus besairiei*. A tiny distal tarsal V is present, adjacent to
the calcaneum part of the astraglocalcaneum. Also present are a very large distal
tarsal VI, and three subequal in size distal tarsals I-III. Distal tarsals I-IV have a
slightly raised rim-like feature that parallels the edge of each bone. This same feature occurs on the distal tarsus bones (distal tarsals) illustrated by Ketchum and
Barrett (2004), and thus appears to be in the Malagasy concretion preservational
regime to be a common feature seen in *Barasaurus besairiei*. Also, a medial groove
is present in a phalanx as shown in Fig. 8.37. Comparable features occur on the

Barasaurus phalanges as illustrated by Ketchum and Barrett (2004). Thus, the identification as *Barasaurus besairiei* of the specimens described in this chapter seems very secure.

Turning to the sacral region (Fig. 8.28), pelvic bones and sacral vertebrae are also preserved on this specimen of *Barasaurus besairiei*. Three right sacral ribs are present on both halves of the concretion (Fig. 8.23). Meckert (1995) noted the presence of three sacral ribs in *Barasaurus besairiei*, and considered this character to be "a diagnostic feature of procolophonids." In *Barasaurus besairiei*, the sacral ribs are "roughly of equivalent length" (Meckert 1995), and this seems to be the case for the specimen considered here.

The right pubis is shown in Fig. 8.38, and the left pubis is shown in Fig. 8.39. In both, an obdurator foramen is clearly visible, and their associated ischia are also preserved (Fig. 8.28). The flat, roughly pentagonal pubis in ventral view (Figs. 8.38 and 8.39) in *Barasaurus besairiei* resembles that of *Procolophon trigoniceps* (a terrestrial procolophonoid from the Lower Triassic of Antarctica and South Africa), and similarly forms the anterior part of the puboischiadic plate (deBraga 2003).

Interestingly, distal caudal vertebrae (but not the tip of the tail), consisting of neural spines, caudal centra, and hemal arches (Fig. 8.40) are preserved on the specimen of *Barasaurus besairiei* described from this second concretion. The shape of the hemal arches demonstrates that this tail fragment belongs to *Barasaurus*, and most likely the individual represented by the pelvic and hind limb remains. Apparently, the tail curled forward strongly enough for part of it to be preserved in the same concretion that preserved the other remains, a fortuitous and fortunate occurrence. This same type of forward curvature of the tail back to the trunk of the animal is also seen in a *Barasaurus besairiei* specimen illustrated by Meckert (1995; his Fig. 24, specimen P5). The caudal vertebrae in the specimen described here are slightly disarticulated but in their proper order within what appears to be a smooth soft-bodied impression of a laterally compressed tail. This flattened tail would have potentially provided propulsive thrust for swimming.

Slightly anterior to and somewhat paralleling the right hind limb, on both dorsal and ventral impressions, a strip of scales (S4) is preserved (Figs. 8.28 and 8.41). Keels are present on several of the scales. Although keels are not present on the scale patches (S1-S3) described from the previous concretion, I consider both of these *Barasaurus*-bearing concretions to be representatives of the same species because there are no keels on the S5 patch in the second concretion. Keeled scales therefore occur only in certain regions of the scale scleritome (similar to the double-pointed scale/osteoderms in *Sclerosaurus*), and in any case the presence or absence of scale keels might have been subject to phenotypic variation with the species.

Again, these are what Meckert (1995) and subsequent workers have referred to as 'fish' or 'fish scales.' Consider once more the Triassic parareptile *Sclerosaurus armatus* from the Buntsandstein of Germany and Switzerland (Sues and Reisz 2008). The preservation of strips of polygonal scales in *Barasaurus besairiei* may indicate that they are directly linked to the dorsal osteoderms in *Sclerosaurus* by common ancestry, and this is likely to be the case considering that both are procolophonoids. These same type of diamond-shaped scales occur on the dorsal side of

Table 8.4 Scale preservation in *Barasaurus besairiei*

Taxon	Number of scales	Scale patch position	Illustration	Specimen	Reference
Barasaurus besairiei	13	Dorsal thorax, right side	Figure 16 of Meckert (1995)	P 7 (National Museum of Natural History, Paris, France)	Meckert 1995 (originally interpreted as fish scales)
Barasaurus besairiei	20	Neck region, ventral side	Figure 22 of Meckert (1995)	CM 47514 (Carnegie Museum, Pittsburgh)	Meckert 1995 (originally interpreted as fish scales)
Barasaurus besairiei	3	Ventral thorax, right side close to spine	Figure 22 of Meckert (1995)	CM 47514	Meckert 1995 (originally interpreted as fish scales)
Barasaurus besairiei	5	Ventral thorax, right side	Figure 22 of Meckert (1995)	CM 47514	Meckert 1995 (originally interpreted as fish scales)
Barasaurus besairiei	4 (patch S1)	Ventral left side, near interclavicle	Figs. 8.8 and 8.10	2 of 10/23/2017	This report
Barasaurus besairiei	4 (patch S2)	Ventral left side, near interclavicle	Fig. 8.8	2 of 10/23/2017	This report
Barasaurus besairiei	14 (patch S3)	Ventral right side, near distal end of right humerus	Figs. 8.8, 8.19, 8.20, 8.21, 8.22 and 8.23	2 of 10/23/2017	This report
Barasaurus besairiei	~15 (patch S4)	Ventral	Figs. 8.28, 8.35, and 8.41	4 of 10/23/2017	This report
Barasaurus besairiei	~ 30 (patch S5)	Dorsal	Figs. 8.28, and 8.42	3 of 10/23/2017	This report

Barasaurus besairiei, as seen in a scale patch (S5) as shown in Figs. 8.28 and 8.42. With these scale patterns, we have confirmation that the pectoral region concretion and the pelvic region concretion described here belong to members of the same species. All available skeletal evidence confirms this conclusion. Table 8.4 summarizes the evidence for scale morphology from these two *Barasaurus* concretions.

The fact that keeled scales are associated with only some of the scales of *Barasaurus besairiei*, and the fact that these occur in a roughly linear series, raises the question as to whether or not there might be a torologous relationship between the keeled scales with pointed tips of *Barasaurus besairiei* and the eight small carinated mucro-anterior plates with beaked anterior margins of *Korifogrammia clementensis* (Figs. 3.22, 3.38, 3.39, 3.40, 3.41, 3.42 and 3.43). This is an interesting possibility that deserves careful evaluation. It is easy to imagine a mechanical benefit to having a keel on a tetrapod's scales or a chiton's plates, as various environmental obstructions or grasping claws or teeth might very well slide harmlessly

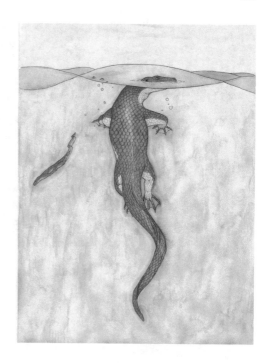

away more easily with a keel than without. This would appear then to be a case of conventional convergent evolution; however, consider that the keel itself runs parallel with the longitudinal morphogenetic field lines and may very well be an expression of these lines as proposed earlier for *Sclerosaurus*. Therefore, the scales/sclerites of *Barasaurus besairiei* and *Korifogrammia clementensis*, respectively, are likely in a torologous rather than merely homoplasic relationship.

Figures 8.43 and 8.44 show reconstructions of *Barasaurus besairiei*. All morphological characteristics described above for *Barasaurus besairiei* suggest that it was indeed an aquatic reptile, but was not as specialized for an aquatic lifestyle as was *Hovasaurus*. Currie (1981) concluded that *Hovasaurus* is "highly specialized for an aquatic existence". The evidence suggesting that *Barasaurus besairiei* was less developed for an aquatic lifestyle is as follows:

1. Metacarpals are less symmetric.
2. The penultimate phalanx is generally not longer than the one(s) before.
3. The radius is neither twisted nor curved.
4. The humerus is relatively thin at its midpoint.
5. Although the rib ends are swollen (for ballast), the ribs are not heavily pachyostitic.
6. Carpus bones are not as relaxed as in *Acerosodontosaurus* and *Hovasaurus*.

Currie (1981) quickly passes over a possible relationship between *Hovasaurus* and the sauropterygians, but he does note the resemblance between *Claudiosaurus*, early nothosaurs and plesiosaurs and the resemblance of all three groups to the

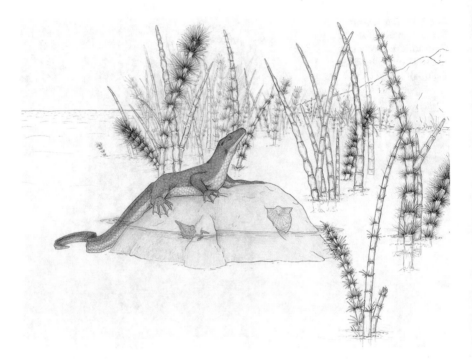

Fig. 8.44 *Barasaurus besairiei*. Shown resting on a rock near shore (Artwork by Kallie Ellen Fellows)

younginiform clade. Currie (1981) bases this resemblance on "the loss of contact between the parietal and postorbital." The *Claudiosaurus* humerus is "shorter and more gracile than the tangasaurid humerus." Currie (1981) concluded that *Claudiosaurus* was closely related to the younginiform clade, but was "not derived from tangasaurid stock."

Carroll (1981) emphasized the similarities between *Claudiosaurus* and the sauropterygians. Merck (1997) held that Sauropterygia and Ichthyopterygia were sister taxa. In what follows I will not be providing a cladistic analysis—the derived state of many Mesozoic marine reptiles renders relationships hard to parse using cladistics. Carroll and Dong (1991) have amply shown the drawbacks of a cladistic approach in the case of hupehsuchians.

Post-cranial material has been shown to be important for sauropterygians (Sachs and Kear 2015). *Barasaurus besairiei* developed, presumably by convergence, traits known in the Sauropterygia and the Ichthyopterygia. The fact that the penultimate phalanx is generally not longer than the one(s) before in *Barasaurus besairiei* resembles the phalanges in ichthyosaurs, where this feature becomes extended in the Ichthyopterygia as the multiplication of phalanges known as hyperphalangy (Fig. 8.45; see Maisch and Matzke 2000). Swollen ends of ribs are seen in early ichthyosaurs such as *Shonisaurus*. Furthermore, as in ichthyosaurs, *Barasaurus besairiei* develops a sclerotic ring around the eyes (Meckert 1995); sclerotic ele-

Fig. 8.45 *Stenopterygius quadricissus*, flipper showing the multiplication of phalanges (hyper-phalangy). Jurassic (Lias), Holzmaden, Germany. Cast. Scale bar in cm (Photograph by Mark McMenamin)

ments are well developed in ichthyosaurs, including large ichthyosaurs such as *Shonisaurus* (Camp 1980).

The somewhat ichthyosaur-like *Hupehsuchus*, bane of the cladists (Carrol and Dong 1991), develops curious rows of bone above its neural spines. These strange accessory bone rows articulate with the neural spines, and represent bone element multiplication that may be related to the hyperphalangy and hyperdactyly of advanced ichthyosaurs. As this involves the sclerotome, not the scleritome, the term torologous to describe a possible relationship between hyperphalangy, hyperdactyly and accessory bone rows above neural spines may not be strictly applicable. In these cases, the morphogenetic torus appears to be experiencing a repeating aspect that may lead to a family of concentric tori. We saw this earlier with the embedding of the dorsal medial spine deep into the dorsal mesenchymal tissue of the coelacanth cf. *Rhabdoderma* sp. as it became surrounded by a concentric torus. Similar consid-erations might apply to cetaceans and the cetacean pelvic remains. We may be able to envision a situation where the coelacanth's set of lobed fins represents an out-growth of an expanded and superimposed partial toroidal morphogenetic field. The lobate fins of the coelacanth are an aspect of the torus expansion, and this explains why most of the coelacanth fins are lobed, even (very oddly) the second dorsal fin that looks like it ought to be evolving into a tetrapod appendage (Ahlberg 1992). This leads us to the Eighth Law of Morphogenetic Evolution:

EXPANSIONS OF OR MATRYOSHKA-NESTINGS OF THE MORPHOGENETIC TORUS CAN LEAD TO CONCENTRIC REPETITION OF BODY LAYERS AND THEIR ASSOCIATED BODY PARTS.

The advanced (parvipelvian) ichthyosaur *Stenopterygius* is known to develop, like *Hupehsuchus*, elements of cartilage that look very much like the rows of bone above *Hupehsuchus's* neural spines. If we consider the above-the-spine bony or cartilaginous elements of *Stenopterygius* and *Hupehsuchus*, respectively, to be both homologous and (in the expanded sense) torologous, then this would lend support to proposals that all Mesozoic marine reptiles should be united in a single clade (Motani et al. 2015). Further support for a shared ancestry between Sauropterygia and the Ichthyopterygia is seen in the key fact that in both groups, the pterygoids extend underneath the braincase.

Barasaurus besairiei never gave rise to a major group of Mesozoic marine reptiles, but it seems that it had the potential to do so. Its pectoral girdle is quite simple and reduced as is usual for early-lineage aquatic reptiles. In descendant marine reptiles, the pectoral girdle reaches a striking state of development. For example, in both the long-necked plesiosaurs such as *Meyerasaurus* and in the short-necked, pliosaur morph that appeared repeatedly by convergent evolution (for example, the polycotylid *Dolichorhynchops* and the plesiosauroid *Elasmosaurus*), we see dramatic evolutionary reinforcement of the pelvic girdle.

Two separate paths were taken in the evolution of the swim dynamics of long-necked plesiosaurs versus short-necked pliosaur morphs. Halstead (1982) noted that pliosaurs had massive muscles for "pulling the limbs down and back," but the muscles involved in the recovery stroke were comparatively weak:

> To begin the swimming stroke, the paddle was held with the upper surface facing forwards at an angle, like a bird's wing in flight; when the limb was pulled downwards, it caused a power stroke with an upwards and forwards thrust. At the end of the power stroke, as the water flowed over the limb, it rotated so that the blade presented minimal resistance to flow. The recovery stroke was completed by the dorsal muscles which pushed the limb forwards and then twisted the blade, ready for the next power stroke.

The powerful stroke of pliosaurs allowed them to pursue their cephalopod prey to depths of over 300 m (Halstead 1982). Long-necked plesiosaur swim dynamics were quite different, as the animal had balanced musculature for both the power and the recovery stroke. This implies an ability to make sharp turns, an asset for a piscivore that was required to spin its body quickly to snap up fish. The long-necked plesiosaur *baüplan* precludes an ability to dive deeply (Halstead 1982), but is well suited for rapid turns that would generally employ all four flippers.

With its nascent development of a robust pectoral girdle as a muscle attachment platform, *Barasaurus besairiei* makes a plausible ancestor for a hypothetical radiation of marine procolophonoids that never happened. Again, we may ask why this never happened. The horrific Permo-Triassic mass extinction wiped out many Paleozoic reptile lineages that might very well have produced fully marine forms. There was thus a tremendous bottleneck at the end Permian that cut short the possibility of returning to the sea for a number of lineages. We know that *Barasaurus* survived into the Triassic. Indeed, a owenettid procolophonoid with affinities to *Barasaurus* (aff. *Barasaurus*) survived until the Middle Triassic, its remains being preserved in the Erfurt Formation of Germany (Martinelli et al. 2016). *Barasaurus* had the favorable Madagascan region as its home, a region that evidently had a refugium aspect that allowed aquatic reptiles to survive the great mass extinction. Why then did the barasaurian lineage not thrive as the Mesozoic got underway?

Currie (1981, his fig. 15a) cleverly demonstrated that the section of caudal vertebrae with very tall neural spines, what Piveteau (1926) had called an 'indeterminate reptile,' was in fact the tail of a *Hovasaurus*, thus making the key point that

Hovasaurus was at least a partly tail propulsion swimmer. If we assume that *Barasaurus besairiei* also had a dorsoventrally flattened tail, as seems a reasonable interpretation of the tail section shown in Fig. 8.40, although its neural spines are not as tall as in *Hovasaurus*, we may be able to infer that *Barasaurus besairiei* was convergent on tail-swimmer ichthyosaurs. It seems likely that both *Barasaurus* and *Hovasaurus* utilized a locomotion strategy that involved both types of swimming in the same animal. It seems plausible that *Barasaurus besairiei* could have been a tail swimmer like *Hovasaurus*, although possibly with less efficiency and with a less dorsal-ventrally flattened tail. *Barasaurus besairiei*'s pectoral girdle morphology also renders it potentially convergent on flipper swimmers such as plesiosaurs.

Discovery of the Eighth Law of Morphogenetic Evolution leads us to consider a key paleontological problem, consideration of which is essential for understanding of the origin of tetrapod limb structure. Recall that the tetrapod forelimb is divided into the stylopod (humerus), zeugopod (radius and ulna), and autopod (digits from the metacarpals onward in a distal direction). The bones of the carpus will be considered here to be part of the zeugopod. A comparable division is seen in the hind limbs, with the femur representing the stylopod and the tibia and fibula representing the zeugopod.

Shubin et al. (1997) considered the enigmatic origins of the tetrapod limbs, and attempted to combine paleontological evidence and emerging genetic evidence to synthesize a new understanding of vertebrate limb patterning. Simply put, Shubin et al. (1997) felt that development of the stylopod was associated with the HoxD-9 gene, that the zeugopod was associated with the HoxD-11 gene, and that the auto-pod was associated with the HoxD-13 gene. The development of a limb is divided accordingly into Phase I, Phase II and Phase III. Phase II would be amply demonstrated by the fin of the problematic early fish *Doliodus*, where the pectoral fin spine serves to delineate the fin axis and the proximal radials of the pectoral fin, distal radials, and cartilages of the 'metapterygial axis' (Maisey et al. 2017) serve as the preaxially (i.e. toward the midline) branching radials as in *Panderichthyes* (Shubin et al. 1997). This is nicely shown in the pelvic fin (Moy-Thomas 1937) of the coel-acanth *Rhabdoderma*. Differentiation of distal fin structure in *Doliodus* is thus involved with HoxD-11.

Shubin et al. (1997) claim that in Phase III, there is a reversal in the polarity of Hox gene expression to allow the digits to form on top of (and not as an extension of) the zeugopod (including the carpus) and stylopod parts. Here the emphasis, and I believe that this emphasis is misplaced in an otherwise important article (Shubin et al. 1997), is on the idea that the HoxD genes are ultimately in control of tetrapod limb development, in other words, the nuclear genome is controlling *baüplan* morphology in a somewhat blueprint-like fashion. Alas, Shubin et al. (1997) are thus beholden to one of the great fallacies of modern biology.

Nevertheless, Shubin et al. (1997) brought to attention something important, namely the very real discontinuity between the zeugopod and the autopod. Consider the stylopod-zeugopod axis, or 'fin axis,' as an elevated morphogenetic field line that lifted up from the surface of the corporeal toroidal morphological field, possibly as a narrow but distending loop. Branching structures are produced as a result to fill the distension gap, with the zeugopod branching off of the stylopod. Hyperphalangy as in the ichthyosaur paddle successively fills the toroidal distension gap by layering row after row of phalanges. But how then to explain the discontinuity between the zeugopod and the autopod in terms of reversed polarity of the Phase II to Phase III Hox genes? Note that gene expression is being controlled by the morphology, not the other way around. Ahlberg (1992) provided supporting evidence for this statement when, in a discussion of coelacanth limb structure, he inferred that "paired-fin structures are being expressed at the posterior dorsal and anal fin sites," and that these median fins developed basal plates comparable to those of the paired (pectoral and anal) fins. Ahlberg (1992) saw this as evidence indicating that the sudden evolution in form was the result of a switch in gene expression. This is no doubt the case in some sense—however, the coelacanth median lobe fins are more likely to be torologous with the paired fins than merely an expression of switched gene expression of the Phase II genes.

What has happened here is distal dislocation in the ventral-lateral limb herniation of the morphogenetic field. This is not something that may be accomplished by a switch in gene expression alone. The toroidal distension has shifted, *mechanically* reorienting the path of Hox gene expression. This allows the autopod to develop in such a way that it becomes sutured to the top of the zeugopod/stylopod complex. Comparable shifts are not unknown in tetrapod limb development. Consider the frameshift (Wagner and Gauthier 1999) that takes place in the embryonic hands of the Australian four-toed earless skink (*Hemiergis perioni*) versus the two-toed earless skink (*Hemiergis quadrilineata*).

We can turn to *Barasaurus besairiei* for an excellent demonstration of this effect. Recall the articulations between the medial centrale and the distal carpals in *Hovasaurus*, *Acerosodontosaurus*, and *Barasaurus besairiei* as shown in Table 8.1. The medial central essentially elongates in the following series: *Acerosodontosaurus, Barasaurus besairiei, Hovasaurus*. This series does not represent an evolutionary series. The varying lengths of the medial central reflect the path of the distal toroidal dislocation. We can confirm this by looking at the *Barasaurus* foot, where the astragalus and the calcaneum felt the tug of toroidal dislocation (being just proximal to the dislocation surface), and have fused together, resulting in the astragalocalcaneum (Figs. 8.32 and 8.33). The elongation of the medial centrale (Figs. 8.9 and 8.10) delineates the track of the dislocation or 'fault line' in the fin axis, caused by *décollement* in matryoshka-nested toroids in accordance with the Ninth Law of Morphogenetic Evolution as follows:

MATRYOSHKA-NESTED MORPHOGENETIC TORI CAN EXPERIENCE DÉCOLLEMENT DISLOCATIONS.

The severed tip of the limb distension reattaches in a new orientation, and this explains the polarity change in Hox gene expression in Phase III, in other words, a reversal in the expression of the HoxD-11 to the HoxD-13 gene.

References

Adams TL, Fiorillo AR (2010) *Platypterygius* Huene, 1922 (Ichthyosauria, Ophthalmosauridae) from the Late Cretaceous of Texas, USA. Palaeontol Electron 14(3):19A:12p

Ahlberg PE (1992) Coelacanth fins and evolution. Nature 358:459

Anderson HM, Cruickshank ARI (1978) The biostratigraphy of the Permian and the Triassic. Part 5. A review of the classification and distribution of Permo-Triassic tetrapods. Palaont afr 21:15–44

Battistini R, Richard-Vindard G (1972) Biogeography and ecology in Madagascar. Junk, The Hague

Bickelmann C et al (2009) The enigmatic diapsid *Acerosodontosaurus piveteaui* (Reptilia: Neodiapsida) from the upper Permian of Madagascar and the paraphyly of "younginiform" reptiles. Can J Earth Sci 46:651–661

Blackburn DG, Flemming AF (2011) Invasing implantation and intimate placental associations in a placentotrophic African lizard, *Trachylepis ivensi* (Scincidae). J Morphology 273:137–159

Boule M (1908) Sur l'existence d'une faune et d'une flore permiennes à Madagascar. C R Acad Sci 147:502–504

Boule M (1910) Sur le Permien de Madagascar. Bull Soc géol Fr 4e sér 10:314–315

deBraga M (2003) The postcranial skeleton, phylogenetic position, and probable lifestyle of the early Triassic reptile *Procolophon trigoniceps*. Can J Earth Sci 40:527–556

deBraga M, Reisz RR (1996) The early Permian reptile *Acleistorhinus pterticus* and its phylogenetic position. J Vetebr Paleontol 16:384–395

Camp CL (1945a) *Prolacerta* and the protorosaurian reptiles. Part 1. Am J Sci 243:17–32

Camp CL (1945b) *Prolacerta* and the protorosaurian reptiles. Part 2. Am J Sci 243:84–101

Camp CL (1980) Large ichthyosaurs from the upper Triassic of Nevada. Paleontographica Abteilung A 170:139–200

Carroll RL (1981) Plesiosaur ancestors from the upper Permian of Madagasar. Phil Trans R Soc London B 293:315–383

Carroll RL, Dong Z-M (1991) *Hupehsuchus*, an enigmatic reptile from the Triassic of China, and the problem of establishing relationships. Phil Trans R Soc London B 331:131–153

Cisneros JC et al (2004) A procolophonoid reptile with temporal fenestration from the middle Triassic of Brazil. Proc R Soc B Biol Sci 271(1547):1541–1546

Clack JA (2002) An early tetrapod from 'Romer's gap. Nature 418(6893):72–76

Clement G (1999) The actinistian (Sarcopterygii) *Piveteauia madagascariensis* Lehman from the lower Triassic of northwestern Madagascar: a redescription on the basis of new material. J Vert Paleo 19(2):234–242

Currie PJ (1980) A new younginid (Reptilia: Eosuchia) from the upper Permian of Madagascar. Canadian. J Earth Sci 17:500–511

Currie PJ (1981) *Hovasaurus boulei*, an aquatic eosuchian from the upper Permian of Madagascar. Palaont afr 21:99–168

Davis K (2012) Lower Permian vertebrates of Oklahoma, Waurika, vol 1. Kieran Davis, Wakefield

Davis K (2013) Lower Permian vertebrates of Oklahoma, Richards Spur, vol 2. Kieran Davis, Wakefield

Estes R (1983) Sauria Terrestria, Amphisbaenia. Handbuch der Paläoherpetologie, Part 10A. Gustav Fisher Verlag, Stuttgart

Fröbish NB et al (2013) Macropredatory ichthyosaur from the middle Triassic and the origin of modern trophic networks. Proc Nat Acad Sci (USA) 110(4):1393–1397

Glaw F, Vences M (2007) A field guide to the amphibians and reptiles of Madagascar, 3rd edn. Vences and Glaw Verlag, Cologne

Gomez S (2002) Crisóstomo Martinez, 1638-1694: the discoverer of trabecular bone. Endocrine 17(1):3–4

Halstead L (1982) The search for the past. Doubleday, Garden City/New York

Haughton S-H (1924) On reptilian remains from the Karoo beds of East Africa. Q J Geol Soc Lond 80:1–11

Hedges SB (2014) The high-level classification of skinks (Reptilia, Squamata, Scincomorpha). Zootaxa 3765:317–338

Holterhoff PF et al (2013) Artinskian (early Permian) conodonts from the Elm Creek limestone, a heterozoan carbonate sequence on the eastern shelf of the Midland Basin, West Texas, USA. New Mex Museum Nat Hist Sci Bull 60:109–119

Houssaye A et al (2014) A new look at ichthyosaur long bone microanatomy and histology: implications for their adaptation to an aquatic life. PLoS One 9(4):e95637. https://doi.org/10.1371/journl.pone.0095637

von Huene F (1940) Osteologie und systematische Stellung von *Mesosaurus*. Palaeontolographica Abteilung A Palaeozoologie-Stratigraphie 92:45–58

Jiang D-Y et al (2016) A large aberrant stem ichthyosauriform indicating early rise and demise of ichthyosauromorphs in the wake of the end-Permian extinction. Sci Rep 6(26232). https://doi.org/10.1038/srep26232

Johnson GD (1988) Abnormal captorhinomorph vertebra from the lower Permian of north-central Texas. J Vetebr Paleontol 8:19A

Ketchum HF, Barrett PM (2004) New reptile material from the lower Triassic of Madagascar: implications for the Permian-Triassic extinction event. Can J Earth Sci 41:1–8

Kiprijanoff W (1883) Studien über die fossilen Reptilien Russlands. III. Theil. Gruppe Thaumatosauria N. aus der Kreide-Formation und dem Moskauer Jura. Mémoires de L'académie impériale de Sciences de St-Petersbourg 31:1–29

Klinkhamer AJ et al (2017) Digital dissection and three-dimensional interactive models of limb musculature in the Australian estuarine crocodile (*Crocodylus porosus*). PLoS One 12(4):e0175079. https://doi.org/10.1371/journal.pone.0175079

Laurin M, Reisz RR (1995) A reevaluation of early amniote phylogeny. Zool J Linnean Soc 113(2):165–223

Lee MSY (2013) Turtle origins: insights from phylogenetic retrofitting and molecular scaffolds. J Evol Biol 26(12):2729–2738

Liebe L, Hurum JH (2012) Gross internal structure and microstructure of plesiosaur limb bones from the late Jurassic, central Spitsbergen. Nor J Geol 92:285–309

Lindgren J et al (2009) Skin of the Cretaceous mosasaur Plotosaurus: implications for aquatic adaptations in giant marine reptiles. Biol Lett 5(4). https://doi.org/10.1098/rsbl.2009.0097

Lindgren J et al (2010) Convergent evolution in aquatic tetrapods: insights from an exceptional fossil mosasaur. PLoS One 5(8):e11998. https://doi.org/10.1371/journal.pone.0011998

Lindgren J et al (2011) Three-dimensionally preserved integument reveals hydrodynamic adaptations in the extinct marine lizard *Ectenosaurus* (Reptilia, Mosasauridae). PLoS One 6(11):e27343. https://doi.org/10.1271/journal.pone.0027343

Maisch MW, Matzke AT (2000) The Ichthyosauria. Stuttgarter Beitr Naturk Ser B 298:1–159

Maisey JG et al (2017) Pectoral morphology in *Doliodus*: bridging the 'acanthodian'-chondrichthyan divide. Am Mus Novit 3875:1–15

Martinelli AG et al (2016) Owenettids and procolophonids from the lower Keuper shed new light on the diversity of pararreptiles in the German middle Triassic. J Paleontol 90(1):92–101

Mayor A (2000) The first fossil hunters: paleontology in Greek and Roman times. Princeton University Press, Princeton

McCoy VE (2015) The formation of concretions and their role in fossilization. Ph.D. Dissertation, Yale University

McMenamin MAS (2005) Microbial influence and environmental convergence in marine (Proterozoic) and lacustrine (Jurassic) depositional settings. Geol Soc America Abstr Prog 37(1):7

McMenamin MAS (2016) Dynamic paleontology. Springer, Cham

McMenamin MAS, Hussey MC (2015) Triassic coprolites from the Luning formation, central Nevada. Geol Soc Am Abstr Prog 47(7):827

McMenamin MAS et al (2016) Ichthyosaur coprolite with nautiloid: new data on the diet of *Shonisaurus*. Geol Soc Am Abstr Prog 48(7). https://doi.org/10.1130/abs/2016AM-284943

Meckert D (1995) The procolophonid *Barasaurus* and phylogeny ,of early amniotes. Ph.D. Dissertation, McGill University, Quebec

Merck J (1997) A phylogenetic analysis of euryapsid reptiles. Ph.D. Dissertation, University of Texas at Austin

Motani R et al (2015) A basal ichthyosauriform with a short snout from the lower Triassic of China. Nature 517:485–488

Moy-Thomas JA (1935) The coelacanth fishes from Madagascar. Geol Mag 72:213–227

Moy-Thomas JA (1937) The Carboniferous coelacanth fishes of great Britain and Ireland. J Zool 107:383–415

Naish D (2014) Skinks, skinks, skinks! Scientific American. https://blogs.scientificamerican.com/tetrapod-zoology/skinks-skinks-skinks/

Naish D (2016) The Madagascan skink *Amphiglossus* eats crabs. Scientific American. https://blogs.scientificamerican.com/tetrapod-zoology/the-madagascan-skink-amphiglossus-eats-crabs/

Perrier de la Bathie H (1919) Au sujet des plus anciennes couches de la série sédimentaire du versant occidental. Bull Acad malgache 4:218–221

Piñeiro G et al (2012) Unusual environmental conditions preserve a Permian mesosaur-bearing Konservat-Lagerstätte from Uruguay. Acta Palaeontol Pol 57(2):299–318

Piveteau J (1926) Paléontologie de Madagascar, XIII. Amphibiens et reptiles permiens. Annales de Paléontologie 15:53–180

Piveteau J (1955) Existence d'un reptile du groupe Procolophonides á Madagascar. Conséquences stratigraphiques et paléontologiques. Comptes rendus hebdomadaires de séances de l'Académie des. Science 241:1325–1327

Pyron RA et al (2013) A phylogeny and revised classification of Squamata, including 4161 species of lizards and snakes. BMC Evol Biol 13:93. https://doi.org/10.1186/1471-2148-13-93

Reisz RR, Scott D (2002) *Owenetta kitchingorum*, n. sp., a small parareptile (Procolophinia: Owenettidae) from the lower Triassic of South Africa. J Vertebr Paleontol 38:224–256

Reisz et al (2011) A new early Permian reptile and its significance in early diapsid evolution. Proc R Soc B 278:3731–3737

Rieppel O (1989) *Helveticosaurus zollingeri* Peyer (Reptilia, Diapsida) skeletal paedomorphosis, functional anatomy, and systematic affinities. Palaeontographica 208:123–152

Robinson JA (1975) The locomotion of plesiosaurs. N Jb Geol Paläont Abh 149:286–332

Rouwet D et al (2016) Cameroon's Lake Nyos gas burst: 30 years later. Eos 97. https://doi.org/10.1029/2016EO055627

Ruta M, Wills MA (2016) Comparable disparity in the appendicular skeleton across the fish-tetrapod transition, and the morphological gap between fish and tetrapod postcrania. Palaeontology 59(2):249–267

Ruta M et al (2011) Amniotes through major biological crises: faunal turnover among parareptiles and the end-Permian mass extinction. Palaeontology 54(5):1117–1137

Sachs S, Kear BP (2015) Postcranium of the paradigm elasmosaurid plesiosaurian *Libonectes morgani* (Welles, 1949). Geol Mag 152(4). https://doi.org/10.1017/S0016756814000636

Scheyer TM et al (2017) A new, exceptionally preserved juvenile specimen of *Eusaurosphargis dalsassoi* (Diapsida) and implications for Mesozoic marine diapsid phylogeny. Sci Rep 7. https://doi.org/10.1038/s41598-017-04514-x

Schlüter T (2008) Geological atlas of Africa: with notes on stratigraphy, tectonics, economic geology and geosites of each country. Springer, Berlin/London

Scott NJ et al (2006) The genera *Boiruna* and *Clelia* (Serpentes: Pseudoboini) in Paraguay and Argentina. Pap Avulsos Zool (São Paulo) 46(9):77–105

Seymour RS et al (1987) Effect of capture on the physiology of *Crocodylus porosus*. In: Webb GJW et al (eds) Wildlife management: crocodiles and alligators. Surrey Beatty, Clipping Norton, pp 253–257

Sharpe PT (2001) Fish scale development: hair today, teeth and scales yesterday? Curr Biol 11(18):R751–R752

Shubin N et al (1997) Fossils, genes and the evolution of animal limbs. Nature 388:639–648

Smith RMH (2000) Sedimentology and taphonomy of Late Permian vertebrate fossil localities in southwestern Madagascar. Palaeontol Afr 36:25–41

Steyer SJ (2002) The first articulated trematosaur 'amphibian' from the lower Triassic of Madagascar: implications for the phylogeny of the group. Palaeontology 14(4):771–793

Sues H-D, Reisz RR (2008) Anatomy and phylogenetic relationships of *Sclerosaurus armatus* (Amniota: Parareptilia) from the Buntsandstein (Triassic) of Europe. J Vertebrate Paleont 28(4):1031–1042

Suess E (1885) Das Antlitz der Erde, v. 1. G. Greytag, Leipzig

Szczygielski T et al (2017) The oldest record of aquatic amniote congenital scoliosis. PLoS One 12(9):e0185338. https://doi.org/10.1371/journal.pone.0185338

Takai F (1976) On *Atherstonia madagascariensis*, a new species of palaeoniscoid fish from Madagascar. Proc Japan Acad 52(1):25–28

Tarlach G (2017) When we left water. Discover 38(6):44–47

Thomson TJ, Droser ML (2015) Swimming reptiles make their mark in the early Triassic: delayed ecological recovery increased the preservation potential of vertebrate swim tracks. Geology 45(7). https://doi.org/10.1130/G36332.1

Tortochaux F (1949) Étude general du Groupe de la Sakamena dans le sud-ouest de Madagascar. Républ Madagascan, Trav Bur Géol 7:1–25

Tsuji LA et al (2013) *Ruhuhuaria reiszi*, a new procolophonoid reptile from the Triassic Ruhuhu Basin of Tanzania. Comptes Rendus Palevol 12(7–8):487–494

Wagner GP, Gauthier JA (1999) 1,2,3=2,3,4: a solution to the problem of the homology of the digits in the avian hand. Proc Nat Acad Sci (USA) 96(9):5111–5116

Witzmann F (2014) Congenital malformations of the vertebral column in ancient amphibians. J Vet Med Ser C Anat Histol Embryol 43:90–102

Yoshida H et al (2015) Early post-mortem formation of carbonate concretions around tusk-shells over week-month timescales. Sci Rep 5:14123. https://doi.org/10.1038/srep14123

Chapter 9
Tetrapteryx

Check for updates

Abstract The primary feathers on the hind limbs of *Microraptor* give us an important clue about the nature of the 'feather scleritome'. The *Microraptor* morphogenetic field hosts four curved projections representing the animal's limbs. This is the case for all tetrapods. Like the bot fly larva, but in reverse, the extra sclerites/enations (maggot spines versus primary feather primordia, respectively) are on the trailing edge of a transverse bulge of the body rather than the leading edge of the bulge (anterior edge larval segment versus posterior edge of limb, respectively). Birds surely passed through an ancestral stage (Tetrapteryx) that developed the hind limbs as wing-like structures and, in accord with Goethe's Law of Compensation, feathering on the hind limbs was reduced as the lineage relied more and more on their forelimbs for powered flight.

Keywords Tetrapteryx · *Dermatobia* · *Archaeopteryx* · *Microraptor* · *Serikornis*

The larva of the human bot fly (*Dermatobia hominis*) is an intriguing if unsettling creature (Sampson et al. 2001; Jacobs and Brown 2006). It causes a disease known as myiasis, the infestation of living human flesh by a fly larva. Cerebral myiasis can be fatal (Rossi and Zucoloto 1973), but the human bot fly larva itself is not known to transmit dangerous pathogens. My favorite title for an article on the subject is as follows: "wiggling subcutaneous lumps" (Lang and Smith 2003). The adult fly has a decidedly 'primitive' appearance, and looks something like a cross between a cicada (superfamily Cicadoidea) and a housefly (*Musca domestica*). Some describe the adult as looking like a bumblebee (Kahn 1999).

The life cycle of *Dermatobia hominis* has an element of hitchhiking. The adult captures a mosquito, deposits its eggs on the underside of the mosquito, and then releases the mosquito. Ticks can also act as the bot fly vector. The mosquito then locates a human and bites for its blood meal. The warmth of the victim's body causes the bot fly eggs to fall off and attach to the person's skin, were they hatch and gain access to subcutaneous tissue by crawling into the tiny hole left by the proboscis of the mosquito. The *Dermatobia* larva then feeds on the pus that its presence generates.

Infection by a human bot fly is something of a badge of honor among tropical biologists and ecologists. Mark W. Moffet (also known as "Dr. Bugs") posted a remarkable video to YouTube of a *Dermatobia hominis* maggot exiting from a

Fig. 9.1 *Dermatobia hominis* larvae, showing short spines deployed most fully on the anterior part of each lumpy segment. Red arrow indicates position of mouth parts (Public domain image by Captain G. Goodman, United States Air Force). Scale in mm

pustule on his left hand at Harvard University. Moffett had acquired the parasite during a trip to Belize. My former colleague at Mount Holyoke College Aaron Ellison attempted to raise a bot fly in his arm, also acquiring the parasite in Belize. As Ellison reported it, the larva began to move around in its chamber near his elbow joint, and its spines irritated a nerve in his joint, causing his arm to spasm ("the spines on its body would give me little electric shocks" in the words of Mark Moffett) and nearly causing Ellison to lose control of the vehicle he was driving at the time. This was the second bot fly Moffett had raised in his own flesh, the first one having infested his scalp while he was a graduate student.

The *Dermatobia* bot fly larva has an interesting scleritome of spines that is somewhat reminiscent of the scleritome of the Early Cambrian kinorhynch-like scalidophoran worm *Eokinorhynchus rarus*. Both species, the extant dipteran larva and the extinct kinorhynch (Zhang et al. 2015), have segmental rings armored with spines. If an isolated specimen of a *Dermatobia hominis* third instar larva were to undergo exceptional soft body preservation in shale, one could be forgiven for mistaking it for a Cambrian Explosion creature from the Burgess Shale or the Chengjiang Biota. There is an interesting difference, however. The segmental rings of *Dermatobia hominis* generally have spines that are placed roughly midway along the segmental band, as seen in the segmental bands at the animal's 'neck.' In *Dermatobia hominis* larvae, the short spines are often deployed most fully on the anterior part of each lumpy segment (Fig. 9.1). Some semi-segments do not completely encircle the larva, and these nevertheless develop the row of spines on their anterior slope. As in *Eokinorhynchus*, in *Dermatobia hominis* larvae there can be larger spines developed on the middle part of the animal during the first larval instar phase. The bands of sclerite spines in the third instar of the bot fly larva are somewhat self-similar from one segmental band to the next as you move from the anterior to the posterior of the maggot. This will be an important observation as we proceed to discussion of Tetrapteryx.

The question "What is a bird?" has a complicated answer. Henry Lappen notes that all birds have feathers (Lappen 2017). But are all feathered animals birds? Lappen (2017) writes:

[Birds] have crest feathers, pin feathers, remiges, retrices, coverts, tail feathers, layers of under feathers and down…The first feathers seen in the fossil record [are] simple quill-like structures coming out of skin follicles in the same way as hair and scales…Later, these simple quills became multi-filamented (as are down feathers) and eventually lengthened, developed barbs and became asymmetrically curved. At which point do we decide that this is a true feather and its owner was a bird?

One of the best acquisitions in my personal library of classic geological literature is a copy of H. H. Swinnerton's *Outlines of Palaeontology, Third Edition* (Swinnerton 1950). This copy once belonged to (and is autographed by) British paleontologist, of Burgess Shale fame, the late H. B. Whittington. In his fig. 315, Swinnerton (1950) illustrates a four-winged reptile referred to in the caption as *Tetrapteryx*. The animal is shown gliding or flying to the viewer's left, with its mouth open and about to clamp down on an insect that appears to be either a damselfly or a dragonfly. The caption, with an assertive scheme of capitalization, reads as follows: "Diagrammatic Restoration of *Tetrapteryx* (after Beebe) illustrating the view that in the Parachute Phase in the Development of Flight among Birds there were large Feathers on the Hind as well as the Fore Limbs."

The image in Swinnerton (1950) is a minor modification (i.e., addition of the insect prey) of the original illustration of the Tetrapteryx concept as introduced by William Beebe in his publication entitled "A Tetrapteryx Stage in the Ancestry of Birds" (Beebe 1915). Beebe (1877–1962) was a famous naturalist of his era, but he is seldom held up as proper role model for aspiring young scientists. There are several reasons for this. First, he never finished his college degree at Columbia University (although he later received honorary doctorates from both Colgate and Tufts). Between bouts of depression, he engaged in a series of troubled romantic relationships. Also, his reputation suffered from "saganization," the view among his peers that he was more a popularizer of science than a serious scientist as is sometimes said of astronomer Carl Sagan. According to Sterling et al. (1997), professional biologists were "reluctant to accord high standing to a successful popularizer." Nevertheless, Beebe accomplished a number of scientific firsts, including seven consecutive deep diving world records in Otis Barton's bathysphere. His description of new species observed from the bathysphere (necessarily done without collecting type specimens) has also raised some eyebrows, especially considering that the putative fish in question (the Abyssal Rainbow Gar; "Bathysphaera intacta"; the Pallid Sailfin, the Three-Starred Anglerfish, and the Five-Lined Constellation Fish) have never been seen again. The latter may have been a misidentified ctenophore. In support of the validity of Beebe's deep sea fish report, however, is the fact that unknown species of snailfish (Liparidae), never previously observed by humans, are periodically captured on camera at great depth by robotic deep sea explorers.

In his remarkable document introducing the Tetrapteryx hypothesis, Beebe (1915) sought to bridge the evolutionary gap between a leaping "lizard" and a flying *Archaeopteryx*:

> But if we arm our imaginations with a prejurassic, parachuting lizard on the one hand, and an *Archaeopteryx* on the other, we still have a hiatus which no logical combining of proportional characters will bridge.

Fig. 9.2 The femoral or pelvic wing as illustrated in Beebe (1915). Beebe's (1915) caption reads in part: "Detail of pelvic wing of white-winged dove... The wing consists of twelve flights and six coverts" (Public domain image)

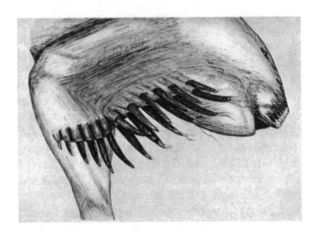

Beebe (1915) goes on to note that flying fish benefit from two pairs of 'wings,' the extended pectoral fins and flared out pelvic fins. Reasoning from this modern example, Beebe (1915) maintains that the ancestors to birds must also have had the functional equivalent of four wings, saying that some "such accessory has always seemed to me necessary if we are to complete our lizard-to-*Archaeopteryx* line of ascent." This comment would seem to indicate that Beebe was beholden to the standard Darwinian model of gradual evolution; however, Beebe was among the first scientists to realize that evolution proceeds in fits and starts, often with a rapid initial diversification followed by relative stasis (Gould 2004).

Beebe's (1915) main line of evidence for Tetrapteryx was "a remarkable development of sprouting quills across the upper part of the hind-leg" of the body of a 4 day old white-winged dove (*Melopelia asiatica*). He called this structure the femoral or pelvic wing (Fig. 9.2) and considered it to be an atavistic feature. Beebe (1915) remarked that the iguana-*Archaeopteryx* "puzzle flashed through my mind and I at once followed up the clue thus given."

Beebe (1915) dismisses any connection to feather-footed and feather-legged domestic birds (primarily pigeons and chickens): "The feathers, which have been bred to great length, sprout from the scaly covering of the tarsus and phalanges and not from the leg proper." Beebe (1915) then proceeds to an interesting literature review of scientific discussion concerning the flight value of feathers on the hind limbs of *Archaeopteryx*. These hind limb feathers are visible only on the Berlin specimen of the famous ancient bird. Vogt (1879) was first to mention the tibial feathers of *Archaeopteryx*: "Le tibia était couvert de plumes dans toute sa longueur. L'Archaeopteryx portait donc des culottes, comme nos faucons, avec les jambes desquels sa jambe a le plus de resemblance, suivant M. Owen. [The tibia was covered with feathers in all its length. *Archaeopteryx* thus developed breeches, like our falcons, with the legs of which its leg has the greatest resemblance, according to Owen]." As an aside, Beebe (1915) quoted Vogt (1879), republishing his comment as a footnote to German military aggression (the year 1915 was in the midst of World War I): "M. Volger se berçait dans l'espérance que S. M. l'empereur

Guillaume achèterait la piecè (Archaeopteryx) pour la conserver à l'Allemagne. Sa Majesté n'entra pas dans ces vues. Ah! si ou lieu d'un oiseau, is s'etait agi d'un canon ou d'un fusil pétrifié! [Mr. Volger cherished the hope that German Emperor William I would buy the piece (*Archaeopteryx*) to preserve it in Germany. His Majesty did not entertain this view. Alas, if in place of a bird, it had been a petrified gun or rifle!]"

Dames (1884) argued that it is not correct that the tibial feathers had the same consistency as those of the wing and the tail, because they are far more delicate and also shorter, about half the length (32 mm versus 65 mm). In spite of this, Dames (1884) continued, if one were to maintain that *Archaeopteryx* also flew with its hind legs, we would then expect that the bones of the hind limbs would also be strengthened and made more robust accordingly.

Beebe (1915) provides a counterpoint view from Abel (1912) in his book on the paleobiology of vertebrates: "Die zweizeilige Befiederung der Unterschenkel spricht dafür, dass diese Federn den Flug der Archaeopteryx als Fallschirmapparate unterstützt haben. [The double fletching of the lower legs speaks to the fact that these feathers supported the flight of *Archaeopteryx* as a parachute apparatus.]" *Befiederung* used here, or "fletching" for feathers, might be considered to be a scientific term equivalent to the presence of zahnreihen in teeth.

Beebe (1915) agreed with Abel's (1912) interpretation, saying that in *Archaeopteryx* or "in our prejurassic Tetrapteryx the function of the pelvic wings would have been merely passive parachutes." Beebe concludes with a restatement of Haeckel's Biogenic Law ("ontogeny recapitulates phylogeny" of recapitulation theory):

And for some unknown reason, Nature makes each squab pass through the Tetrapteryx stage. The line of feathers along the leg of the young bird reproduces on this diminutive, useless scale the glory that was once theirs. No fossil bird of the ages prior to Archaeopteryx may come to light, but the memory of Tetrapteryx lingers in every dove-cote.

Interestingly, the year before publication of Beebe's (1915) article on Tetrapteryx, the ground-breaking animated film *Gertie the Dinosaur* (1914) was produced by Windsor McCay. This was the earliest animated film to feature a dinosaur in the leading role. It was also the first film to use key frames, registration marks and other innovations. The show was originally part of a vaudeville act—McCay would appear in person, give vocal commands to the dinosaur, and she would obey! This original version was replaced by a live action sequence that was shown in movie theatres. *Gertie* was inspiring to Walt Disney and other animated film makers, and this inspiration is especially clear in Disney's *Fantasia* (1940).

Gertie the Dinosaur included a cameo appearance by a four-winged reptile (Fig. 9.3). The next four text frames (this is a silent film) has McCay addressing Gertie: "Did you see that four-winged lizard?" "Sure?" "Are you in the habit of seeing things?" "Are you fibbing to me?" McCay spoofs on the improbability of the whole concept, apparently unaware that Tetrapteryx was going to be launched the following year. This may be an example of paleoartistic anticipation of subsequent scientific interpretation (and is not an isolated case of this phenomenon). The specter of a four-winged early flying reptile retains its appeal to this day, and Beebe's

Fig. 9.3 Four-winged lizard as rendered in an early animation by Windsor McCay for the 1914 film *Gertie the Dinosaur* (Sketch/reconstruction of four-winged lizard in flight by Ariel Samantha Kimberley)

Tetrapteryx appears on the cover of Jonathan Kane, Emily Willoughby and T. Michael Keesey's book "God's Word or Human Reason? An Inside Perspective on Creationism" (Kane et al. 2016), a book that features writings by five former creationists who have embraced theistic evolution.

Beebe's (1915) Tetrapteryx concept lapsed into disfavor, largely due to its rejection by Gerhard Heilmann (Heilmann 1926). Heilmann (1926) considered Beebe's Tetrapteryx concept at length, but ultimately rejected it because he (Heilmann) could not find additional evidence for hind wings in juveniles of numerous bird species. Heilmann claimed that hatchlings' thighs showed permanent feathers but "no atavism." He argued that if this truly represented an atavism, it should appear ("like a glimpse") and then vanish again as the bird reached maturity. Heilmann's (1926) conclusions led the entire ornithological community to reject the Tetrapteryx concept (Welker 1975).

Reception of the Tetrapteryx hypothesis underwent a dramatic change for the better with the discovery and description of the Cretaceous paravian dinosaur *Microraptor* in 2000. *Microraptor* is reconstructed as having been able to fly with four wings (Chatterjee and Templin 2007). Chatterjee and Templin (2007) noted that the small maniraptoran dinosaur *Pedopenna* has primaries on its feet, and this taken together with the hind limb feathers of *Archaeopteryx*, concluded that the Tetrapteryx hypothesis has merit and deserves to be reevaluated.

The discovery and description of *Microraptor* has a checkered history. The first specimen to come to the attention of the scientific community was a fraudulent specimen consisting of parts of fossils of three different animals (*Yanornis*, *Microraptor*, and an unidentified third feathered dinosaur) that had been glued

Fig. 9.4 *Microraptor* model. Painted plastic, made in China, marketed by Safari Ltd. (Chinese manufacture) of Miama, Florida in 2006 (Photograph by Mark McMenamin)

together in China and offered for sale in the United States (Olson 2000). The tail part of the patchwork fraud is now known to be the piece that belonged to *Microraptor*. Fortunately, the remainder of the specimen from which the tail was removed was eventually located, and given the name *Microraptor zhaoianus* by Xu et al. (2000). The genus name is now well enough established to have been given the honor of being fashioned into a plastic toy model (Figs. 9.4 and 9.5). This model, an Early Cretaceous (118-123 million years ago) *Microraptor*, was marketed in *circa* 2006 by Safari Ltd. of Miama, Florida. Made in China, it comes with a small hangtag booklet in five languages (McMenamin and Lafreniere 2015).

Based on the relatively advanced skeletal anatomy of *Microraptor*, the small theropod dinosaur may very well have been capable of powered flight (Agnolín and Novas 2013). As such, *Microraptor* supports the hypothesis that other, larger, flightless members of the Dromaeosauridae such as *Deinonychus*, were descended from toothy ancestors that could fly. Flying dromaeosaurs such as *Microraptor* evidently evolved the ability to fly independently from the true bird lineage (Dececchi et al. 2016).

The question of flight capability in four-winged dinosaurs is a topic of considerable debate. The recent description of *Serikornis sungei*, a small four-winged theropod (Fig. 9.6) from the Upper Jurassic Tiaojishan Formation of the Liaoning Province, China, has reignited the controversy (Lefèvre et al. 2017). Lefèvre et al. (2017) place *Serikornis* in a group of early paravians, and classify the genus outside of the eumaniraptoran clade that produced birds. Nevertheless, *Serikornis* is covered with feathers, earning it its nickname "Silky" (*serikos* is the ancient Greek

Fig. 9.5 *Microraptor*
model, detail of head and
neck. Painted plastic,
marketed by Safari Ltd.
(Chinese manufacture) of
Miama, Florida in 2006
(Photograph by Mark
McMenamin)

Fig. 9.6 *Serikornis sungei*, a four-winged paravian theropod from the Upper Jurassic Tiaojishan
Formation of the Liaoning Province, China. The small dinosaur is approximately the size of a
modern pheasant, and is depicted here feeding on a lactrodectid spider of the family Theridiidae
(Artwork by Emily Willoughby, used here as per Creative Commons BY-SA 4.0)

word for silk). As quoted in Pickrell (2017), senior author Ulysse Lefèvre of the Royal Belgian Institute of Natural Sciences in Brussels, claims that the "feathering of *Serikornis* shows for the first time a complete absence of barbules—that is, the microstructures that allow feathers to resist air pressure during wing beats... [the] plumage is composed of four wings, as with many theropod dinosaurs from China, but it did not allow 'Silky' to take off from the ground or from a tree." Nevertheless, *Serikornis* bears on all four of its limbs pennaceous, elongate feathers stiffened by central vanes.

Michael Benton of the University of Bristol is not convinced that *Serikornis* was a non-flier. As quoted in Pickrell (2017), Benton notes that the wings on the hind limbs would be a serious hindrance "for a ground-runner. The long feathers on the thigh and calf would be like very elaborate bell-bottomed trousers, rubbing and catching as the animal walked or ran." Benton advocates for gliding capability in *Serikornis*; however, Lefèvre, defending his 'no flight' hypothesis, only allows that the plumage in *Serikornis* might be used as a parachute to slow descent to the ground after leaping from the branch of a tree. I agree with Lefèvre that flight in *Serikornis* (and other four-winged dinosaurs) is indeed "a tricky issue and requires more time and fossils" before we more fully understand early dinosaurian flight.

The primary feathers on the hind limbs of *Microraptor* give us an important clue about the nature of the feather 'scleritome'. The *Microraptor* morphogenetic field hosts four projections representing the animal's limbs. This is the case for all tetrapods. Like the bot fly larva, but in reverse, the extra sclerites/enations (maggot spines versus primary feather primordia, respectively) are on the trailing edge of the bulge rather than the leading edge of the bulge (anterior edge larval segment versus posterior edge of limb, respectively). Bluntly pointed osteoderms in the tail of the Pleistocene armored mammal *Glyptodon* follow this same 'trailing-edge-of-the-toroidal ring' pattern.

Birds surely did pass through an ancestral stage (Tetrapteryx) that developed the hind limbs as wing-like structures. Unrelated feathered tetrapods may have done the same by torologous convergent evolution. At least four small dinosaur genera are known to have expressed this trait (*Anchiornis*, *Aurornis*, *Microraptor* and *Serikornis*). Due to Goethe's Law of Compensation, feathering on the hind limbs was reduced as the lineage relied more and more on their forelimbs for powered flight. Loss of avian dentition may be as much due to Goethe's Law of Compensation as it is to weight reduction. Morse's Law of Digital Reduction (Shubin et al. 1997), which surely applies to birds, may be seen here as a corollary of Goethe's Law of Compensation.

References

Abel O (1912) Grundzüge der Palaeobiologie der Wirbeltiere. E. Schweizerbart, Stuttgart

Agnolín FL, Novas FE (2013) Avian ancestors. A review of the phylogenetic relationships of the theropods Unenlagiidae, *Microraptoria, Anchiornis* and Scansoriopterygidae. SpringerBriefs Earth Syst Sci. https://doi.org/10.1007/978-94-007-5637-3

Beebe W (1915) A tetrapteryx stage in the ancestry of birds. Zoologica 2(2):37–52

Chatterjee S, Templin RJ (2007) Biplane wing planform and flight performance of the feathered dinosaur *Microraptor gui.* Proc Natl Acad Sci USA 104(5):1576–1580

Dames WB (1884) Ueber Archaeopteryx. Paleontol Abh 2:39–41

Dececchi TA et al (2016) The wings before the bird: an evaluation of flapping-based locomotory hypotheses in bird antecedents. PeerJ 4:e2159. https://doi.org/10.7717/peerj.2159

Gould CG (2004) The remarkable life of William Beebe. Island Press, Washington, DC

Heilmann G (1926) The origin of birds. Dover, New York

Jacobs B, Brown D (2006) Cutaneous furuncular myiasis: human infestation by the botfly. Can J Plast Surg 14(1):31–32

Kahn DG (1999) Myiasis secondary to *Dermatobia hominis* (Human Botfly) presenting as a long-standing breast mass. Arch Pathol Lab Med 123:829–831

Kane J et al (2016) God's word or human reason? An inside perspective on creationism. Inkwater Press, Portland

Lang T, Smith DS (2003) Wiggling subcutaneous lumps. Clin Infect Dis 37:2087–2088

Lappen H (2017) What is a bird?—easy question, complicated answer. Daily Hampshire Gazette (Northampton, Massachusetts) 231(292):C7

Lefèvre U et al (2017) A new Jurassic theropod from China documents a transitional step in the macrostructure of feathers. Sci Nat 104:74

McMenamin MAS, Lafreniere MM (2015) A concise field guide to miniature models of dinosaurs and other extinct monsters. Meanma Press, South Hadley

Olson SL (2000) Countdown to Piltdown at *National Geographic*: the rise and fall of *Archaeoraptor.* Backbone 13(2):1–3

Pickrell J (2017) New feathered dinosaur had four wings but couldn't fly. National Geographic. http://news.nationalgeographic.com/2017/08/feathered-dinosaur-four-wings-species-serikornis-science/?utm_source=NatGeocom&utm_medium=Email&utm_content=wildscience_20170916&utm_campaign=Content&utm_rd=927031478. Accessed 16 Sept 2017

Rossi MA, Zucoloto S (1973) Fatal cerebral myiasis caused by the tropical warble fly, *Dermatobia hominis.* Med Vet Entomol 15:22–27

Sampson CE et al (2001) Botfly myiasis: case report and brief review. Ann Plast Surg 46:150–152

Shubin N et al (1997) Fossils, genes and the evolution of animal limbs. Nature 388:639–648

Sterling KB et al (1997) Biographical dictionary of American and Canadian naturalists and environmentalists. Greenwood Press, Westport

Swinnerton HH (1950) Outlines of palaeontology, 3rd edn. Edward Arnold, London

Vogt MC (1879) L'Archaeopteryx macrura.—Un intermédiaire entre les oiseaux et les reptiles. La Revue Scientifique, Second Series 9(11):241–245

Welker RH (1975) Natural man: the life of William Beebe. Indiana University Press, Bloomington and Indianapolis

Xu X et al (2000) The smallest known non-avian theropod dinosaur. Nature 408:705–708

Zhang H et al (2015) Armored kinorhynch-like scalidophoran animals from the Early Cambrian. Sci Rep. https://doi.org/10.1038/srep16521

Chapter 10
Zealanditherians

Abstract Marsupials first appear in Cretaceous North America, their ancestors having arrived from Asia during an eastward migration of Mesozoic metatherians. By the end of the Mesozoic, the North American metatherians had developed into large (over one meter long) animals with a powerful bite force, partly a function of hypertrophied premolars (p3) in some species. The inflated premolar is associated with reappearance of zahnreihen in a Cretaceous metatherian mammal, *Didelphodon coyi*. The metatherian migration begun in Asia continued through North America, to South America, to Antarctica and on to Australia where marsupials underwent a well known adaptive radiation in 'splendid isolation'. Until recently it was thought that terrestrial mammaliaforms never reached New Zealand, as New Zealand had tectonically rifted away from Antarctica at 82 million years ago, supposedly before marsupials had reached Antarctica. Recent discoveries from limited exposures of Miocene strata in New Zealand near Otago show that mammaliaforms had indeed colonized and diversified in New Zealand apart from their ancestors in the rest of the Gondwanan continental diaspora. These zealanditherians inhabited the newly characterized continent Zealandia, and were apparently driven to extinction by habitat loss when most of Zealandia was submerged by the sea.

Keywords Zealanditherians · Zealandia · *Didelphodon* · Hypertrophied premolars · *Repenomamus* · Zahnreihen · *Thylacoleo* · *Thylacinus* · *Thylacosmilus* · Saint Bathans mammal

Several types of Mesozoic mammals are thought to have preyed upon dinosaurs. One of these is *Repenomamus*, a relatively large Cretaceous mammal discovered in the Yixian Formation, Liaoning Province, northeast China. *Repenomamus* belongs to the Order Gobiconodonta, an Asian mammaliaform order only distantly related to modern mammals. A specimen of *Repenomamus robustus* was found with the masticated remains of a young *Psittacosaurus* preserved in its abdominal cavity. This 130 million year old fossil association has been interpreted as evidence that *R. robustus* had either scavenged or preyed upon small dinosaurs (Hu et al. 2005).

M.A.S. McMenamin, *Deep Time Analysis*, Springer Geology,
https://doi.org/10.1007/978-3-319-74256-4_10

Another Mesozoic mammal thought capable of subduing and consuming non-avian dinosaurs (Wilson et al. 2016) is the marsupialiform metatherian *Didelphodon*. Also known as a marsupicarnivoran, *Didelphodon* (like *Repenomamus*) was one of the largest types of mammals in the Mesozoic. *Didelphodon* reached up to 5 kg body mass. This metatherian (marsupials and related mammals) predator is thought to be both omnivorous and semiaquatic. An otter shape is implied by a 30% complete skeleton of *Didelphodon* that was found in the Hell Creek Formation of Harding County, South Dakota, a short distance away from a *Triceratops* skeleton. *Didelphedon* and its family Stagodontidae represent perhaps the most interesting group of marsupials to fall victim to the end-Cretaceous mass extinction.

Marsupials are generally thought to be less active and less aggressive than their placental relatives (both represent types of therian mammals), so it seems remarkable that a medium-sized marsupial predator could thrive during the Mesozoic. Three species are currently known in Family Stagodontidae: *Didelphodon vorax* Marsh, 1889, *Didelphodon padanicus* (Cope 1892) and *Didelphodon coyi* Fox and Naylor, 1986. Note that two of the three species were named by Othniel Charles Marsh (1831–1899) and Edward Drinker Cope (1840–1897), best known for their Bone Wars rivalry in the discovery and description of the Mesozoic paleontological riches of the western United States. Although their conflict over dinosaur fossils receives the most attention, they also sparred over mammalian fossil material including *Didelphodon*. Cope (1892) begins an important paper on the subject on a contentious note:

> In 1881 I had the pleasure of announcing the existence of Mammalia in the Laramie formation [of Late Cretaceous age], and described the new genus and species of Multituberculata, *Meniscoëssus conquistus*. Since then Prof. Marsh has described several species from the same formation, exaggerating the number very considerably, as has been precisely shown by Prof. Osborn.

Kielan-Jaworowska et al. (2004) remark that the "heated exchanges [between Marsh and Osborn] on the subject of Mesozoic mammals bring a breath of entertainment to the subject."

Since 1892, the number of Mesozoic mammal species has grown considerably (Kielan-Jaworowska et al. 2004). *Didelphodon* is thought to have consumed a durophagous diet (perhaps including mollusks and other shellfish) because of the impressive enlargement of its posteriormost premolar (p3). In *Didelphodon vorax*, this tooth forms an onion-dome shape, and thus strongly resembles one of the teeth of the Late Cretaceous shell-cracking mosasaur *Globidens phosphaticus* from Angola (Polcyn et al. 2010). The *Didelphodon* premolar p3 is the most robust and most preservable part of the animal. Just such a single tooth records the genus *Didelphodon* in the Upper Cretaceous (Campanian-Maastrichtian) fossil assemblage of the Williams Fork Formation, western Colorado (Brand et al. 2017).

The species *Didelphodon coyi* Fox and Naylor, 1986 is especially noteworthy because it appears to be one of the very few types of mammals that shows the clear presence of control by zahnreihen in the dentition of its mandible. Figure 10.1 shows the occlusal view of the holotype, a right dentary, of *Didelphodon coyi*.

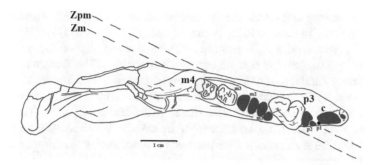

Fig. 10.1 *Didelphodon coyi*, holotype, occlusal view of the right dentary. Molars **m4** and **m3**, plus the huge premolar **p3**, are represented by preserved teeth. The upper dashed line represents the *premolar zahnreihe* (**Zpm**, consisting of: tiny **p1** [one root], **p2** [two roots], and huge **p3**). The lower dashed line represents the *molar zahnreihe* (**Zm**, consisting of: **m1** [two roots], **m2** [two roots], **m3** and **m4**). A large canine alveolus (**c**) and incisor alveolus (**i**) are also visible (Redrawn and modified from Fox and Naylor (1986)). Scale bar 1 cm

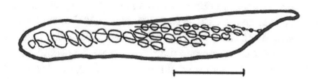

Fig. 10.2 *Captorhinus aguti*, a Permian terrestrial reptile, dorsal view of right dentary. Solid lines connect zahnreihen (Redrawn from de Ricqlès and Bolt (1983)). Scale bar 1 cm

The zahnreihen are shown with dashed lines in Fig. 10.1. The premolar zahnreihe begins with a very small p1, moves to a moderate p2, and ends with an enormous subtriangular p3. There is no special enlargement of the teeth in the molar row in *D. coyi*. The molar row seems to slide anteriorly off the edge of the jaw as it follows its zahnreihe. An alternate explanation, that the pattern merely represents tooth crowding due to the enormous size of p3, seems less likely because the teeth are clearly deployed in rows rather than being more randomly displaced by the big premolar.

This is very unusual for mammalian teeth, and is more characteristic for reptiles such as *Omphalosaurus* (a Triassic ichthyosauroid with crushing dentition comparable to that of the mosasaur *Globidens alabamaensis* and the pycnodont fishes; Sander and Faber 2003) and the Permian terrestrial species *Captorhinus aguti*, where as many as eight zahnreihen may be seen running off the side of the jaw (Fig. 10.2; de Ricqlès and Bolt 1983). Curiously, however, the multiple rows of teeth are not always expressed in *Captorhinus* and single rows can occur (Kissel et al. 2002) as in *Captorhinus magnus*.

Fox and Naylor (1986) analyzed *D. coyi's* unusual jaw morphology by describing how, if you hold the mandible in a horizontal position with its anterior end pointed toward you (i.e., the orientation that it would have if you were looking the animal in the face) that the "coronoid process slopes inward, towards the midline." This differs from the situation with two other marsupials, the common opossum

(*Didelphis marsupialis*) and the Tasmanian devil (*Sarcophilus harrisi*). The coronoid process in these animals is almost vertical. Fox and Naylor (1986) infer that this "peculiar orientation probably reflects unusual size of the mandibular adductors" in *Didelphodon coyi*, an interesting suggestion. The Tasmanian devil, a living marsupicarnivoran, has a tremendously powerful bite for its size. Thus *Didelphodon coyi* may have had one of the strongest bite forces in the entire history of jawed animals (a trait it may have shared with the Australian Plio-Pleistocene marsupial *Thylacoleo carnifex*, known for its massive p3 premolar). Perhaps its powerful bite helps to explain *Didelphodon's* survival during the Mesozoic in the shadow of the non-avian dinosaurs.

The placement of the post-canine teeth in the dentary of *Didelphodon coyi* clearly shows a pair of zahnreihen (Fig. 10.1). Fox and Naylor (1986) do not address the concept of zahnreihe, and claim that the less derived condition in "primitive opossum-like marsupials" is a single row of teeth that parallels the jaw line. They thus see the odd jaw tooth configuration of *Didelphodon coyi* as a derived feature, when in fact may represent the less derived state. Whether or not this represents a reversion to an earlier state, in other words, an evolutionary atavism, is not yet certain but may very well be the case. Fox and Naylor (1986) attribute the offset in the premolar and molar tooth rows to the fact that "the dentigerous parts of the jaw are short in *D. coyi*," and the overlapping tooth rows have the overall effect of decreasing "the total length of the tooth row" without loss of important teeth. This may be true, but we must also note that we see in *Didelphodon coyi* the reestablishment of zahnreihen, a remarkable occurrence in a mammal of any type, as a result of modification to its toroidal morphogenetic field. This would thus be comparable to the reestablishment of a multielement scleritome in the scaly-foot gastropod (*Chrysomallon* spp.; Chen et al. [2015]) and represent an apparent violation (by atavism) of the Second Law of Morphogenetic Evolution.

In the case of *D. coyi*, evolutionary pressure to shorten the jaw and strengthen the bite led to an atavistic reappearance of two adjacent zahnreihen, the premolar row and the molar row. This may indicate that the single row of teeth characteristic for most mammalian jaws in fact represents a superimposition of two or more once formerly separate, relict zahnreihen. This also suggests that toroidal expansions, or dilations in the morphogenetic field lines, may be associated with atavisms such as the reappearance of zahnreihen in *Didelphodon coyi* (tooth enamel, largely ectodermal = scleritome) or the reappearance of the median dorsal plate of the early fish *Guiyu* as the otico-occipital shield in the bizarre Triassic coelacanth *Foreyia* (bone, largely mesodermal = sclerotome).

Figure 10.1 merits even closer inspection. The molar zahnreihe (m1, m2, m3, m4) in *D. coyi* appears to show a normal progression in tooth size, with m1 slightly smaller (judging from its alveolus) and the other three molars of roughly similar size. The premolar zahnreihe, however, is a quite different story. The first premolar (p1) is tiny and has only a single root. The second premolar (p2) is of typical size. But the third premolar (p3) is enlarged all out of proportion in comparison the other teeth, and was presumably used for crushing shell or bone.

This suggests that the *Didelphodon coyi* premolar zahnreihe (Zpm in Fig. 10.1) has some torologically unusual properties. A massive dilation in the morphogenetic field lines running diagonal to the jaw line induces the hypertrophy in the durophagy-enabled premolar p3. The placement of the field line has the effect of greatly diminishing the size of p1. The size of m1 is slightly compressed (right where you would expect, on its anteriolabial corner) by the bulge in the premolar zahnreihe and the expansion of p3.

Scott and Fox (2015) report a fragmentary specimen they refer to *Didelphedon* sp. from the Dinosaur Park Formation (near Steveville), Late Cretaceous (Judithian), southeast Alberta, Canada. It is an edentulous left dentary. In spite of its incomplete nature, it shows parts of two zahnreihe, the molar (alveoli for m1-m3) and premolar (alveolus for a gigantic p3, a diagnostic dental characteristic for *Didelphodon*). This is the oldest occurrence of *Didelphodon*, and its mandible morphology with pre-served molar and premolar zahnreihe make this specimen of *Didelphedon* sp. quite comparable to *D. coyi*, known from the immediately overlying Edmontonian Horeshoe Canyon Formation.

A fragmentary, edentulous dentary of *Didelphodon vorax* is shown in Figs. 10.3, 10.4, 10.5, 10.6, 10.7, 10.8 and 10.9. The specimen is from the Late Cretaceous Hell Creek Formation of Montana. *Didelphodon* has been known from the Hell Creek

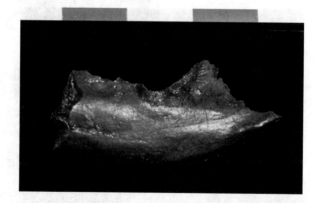

Fig. 10.3 *Didelphodon vorax*. Lingual side of jaw fragment, mandibular body. Sample 1 of 7/7/2017. An unnamed foramen on the myohyloid line is visible beneath the anterior part of the alveolus for m3, close to a position beneath the junction of m2 and m3. Scale bar in cm (Photograph by Mark McMenamin)

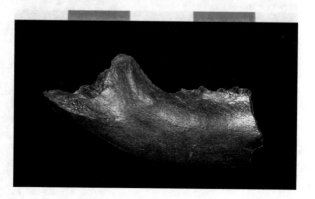

Fig. 10.4 *Didelphodon vorax*. Labial side of jaw fragment, mandibular body, showing partial masseteric fossa on left (posterior) end, and just above this, the lower portion of the coronoid process. Scale bar in cm (Photograph by Mark McMenamin)

Fig. 10.5 *Didelphodon vorax*. Occlusal view of jaw fragment. From right to left, the alveoli for molars m1 (partial), m2, m3 and m4 are visible. The molar roots for m4 are deeply set in a huge alveolar cavity. Scale bar in cm (Photograph by Mark McMenamin)

Fig. 10.6 *Didelphodon vorax*. Underside of jaw fragment showing flattened region at the posterior end of the mandibular body. Scale bar in cm (Photograph by Mark McMenamin)

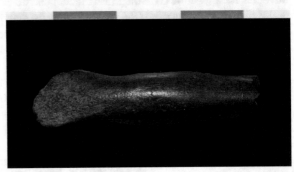

Fig. 10.7 *Didelphodon vorax*. View of jaw fragment posterior to the coronoid process showing a crest in the mandibular canal that runs diagonal to the line of the jaw. The crest tilts to the right and is just above a horizontal surface that represents the flattened region at the posterior end of the mandibular body. Scale bar in cm (Photograph by Mark McMenamin)

Fig. 10.8 *Didelphodon vorax*. View of jaw fragment at anterior end in the direction of the symphysis, showing mineral deposits (lighter in color; ?calcite) in the mandibular canal. The downward pointing triangular marking is a partial alveolus (posterior edge) of molar m1. Scale bar in cm (Photograph by Mark McMenamin)

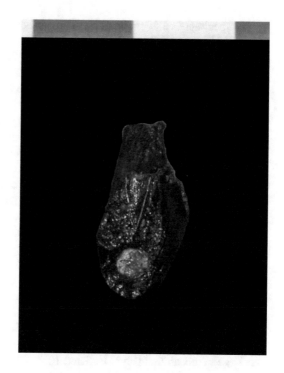

Fig. 10.9 *Didelphodon vorax*. Reconstruction of entire skull. Note the huge m4 on the mandible (Artwork by Becky Wah. Scale bar 1 cm)

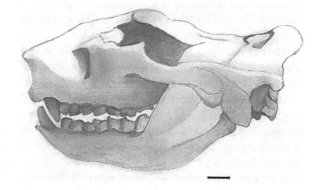

Formation for some time (Simpson 1927; Clemens 1968; Fox and Naylor 1986), and this specimen confirms some important generalizations that have been made concerning the dentition of this species. The Hell Creek Formation has recently produced a complete specimen of a semi-aquatic *D. vorax* that died in its burrow. Description of this specimen is in preparation by Kraig Derstler, Robert Bakker and coauthors. The fluvial deposits that encase the specimen also host a *Triceratops* skeleton only 12 m away from the complete *Didelphodon vorax*.

The specimen is a fragment (3.6 cm greatest dimension) of a very robust right jaw (mandibular ramus). The muscle insertion (masseteric fossa) is very clear and forms a parabolic arc on the labial side of the jaw (Fig. 10.4). The break at the posterior part of the jaw exposes the mandibular canal (Beck 2016). There is a crest in the mandibular canal that runs diagonal to the line of the jaw (and is roughly parallel to the molar zahnreihe). This mandibular canal crest is approximately 8 mm long and curves slightly at its anterior end in the direction of the line of the jaw. This crest is roughly coincident with a flattening that occurs on the ventral posterior part of the dentary as preserved here. The coronoid process has a vertical orientation, as indicated by its remaining base. The mandibular canal is visible on the break on the other (anterior) end of the jaw. It is circular and partly occluded with mineral deposits.

Alveoli for three molars are preserved on the jaw, m2, m3 and m4. A partial root socket at the anterior broken edge of the jaw represents the alveolus for m1 (Figs. 10.5 and 10.8). Fragments of tooth roots are preserved in both root channels for m4 and possibly also in the base of the other root sockets. The underside of the dentary (Fig. 10.6) is smooth and bulbous and flattens posteriorly. An unnamed foramen is preserved on the lingual side of the jaw (Fig. 10.3). A scratch or bone suture runs obliquely posterior form the foramen, continues beneath the jaw and runs up the labial side.

The jaw is notable for two reasons. First, judging by its alveolus, and even accounting for the fact that there has been some damage around its edges, the m4 was large in this animal (Figs. 10.5 and 10.9). Clemens (1968) in his reconstruction of the jaw of a Hell Creek, Montana specimen of *D. vorax* shows m4 as larger than m3, and in fact the largest of the molars and premolars, and I concur with this reconstruction. *D. vorax* evidently, in mechanical terms, more often placed its main bite force over m4 than over its premolars.

The expanded base of the alveolar cavity indicates that bone morphogenetic protein-2 (BMP-2; Wise et al. 2007) was very active at the base of the developing tooth crypt (the bony structure that surrounds developing teeth). The molar row in this jaw extends further to the posterior than is the case in *Didelphodon coyi*, curving slightly lingually, and the edge of m4 runs right up against the coronoid process in an almost primate (hominin) fashion. The holotype of the small Cretaceous metatherian *Alphadon eatoni* shows the m4 (in crypt) in a similar position, although the erupting m4 is of modest size (Cifelli et al. 1996; Kielan-Jaworowska et al. 2004).

A case of marsupial "impacted" molar is seen in the recently extinct thylacine (*Thylacinus cynocephalus*, the Tasmanian tiger; Clarkson et al. 2017, their Fig. 2), but this is a fragment of the maxillary, not the dentary. In the case of *Thylacinus cynocephalus* the last upper molar, still with a triangular aspect (occlusal surface), is turned sideways to fit into the back of the maxillary. This may seem an odd configuration, but it crops up (presumably by homoplasy) in the maxillaries of unrelated (placental) predatory mammals such as the Pleistocene felid *Smilodon californicus*. A curious archaeological specimen of part of the upper jaw of a Tasmanian tiger, coated with red pigment, is the earliest known evidence of interaction between marsupials and humans (Clarkson et al. 2017).

Didelphodon vorax's enlargement of m4 suggests a crushing bite for its body size. The bite force was deployed more posteriorly in *D. vorax* rather than anteriorly (over the premolars) as in *D. coyi*. As Fox and Naylor (1986) put it:

> [The] muscle mass would have inserted more anteriorly relative to mandibular length in *D. coyi*, closer to the crushing premolars (thereby increasing the mechanical advantage in adduction and the forces exerted by these teeth), a point of difference, for example, from *D. vorax*, which has been interpreted as having posteriorly inserting adductors and hence poor leverage on the mandible and crushing premolars (Clemens 1968), or *Sarcophilus harrisi* [another carnivorous marsupial], in which the molars, not the premolars, are the main crushing teeth.

This moving locus, from one species to the next, of the main bite force in the metatherian jaw is a repeated theme among ancient marsupials. The posterior location of the presumed maximum bite force in *D. vorax* links it to its ancestor *Eodelphis*.

In an analysis of possible durophagy in stagodontid metatherians, Brannick and Wilson (2017) analyzed the respective biting forces of *Eodelphis* and *Didelphodon*. Brannick and Wilson (2017) made estimates of mandibular bending strength (using beam theory) at six interdental gap locations in *Didelphodon*, *Eodelphis cutleri* and *Eodelphis browni*. Of the *Eodelphis* species, *E. cutleri* has larger premolars and a stronger jaw, leading Brannick and Wilson (2017) to hypothesize that *E. cutleri* gave rise to *Didelphodon*, a stagodontid with an even more robust jaw. Regarding respective bite strengths, Brannick and Wilson (2017) conclude that:

> The mandibular bending strength profiles of *Eodelphis* differ from those of *Didelphodon* . . . the jaw of *Eodelphis* is better suited to handle those forces posteriorly than anteriorly … In contrast, [in] jaws of both *D. vorax* and *D. coyi* . . . their labiolingual bending strength varies little from the symphysis to below the ultimate molar . . . The low labiolingual bending strength values anteriorly in the jaw of *Eodelphis* suggest that, among the morphological changes associated with durophagy in stagodontids, the inflated premolar morphology evolved prior to the broad anterior mandibular morphology for large labiolingual loads.

Brannick and Wilson (2017) assert that the inflated premolar came first in the stagodontid lineage, whereas the exceptional bite forces came later with the advent of genus *Didelphodon*. The morphology of *D. vorax*, with its large molar set far back in the jaw, strongly suggests that in this species (as in *Eodelphis*; Brannick and Wilson 2017) that there was still a peak in labiolingual bending strength posteriorly and this may be reflected in the development of its large ultimate molar. *D. coyi* apparently shifted maximum bite force anteriorly, and its premolar p3 inflates even more to form a mid-jaw massive crushing tooth. The large labiolingual load at mid-jaw may be the factor that causes the reappearance of zahnreihen in *D. coyi*, in other words, this may be thought of as a mechanical stress-induced atavism. A single jaw field line splits and distends in a localized case of toroidal inflation or expansion ('herniation' of McMenamin 2015), thus facilitating the hypertrophy of premolar p3.

The fact that marsupials first appeared in North America seems odd, because the only marsupial native to the United States today (and hence the continent's only metatherian north of the Mexican border) is the Virginia or North American Opossum (*Didelphis virginiana*). This resourceful gray to whitish-colored metatherian (*wapathemwa* = "white creature" in Algonquian) colonized North America from

South America, thus returning its metatherian kind to a North American homeland. The species continues to expand its range northward, sometimes with human assistance. During the Great Depression *Didelphis virginiana* was taken westward as a potential food source, and has since colonized much of California and coastal Oregon and Washington. The animal continues to work its way north, and has reached both western and eastern Canada. This is an impressive feat—an archaic metatherian holding its own against the aggressive and adept placental fauna of the north. *Didelphis virginiana* was one of the few hardy northbound colonists that travelled across the newly established Isthmus of Panama during the Great American Faunal Interchange at 3 million years ago.

Since the beginnings of evolutionary theorizing, it seemed strange that marsupials were found in South America and Australia, but except for *Didelphis virginiana* they were largely absent from Northern Hemisphere mammalian assemblages. Even more perturbing was the following question: Why were there so many marsupials in Australia but none in relatively nearby New Zealand?

The remarkable marsupial evolutionary experiment in Australia produced a long list of cases of convergent evolution with regard to the placental mammalian faunas to the north. The respective ecological counterparts partitioned their environments in much the same ways. Antelope was matched by kangaroos (*Macropus*, etc.). Sloths were matched by the koala. The marsupial 'mole' was matched by the placental mole. Marmots were matched by wombats. Squirrels were matched by phalangers (*Petaurus,* the sugar glider). Anteaters had their counterpart in numbats. The wolverine saw its counterpart in the Tasmanian devil. Rhinoceros found a counterpart in the extinct, lumbering *Diprotodon* (the largest marsupial that ever lived).

Tapirs and perhaps even proboscideans may also have had a match if the diprotodontid *Palorchestes* had a trunk as some researchers have suggested. And of course canids found their counterpart in the thylacine (*Thylacinus cynocephalus*), and felids in the marsupial 'lion' (*Thylacoleo carnifex*), actually a predatory diprotodont. The enormous premolars (carnassial premolars) in Plio-Pleistocene *Thylacoleo carnifex* recall the enlarged p3 in *Didelphodon vorax*. In *Thylacoleo carnifex*, p3 gets ridiculously elongated along the Zpm zahnreihe field line. These teeth in both animals were probably used to shear and penetrate bone. An entire, intact skeleton of *Thylacoleo carnifex* was discovered in 2002 in a cave in the Nullarbor region on the coast of Western Australia, with its enormous premolars and associated dentition quite visible and intact (Cremin 2007).

South America produced what most certainly be the most impressive case of marsupial homoplasy with the appearance of *Thylacosmilus atrox*, the saber-toothed marsupial (Figs. 10.10 and 10.11). The superficial similarities between *Thylacosmilus* and the placental felid *Smilodon* are striking. *Thylacosmilus* went extinct some time during the million year interval between 3.6 and 2.6 million years ago. This coincides with the time of the Great American Faunal Interchange; however, saber-tooth felids are not known from South America until 781,000–126,000 years ago (Prevosti et al. 2013). Numerous other types of metatherian carnivores such as *Arctodictis sinclairi* are also known from South America (Forasiepi 2009). With a certain irony, the metatherian predator *Thylacoleo* has the

Fig. 10.10 *Thylacosmilus*, the saber-toothed metatherian from Miocene strata of Argentina. Photograph of cast of complete skull. Note the interestingly elongate masseteric fossa, 5 cm in length. Scale in cm (Photograph by Mark McMenamin)

Fig. 10.11 *Thylacosmilus*, reconstruction of head region (Public domain artwork by Dibrd)

Fig. 10.12 *Sinodelphys szalayi* from the Yixian Formation of Liaoning, China. Length of metatherian 15 cm (Photograph by Ghedo Ghedo and used here per CCBY-SA3.0 license)

strongest bite of any known mammal, but greatly reduced canine teeth, whereas the sabertooth metatherian *Thylacosmilus* has enormously enlarged canines that constitute its dirk teeth. The somewhat similar lifestyles contrast strongly with dramatically different dentition.

Initial attempts to explain the paleobiogeographic discrepancies of the therian (that is, marsupial and placental) mammals relied on such solutions as land bridges, floating islands adrift at sea, and even Atlantis-like sinking continents. Ironically, the oldest true marsupials are known from North America. They are approximately 100 million years old (Cretaceous). Somewhat younger fossil marsupials (70 million years old) are known from South America.

The metatherian ancestors of marsupials, however, are first known from Early Cretaceous fossils in China (Luo et al. 2003). The oldest of these is the Early Cretaceous (125 million years old) *Sinodelphys szalayi* of the Yixian Formation in Liaoning, China (Luo et al. 2003). The only confidently known fossil shows the exceptional preservation characteristic for the Yixian Formation, and the holotype of *Sinodelphys szalayi* at first glance resembles an opossum with very short or no hairs on its face (Fig. 10.12). Foot bones of a mammal resembling *Sinodelphys* have been found in the gut contents of the Cretaceous paravian dinosaur *Microraptor zhaoianus* (Larsson et al. 2010).

Interestingly, the basal or primitive eutherian mammal, *Achristatherium yanensis*, also occurs in Cretaceous rocks of Liaoning (Hu et al. 2010). Eutherians represent the group that generated placental mammals, whereas metatherians developed the marsupials. The two groups are distinguished by their dentition. Eutherians never have more than three pairs of molars (m1-m3), whereas metatherians have four pairs of molars (m1-m4). *Achristatherium yanensis* has vestigial remains of the septomaxilla, an archaic feature known from its ancestral therapsids (Hillenius 2000; Hu et al. 2010).

Before the breakup of Pangea was complete, early metatherians were able to expand their range from Asia into North America. It was in North America where true marsupials first evolved. Marsupials are partly defined by the presence of pouches for the young, and as pouches consist of soft tissue they do not fossilize very well. This makes it somewhat less than perfectly straightforward to recognize true marsupials from fossil material. Nevertheless, what is considered by many to be the earliest crown group marsupial (and everyone agrees that it is a metatherian) is from Paleocene strata of Montana (*Peradectes minor*; Clemens 2006; O'Leary et al. 2013), deposited shortly after the end-Cretaceous mass extinction event. *Peradectes* is also known from the Nacimiento Formation, San Juan Basin, New Mexico (Williamson and Taylor 2011). The San Juan Basin has also produced the early metatherian *Swaindelphys*. Paleocene strata of Montana have also produced the eutherian mammal *Purgatorius*, considered to be the first proto-primate (O'Leary et al. 2013), although others consider the earliest fossil primates to be African (*Altiatlasius*; Williams et al. 2010).

This is a somewhat troubling discrepancy in our understanding of early mammalian evolution, as it would seem to be a rather key point to determine the geographic origins of our own Order Primates. As the breakup of Pangea proceeded throughout the Cenozoic, there was a major difference between the paleobiogeography of North America and the paleobiogeography of Africa. Crown marsupials, for example, originate in North America and then spread to Gondwanan continents via South America, but never reached Africa because the South Atlantic Ocean had already opened, creating an impassible barrier for metatherians on their southward trek.

This also makes the locus of primate origins all the more puzzling. If primates originated in North America (Montana, New Mexico, etc.) how did they get to Africa to initiate the Cenozoic primate adaptive radiation that eventually led to our species? Compounding the mystery, Forasiepi (2009) notes that during the latest Eocene to earliest Oligocene there "was sporadic land relations" between South America and Africa, bringing the first rodents and the first primates to the Americas. George Gaylord Simpson (1980) called these immigrants the "Old Island Hoppers." These pioneering rodents and primates, possibly carried across a (much narrower) South Atlantic by floating rafts of plant debris, lead to the radiation of the New World monkeys (Platyrrhini) and broke the general pattern of mammalian migration by land from North America, to South America, to Antarctica and finally to Australia. We can then include primates in the list of mammals able to survive inadvertent transoceanic journeys and able to diversify on the opposite coast! Equatorial currents in the early South Atlantic presumably flowed from east to west (as they do today), and for some reason the return route (assuming that there was an ancient equivalent to the Gulf Stream) was never able to carry metatherians across the ocean from west to east to colonize Africa, or, once having arrived, they were unable to survive for some reason.

Cretaceous marsupials found in South America appeared to provide support for the southern migration route hypothesis. Then in 1981, scientists from the University of California at Riverside (M. Woodburne and B. Daly) discovered an approximately 40 million year old jaw fragment that turned out to be the fossil of an Antarctic marsupial (Woodburne and Zinsmeister 1982). This provided compelling evidence supporting the hypothesis of a southern migration route. The fossil portrayed the marsupial clade midway in its journey from South America to Australia.

It is now evident that the fossil is a specimen of a polydolopine polydolopimorphian, the most common type of Antarctic fossil marsupials in that time interval (Goin et al. 2007).

Before the acceptance of continental drift and plate tectonics (before about 1960), the prevailing hypothesis was that early marsupials had taken a northern route, migrating from midcontinent North America, to Alaska, and then southward to Asia and Australia where today there is almost continuous land or at least relatively short-distance island hopping. A major problem with this hypothesis at the time was that there were no marsupial fossils of appropriate age known from Alaska or Asia.

The plate tectonics revolution in the earth sciences provided a solution to this conundrum. A quick review of the chronology of supercontinental breakup is in order here. At approximately 200 million years ago, Pangea split into a northern half and a southern half. These halves were Laurasia and Gondwana, respectively. At approximately 150 million years ago, Gondwana began splitting into Africa, Australia, India, Madagascar, South America and New Zealand with Antarctica as the centerpiece from which the other southern continents moved away. New Zealand broke off from Gondwana at about 82 million years ago. After having been in continuous contact since the times of Rodinia, Australia was the last modern continent to break away from Antarctica, with the split occurring approximately 40 million years ago during the Eocene.

At the time of the Cretaceous-Paleogene boundary, and hence the end-Cretaceous mass extinction, Australia was thousands of kilometers south of its present location. It is now clear that marsupials would have had a very difficult time migrating to Australia through Asia, even though, with some irony, the oldest metatherians known are from Asia. The only plausible alternative hypothesis is that marsupials migrated from South America, then to Antarctica, and then on to New Zealand. New Zealand had already broken off from Gondwana by 82 million years ago. This made it impossible for marsupials to reach the New Zealand islands via land, as the marsupial clade had not yet arrived in Australia by that time. The early tectonic departure from the Gondwanan continental plexus explains a number of paleobiogeographic oddities associated with New Zealand. The preponderance of birds on the islands, including flightless forms such as the kiwi (*Apteryx australis*) and the extinct moas (Order Dinornithiformes), has led some to refer to New Zealand as a "bird world." Ecologist Jared Diamond (1990) sees New Zealand, because of its geographically remote status, as being of exceptional scientific interest and "as close as you will get to the opportunity to study life on another planet."

There is also an island refugium aspect to the New Zealand biota. This refugium aspect partly explains the survival of the last living sphenodontid reptile (the only surviving member of Order Rhynchocephalia, superorder Lepidosauria, and clade Lepidosauromorpha), the Tuatara (*Sphenodon* sp.) of New Zealand. Sphenodontids have a fossil record reaching back to the Triassic, and are diapsid reptiles related to lizards, snakes and the tangasaurids such as *Hovasaurus*. Interestingly, the sphenodontid lineage produced an aquatic genus (*Pleurosaurus*) in the Jurassic (Dupret 2004).

In spite of how it may seem from its current geographic location, New Zealand is now thought to have rifted away from the coast of Antarctica rather than the coast of western Australia. New Zealand has always been marsupial free (not counting marsupials imported by humans) and also completely free of nonvolant (non-flying) mammals because of its 82 million years ago separation from the rest of Gondwana. Bats, a placental mammal type allied to rodents and primates, were thought to be the only naturally occurring mammals on New Zealand for its entire history. New Zealand bats are another odd group—bats of the Mystacinidae (short-tailed bats such as *Mystacina tuberculata*) spend a lot of their time foraging for food on the ground (Arkins 1999).

Before continuing further here, it is important to note that the marsupials were not the only strange mammals to reach Australia. The egg-laying monotremes ('prototherian' mammals) are represented today only by the echidna (*Zaglossus*) and the platypus (*Ornithorhynchus*; Ferner et al. 2017). There exist numerous intriguing differences between monotremes and therian mammals (marsupials and placentals). One of the most important for interpretation of the fossil record is that they bear in their shoulder girdle, as do many reptiles, both and interclavicle (Fig. 8.14) and coracoids. The fact that the male platypus has a powerful venom (β-defensins) in spurs on its hind limbs has led to the proposal that the first mammals were venomous (Hurum et al. 2006).

The fossil record of monotremes is also very intriguing. The earliest monotreme fossils add to the growing list of large Mesozoic mammals. The Cretaceous (middle Albian) Australian monotreme *Steropodon galmani* occurs in the Griman Creek Formation of Lightning Ridge, New South Wales, Australia and is apparently the oldest known monotreme fossil (Archer et al. 1985).

Lightning Ridge has also produced the early mammal *Kollikodon ritchiei*. Originally thought to be a monotreme (Flannery et al. 1995) like *Steropodon*, it has been reclassified as a non-monotreme australosphenidian (Pain et al. 2016). Both *Steropodon* and *Kollikodon* are preserved in opal, an unusual and strikingly beautiful mode of preservation. The naming of genus *Kollikodon* is an amusing story. The mammal has odd-looking teeth in occlusal view that look like hot cross buns, leading M. Archer to propose the name "Hotcrossbunodon" for the animal. This name was voted down by Archer's colleagues, nevertheless, a trace of the nickname remains as the genus name is derived from the ancient Greek *kollix* (κολλιξ) for muffin or bun (Long et al. 2002).

Fossil monotremes are also known from the Paleocene, Oligocene and Miocene Series. Giant fossil echidnas also appear in the Australian fossil record (Griffiths et al. 1991), and include the genera *Megalibgwilia* and the gigantic *Zaglossus hacketti*. *Z. hacketti* is the largest known monotreme of all time. This Pleistocene giant was over a meter long and weighed perhaps 30 kg (Long et al. 2002). The giant platypus *Odurodon* from the Oligocene-Miocene of Australia (Asahara et al. 2016) is remarkable because it retains its molar teeth as an adult (Archer et al. 1992); modern monotremes are toothless in the adult stage. Even more remarkable is the fact that a Paleocene giant platypus (*Monotrematum sudamerianum*) was discovered in 1992 at Punta Peligro, Argentina (Pascual et al. 1992). The similarities between *Obdurodon* and *Monotrematum*, particularly tooth crown morphology consisting of a pair of V-shaped lobes, are so strong that there are proposals to synonymize the two genera.

It is not clear whether or not the platypus lineage first evolved in South America and then migrated to Australia, or whether the lineage first evolved in Australia and then migrated to South America. The fact that the South American platypus is older (Paleocene) would support the former hypothesis, whereas the fact that the oldest monotreme is Australian (Cretaceous; *Steropodon*) would support the latter. Either way, the paleobiogeographic distribution of the Ornithorhynchidae indicates that Eocene platypus fossils await discovery in Antarctica. I predict that the first Antarctic monotreme fossils will be found in the Eocene of the James Ross Basin, Antarctica, probably within the estuarine or deltaic facies of the La Meseta Formation (Marenssi 2006). A lively mammalian fauna has already been recovered from the La Meseta Formation (including polydolopimorphids, microbiotherians [Goin et al. 2007], didelphimorphids, marsupials, gondwanatherians, xenarthrids, litopterns, astrapotherians, and cetaceans; Gelfo et al. 2015); continuing research is likely to add monotremes to the faunal list.

We now arrive at the curious case of the Saint Bathans mammal. Recall that the common consensus has been that (apart from bats) no mammals ever reached New Zealand because it had separated from the rest of Gondwana (or more specifically, Antarctica) at an early date (Cretaceous). A fascinating report (Worthy et al. 2006) promises to overturn this conclusion with the description of what the authors claim is the discovery of a nonvolant mammal of Miocene age from the Manuherikia Group on South Island, New Zealand. The locality is in Central Otago near Saint Bathans.

The fossils of interest from St. Bathans are fragmentary, consisting of two partial mandibles and part of a femur. Worthy et al. (2006) interpreted these fossils as representing "at least one mouse-sized mammal of unknown relationships." In other words, Worthy et al. (2006) are not certain that the mandibles and the fragmentary femur belong to the same species, but they make this assumption "on the basis of parsimony," calling the creature the "first nonvolant terrestrial mammal from New Zealand."

The most complete of the two mandibles is an edentulous fragment with alveoli for an incisor, a canine tooth, and two premolars (p1 and p2). There is a fragment of tooth root material remaining in the posterior root of the anterior (p1) premolar. A fairly large mental foramen is visible on the labial (lateral) sides of the jawbone. The jaw developed a massively fused symphysis that unites the two halves of the mandible, and the symphysial suture (i.e., the juncture between the right and left halves of the jawbone) is rather long.

The most interesting aspect of the jaw is the alveoli for the incisors. These are oriented almost horizontally, as if they were the insertion points for a pair of tiny tusks ("procumbent medial incisor"; Worthy et al. 2006) that projected directly forward from the jaw on either side of the symphysis. As such, these incisors angled forward much more acute angle than in, say the Australian marsupicarnivoran *Wakaleo alcootaensis* where the angle of tilt ranges from 54° to 27° (Yates 2015).

The femur fragment consists of dense bone indicating that the animal was adult when it died. The right femur fragment is from its proximal end, and "preserves the head, greater trochanter, and proximal edge of the crest that led to the lesser trochanter" (Worthy et al. 2006). Worthy et al. (2006) continue their analysis of the tiny bone, infer a "somewhat sprawling femoral posture," and conclude that "if we are correct in interpreting that the femoral head projects slightly dorsomedially with respect to the shaft of the femur, and that the lesser trochanter had its origin ventrally, then the femur did not project out to the side in the extreme abducted posture of monotremes." In other words, the St. Bathans mammal is not a monotreme but represents a different and presumably even more archaic type of terrestrial mammal.

Exactly what sort of mammaliaform or theriiform creature the St. Bathans mammal represents is not entirely clear, but the possibility exists that it represents a unique group that is "more derived than morganucodontans and more primitive than multituberculates and trechnotheres and which is not a monotreme [n]or an eutriconodontan" (Worthy et al. 2006). Some have speculated that the animal might have been some type of aberrant, ground-dwelling or flightless bat species native to New Zealand, but greatly procumbent incisors is not typical for a bat. On July 25, 2017, I emailed Trevor Worthy at Flinders University asking for a brief update on the status of the St. Bathans mammal. He graciously replied with what I consider to be the latest word on this critical paleontological project: "I believe we have evidence of two very different non-bat mammals, from teeth and post cranial bits. In addition, there are minimally 5 kinds of bats in 3 families." Although the fossil record of interest is scrappy and the fossils fragmentary, rather than being an isolated "bird world" ancient New Zealand might better be considered as an isolated "bird-bat-archaic mammal world."

If we accept that the Saint Bathans mammal (or mammals) represents an archaic terrestrial mammal endemic to New Zealand (Campbell and Hutching 2007), several fascinating conclusions and new questions soon come to light. First, we must conclude that some type of mammaliaform animal had reached Antarctica by the Cretaceous (before 82 million years ago) in order to colonize New Zealand. This seems to be a plausible conclusion, as the land connections were available up until that time.

If mammaliaforms had reached New Zealand, why did they subsequently go extinct? The refugium aspect of an isolated "land of birds" (Worthy et al. 2006) would seem to be an ideal place for the survival (as in the case of the tuatara *Sphenodon*) and perhaps diversification of an indigenous and unusual group of archaic mammals. The prospect provides ample fuel for the paleontological imagination. For example, one can imagine a diversification of these mammaliaforms along the lines of the therian mammals or, to a more limited extent, the monotremes.

I propose that we refer to this clade of mammaliaforms (Fig. 10.13) as the *zealanditherians*. As is the case for the mostly northern hemisphere haramiyidan basal mammal or mammaliaform group, the zealanditherians are an extinct group. Haramiyidians, by the way, are the longest lived mammalian group, surviving from the late Triassic to the end of the Cretaceous (Luo et al. 2015). The haramiyidians have recently produced two volant (flying or gliding) mammaliaforms, 160 million years in age (Late Jurassic; Oxfordian age; Meng et al. 2017; Luo et al. 2017), *Maiopatagium furculiferum* and *Vilevolodon diplomylos*.

Zealanditherians, like their distant mammaliaform relatives, had the potential to radiate into a diverse suite of organisms, perhaps every bit as diverse as the

Fig. 10.13 Zealanditherian currently known as the St. Bathans mammal. Length of animal approximately 5 cm. Digital artwork, entitled "New Zealand's Precious Darling," by Dylan Bajda (Used here as per CC BY-SA 4.0 license)

marsupials of Australia, or the more than ten orders of placental mammals that appeared in rapid succession in the northern hemisphere after the extinction of the nonavian dinosaurs at the end of the Cretaceous. Had evolutionary diversification been allowed to take its course, even the production of volant zealanditherians would not have been out of the question. But why did zealanditherians go extinct instead of engendering an Australia-style adaptive radiation? New Zealand never went into a deep freeze as was the case for Antarctica, a circumstance that led (lamentably in my opinion) to the demise of the indigenous Antarctic terrestrial mammals (including the hypothetical Antarctic monotremes [possibly including the genus *Obdurodon*] that I predict will eventually be found in Antarctica). Furthermore, unlike aquatic *Barasaurus* and its Mesozoic exclusion from a full radiation due to competition with presumably more aggressive marine tetrapods, there were no land mammals in New Zealand to compete with the zealanditherians.

As New Zealand was still attached to Gondwana (East Antarctica) during the Early Cretaceous, and since this part of Gondwana at the time was filled with Mesozoic terrestrial mammals, it is highly plausible that these creatures would also have lived in what is now New Zealand. More than 50 million years after the departure of New Zealand from Antarctica, however, there was a geological event of great significance for New Zealand mammaliaforms.

In an event of biblical proportions, New Zealand was almost completely submerged by the sea during the Oligocene at 25–30 million years ago (Worthy et al. 2006). This traumatic flooding and loss of land area led to massive habitat losses, and not only did this cut short the possibility of a zealanditherian adaptive radiation along the lines of marsupials in Australia, it also led to the extinction of the entire zealanditherian clade. Vertebrates on islands are prone to extinction anyway (even without the arrival of humans accelerating the process) because of relatively small population sizes and because of the limited extent of land surface area, leaving the entire island vulnerable to sudden environmental changes (Grayson and Meltzer 2003). Thus the Antarctic mammals were destroyed by ice, whereas the zealanditherians were destroyed by flood.

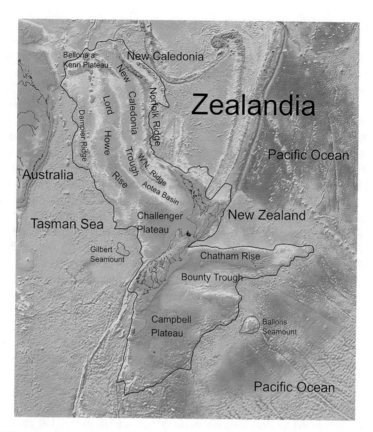

Fig. 10.14 Continent of Zealandia, topographic map based on data in Mortimer and Campbell (2014) (Cartography by Ulrich Lange, used here per CC0 [public domain image])

The tragedy of the zealanditherians is actually far worse than loss of approximately 80% of the present day land area of the New Zealand islands. For the New Zealand North and South islands are merely the topographic high ground or peak of a recently discovered and described continent, the last continent to be 'discovered' on earth, the hidden continent Zealandia (Mortimer and Campbell 2014; Mortimer et al. 2017). Most of Zealandia is sunken to a depth of approximately 1000 m, and the bulk of the submergence occurred about 80 million years ago.

The existence of Zealandia, a real-life Atlantis that actually did for the most part subside beneath the waves of the ocean, was inferred (and named) by Bruce Luyendyk, one of my professors when I was a graduate student at the University of California at Santa Barbara (Luyendyk 1995). Luyendyk (1995) had realized that there was an enormous patch of continuous continental crust linking the Campbell Plateau, the Chatham Rise, the Lord Howe Rise and New Zealand, which he named Zealandia.

Mortimer et al. (2017) propose the following criteria for recognition of a new continent. First, it must be topographically elevated relative to surrounding ocean floor crust. Second, it should contain a diverse suite of sedimentary, metamorphic and igneous rock types. Third, it must have a thicker crust than typical for oceanic crust. Fourth, it must have a lower seismic velocity than oceanic crust. Finally, it must have "well-defined limits around a large enough area to be considered a continent rather than a microcontinent or continental fragment" (Mortimer et al. 2017).

Zealandia has the form of an inverted 'T' with the New Zealand islands and a major fault zone (Alpine Fault) separating the cross bar of the 'T' from the inverted upright (Fig. 10.14). A massive 7.8 magnitude earthquake (the Kaikoura earthquake) in November 2016 on the Alpine Fault caused two deaths and obliterated long stretches of coastal roads in New Zealand. New Zealand is bracing for an even larger earthquake on the Alpine Fault, a major fault with a dangerous bend in its center. The fault trace of the Alpine Fault (a right-lateral strike slip fault comparable to California's San Andreas Fault) roughly parallels the coastlines of the New Zealand islands (Sutherland et al. 2017; Zachariasen et al. 2006).

At 4,920,000 square kilometers, Zealandia is more than half the area of Australia (Mortimer and Campbell 2014). Depending on how much of the continent was exposed land surface in the times after the tectonic separation from Antarctica during the Cretaceous, there may well have been an adaptive radiation of zealanditherians in an extensive yet isolated terrestrial setting. If so, their fossils may mostly reside beneath the waves of the sea. New Zealand's sedimentary record is very sparse in terms of Cenozoic strata, and that adds great significance to the Manuherikia Group sediments (Bannockburn Formation) that produced the Saint Bathans mammal. These sedimentary rocks play host to the last remaining scraps of the zealanditherian lineage. More fossils no doubt occur beneath the waves, but these will be very difficult or impossible to access and study.

Was Zealandia ever completely submerged, and was this the reason for the demise of the zealanditherians? The survival of the tuatara indicates that this was never the case, thus the zealanditherian extinction event must be the result of extreme habitat loss. Extinction evidently did not occur by means of wholesale drowning of the entire population.

References

Archer M et al (1985) First Mesozoic mammal from Australia—an early Cretaceous monotreme. Nature 318:363–366

Archer M et al (1992) Description of the skull and non-vestigial dentition of a Miocene platypus (*Obdurodon dicksoni* n. Sp.) from Riversleigh, Australia, and the problem of monotreme origins. In: Augee ML (ed) Platypus and echidnas. Royal Zoological Society, New South Wales, pp 15–27

Arkins AM (1999) Diet and nectarivorous foraging behavior of the short-tailed bat (*Mystacina tuberculata*). J Zool 247(2):183–187

Asahara M et al (2016) Comparative cranial morphology in living and extinct platypuses: feeding behavior, electroreception, and loss of teeth. Sci Adv 2(10):e1601329. https://doi.org/10.1126/sciadv.1601329

Beck RMD (2016) The skull of *Epidolops ameghinoi* from the early Eocene Itaboraí Fauna, southeastern Brazil, and the affinities of the extinct marsupialiform order Polydolopimorphia. J Mammal Evol. https://doi.org/10.1007/s19914-016-9357-6

Brand NA et al (2017) The microinvertebrate fossil assemblage of the upper Cretaceous (Campanian-Maastrichtian) Williams Fork formation, western Colorado. Society of Vertebrate Paleontology 77th Annual Meeting Program & Abstracts, Calgary, Canada, 85

Brannick AL, Wilson GP (2017) Exploring the evolution of durophagy in stagodontid metatherians using relative mandibula bending strength. Society of Vertebrate Paleontology 77th Annual Meeting Program & Abstracts, Calgary, Canada, 86

Campbell H, Hutching G (2007) In search of ancient New Zealand. Penguin Books, Auckland/New Zealand

Chen C et al (2015) How the mollusc got its scales: convergent evolution of the molluscan scleritome. Biol J Linnean Soc 114:949–954

Cifelli RL et al (1996) Origin of marsupial pattern of tooth replacement: fossil evidence revealed by high resolution X-ray CT. Nature 379:715–718

Clarkson C et al (2017) Human occupation of northern Australia by 65,000 years ago. Nature 547:306–310

Clemens WA (1968) A mandible of *Didelphodon vorax* (Marsupialia, Mammalia). Los Angeles County Museum, Contributions in. Science 133:1–11

Clemens WA (2006) Early Paleocene (Puercan) peradectid marsupials from northeastern Montana, North American western interior. Palaeontogr Abt A 277:19–31

Cope ED (1892) On a new genus of Mammalia from the Laramie formation. Amer Nat 26:758–762

Cremin A (2007) The world encyclopedia of archaeology. Firefly Books, Buffalo/New York

Diamond JM (1990) New Zealand as an archipelago: an international perspective. In: Towns DR et al (eds) Ecological restoration of New Zealand islands. Department of Conservation Wellington, New Zealand, pp 3–8

Dupret V (2004) The pleurosaurs: anatomy and phylogeny. Rev Paléobiol 9:61–80

Ferner K et al (2017) Comparative anatomy of neonates of the three major mammalian groups (monotremes, marsupials, placentals) and implications for the ancestral mammalian neonate morphotype. J Anat. https://doi.org/10.1111/joa.12689

Flannery TF, Archer M, Rich TH, Jones R (1995) A new family of monotremes feom the Creataceous of Australia. Nature 377(6548):418–420

Forasiepi AM (2009) Osteology of *Arctodictis sinclairi* (Mammalia, Metatheria, Sparassodonta) and phylogeny of Cenozoic metatherian carnivores from South America. Monogr Mus Argentino Cienc Nat, n s 6:1–174

Fox RC, Naylor BG (1986) A new species of *Didelphodon* marsh (Marsupialia) from the upper Cretaceous of Alberta, Canada: paleobiology and phylogeny. N Jb Geol Paläont (Abh) 172(3):357–380

Gelfo JN et al (2015) The oldest mammals from Antarctica, early Eocene of the La Meseta formation, Seymour Island. Palaeontology 58(1):101–110

Goin FJ et al (2007) New marsupial (Mammalia) from the Eocene of Antarctica, and the origins and affinities of the microbiotheria. Rev Asoc Geol Argent 62(4):597–603

Grayson DK, Meltzer DJ (2003) A requiem for North American overkill. J Archaeol Sci 30:585–593

Griffiths M et al (1991) Observations on the skulls of fossil and extant echidnas (Monotremata: Tachyglossidae). Aust Mammal 14:87–101

Hillenius WJ (2000) Septomaxilla of nonmammalian synapsids: soft-tissue correlates and a new functional interpretation. J Morphol 245:29–50

Hu Y et al (2005) Large Mesozoic mammals fed on young dinosaurs. Nature 433:149–152

Hu Y et al (2010) New basal eutherian mammal from the early Cretaceous Jehol biota, Laioning, China. Proc R Soc B 277:229–236

Hurum JH et al (2006) Were mammals originally venomous? Acta Pal Polonica 51(1):1–11

Kielan-Jaworowska Z et al (2004) Mammals from the age of dinosaurs: origins, evolution, and structure. Columbia Univ Press, New York

Kissel RA et al (2002) *Captorhinus magnus*, a new captorhinid (Amniota: Eureptilia) from the lower Permian of Oklahoma, with new evidence on the homology of the astragalus. Can J Earth Sci 39:1363–1372

Larsson H et al (2010) The winged non-avian dinosaur *Microraptor* fed on mammals: implications for the Jehol biota ecosystem. Soc Vert Paleont Prog Abstr 70:114A

Long JA et al (2002) Prehistoric mammals of Australia and new Guinea: one hundred million years of evolution. Johns Hopkins University Press, Baltimore/Maryland

Luo Z-X et al (2003) An early Cretaceous tribosphenic mammal and metatherian evolution. Science 302(5652):1934–1940

Luo Z-X et al (2015) Mandibular and dental characteristics of late Triassic mammalia-form *Haramiyavia* and their ramifications for basal mammal evolution. Proc Nat Acad Sci (USA):201519387. https://doi.org/10.1073/pnas.1519387112

Luo Z-X et al (2017) New evidence for mammaliaform ear evolution and feeding adaptation in a Jurassic ecosystem. Nature. https://doi.org/10.1038/nature23483

Luyendyk B (1995) Hypothesis for Cretaceous rifting of east Gondwana caused by subducted slab capture. Geology 23:373–376

Marenssi SA (2006) Eustatically controlled sedimentation recorded by Eocene strata of the James Ross Basin, Antarctica. Geol Soc London Spec Pub 258:125–133

Marsh OC (1889) Discovery of Cretaceous Mammalia. Amer J Sci 38(3):81–92

McMenamin MAS (2015) Paramphibia: a new class of tetrapods. Meanma Press, South Hadley/Massachusetts

Meng Q-J et al (2017) New gliding mammaliaforms from the Jurassic. Nature. https://doi.org/10.1038/nature23478

Mortimer N, Campbell H (2014) Zealandia: our continent revealed. Penguin Books, Auckland/New Zealand

Mortimer N et al (2017) Zealandia: earth's hidden continent. GSA Today 27(3–4):28–35

O'Leary MA et al (2013) The placental mammal ancestor and the post-K-Pg radiation of placentals. Science 339(6120):662–667

Pain R et al (2016) The upper dentition and relationships of the enigmatic Australian Cretaceous mammal *Kollikodon ritchiei*. Mem Mus Victoria 74:97–105

Pascual R et al (1992) First discovery of monotremes in South America. Nature 356:704–706

Polcyn MJ et al (2010) The North African mosasaur *Globidens phosphaticus* from the Maastrichtian of Angola. Historical Biol 22(1–3):175–185

Prevosti FJ et al (2013) The evolution of the Cenozoic terrestrial mammalian predator guild in South America: competition or replacement? J Mammalian Evol 20:3–21

de Ricqlès A, Bolt JR (1983) Jaw growth and tooth replacement in *Captorhinus aguti* (Reptilia: Captorhinomorpha): a morphological and histological analysis. J Vert Paleont 3(1):7–24

Sander PM, Faber C (2003) The Triassic marine reptile *Omphalosaurus*: osteology, jaw anatomy, and evidence for ichthyosaurian affinities. J Vert Paleont 23(4):799–816

Scott CS, Fox RC (2015) Review of Stagodontidae (Mammalia, Marsupialia) from the Judithian (late Cretaceous) Belly River group of southeastern Alberta, Canada. Can J Earth Sci 52:682–695

Simpson GG (1927) Mammalian fauna of the Hell Creek formation of Montana. Amer Mus Novitates 267:1–7

Simpson GG (1980) Splendid isolation. The curious history of South American mammals. Yale Univerity Press, New Haven

Sutherland R et al (2017) Extreme hydrothermal conditions at an active plate-bounding fault. Nature 546:137–140

Williams BA et al (2010) New perspectives on anthropoid origins. Proc Nat Acad Sci (USA) 107(11):4797–4804

Williamson T, Taylor L (2011) New species of *Peradectes* and *Swaindelphys* (Mammalia; Metatheria) from the early Paleocene (Torrejonian) Nacimiento formation, San Juan Basin New Mexico. Palaeontol Electron 14(3.23A):1–16

Wilson GP et al (2016) A large carnivorous mammal from the late Cretaceous and the North American origin of marsupials. Nat Commun 7:13734. https://doi.org/10.1038/ncomms13734

Wise GE et al (2007) Bone formation as a potential motive force of tooth eruption in the rat molar. Clin Anat 20(6):632–639

Woodburne MO, Zinsmeister WJ (1982) Fossil land mammal from Antarctica. Science 218:284–286

Worthy TH et al (2006) Miocene mammal reveals a Mesozoic ghost lineage on insular New Zealand, southwest Pacific. Proc Nat Acad Sci (USA) 103(51):19419–19423

Yates AM (2015) New craniodental remains of *Wakaleo alcootaensis* (Diprotodontia: Thylacoleonidae) a carnivorous marsupial from the late Miocene Alcoota local fauna of Northern Territory, Australia. Peer J. https://doi.org/10.7717/peerj.1408

Zachariasen J et al (2006) Timing of late Holocene surface rupture of the Wairau fault, Marlborough, New Zealand. N Z J Geol Geophys 49:159–174

Chapter 11
Bifaces to the Ends of the Earth

Abstract The date of arrival of humans in the Americas is a long-standing puzzle in archaeology. A vein quartz biface from southern New England (South Hadley, Massachusetts) represents the oldest evidence for human occupation in New England, one of the last places in the United States to be colonized by *Homo sapiens*. The biface, which was resharpened and reused by a member of a subsequent lithic culture, is dated here to approximately 18,000 years ago, which means that pre-Clovis hunters arrived in Massachusetts just as the most recent glacial maximum was melting back to the north.

Keywords Paleoindians · Clovis culture · Connecticut Valley · Beringia · Overkill hypothesis · Cartifact · Geofact · Projectile point · Vein quartz · Loess

While decrying the prevailing anomie in contemporary Western culture, A. James Gregor (2017) of the University of California, Berkeley noted that a professional anthropologist had recently argued that:

> no distinction in truth could be made between the Zuni Indian tribe's claim that the indigenous population of North America spontaneously rose out of the earth and the claim…that Native Americans migrated over the prehistoric land bridge that once connected North America with Asia. [In this view,] both characterizations of the origins of Native Americans are equally 'true'.

Whether or not humans can spontaneously arise out of rocks and soil is an interesting question that appears to be beyond the purview of natural science. Natural science may, however, address the question of the migration routes taken by the first human colonists of North America.

Once maritime explorers charted the general dimensions of the Americas, it became generally assumed that *Homo sapiens* was a relative newcomer to the New World. Early estimates made in the seventeenth century concluded that the ancestors of the "Indians" could not have reached the Americas more than a few thousand years ago. These estimates were self-serving, and were used as justification to displace native populations in the Americas, who after all, so the thinking went, had not been there very long in any case. With this restrictive time framework, various theories were proposed to explain the origin of native American populations.

The Mediterranean region, the Lost City of Atlantis, and even islands across the Pacific were all proposed as the source of America's first men and women.

The first inhabitants of America have been called "archaeology's greatest mystery" (Adovasio and Page 2002) by James Adovasio, who has vigorously advanced the concept of very early occupation (14,000 or more years ago) for the Meadowcroft Rockshelter in southwest Pennsylvania (Cremin 2007). One very popular idea regarding *Homo sapiens'* first appearance in the New World, based on geological, climatological, ecological and archeological evidence, is the concept that people walked from Asia to Alaska across an intercontinental strip of tundra known as Beringia. Beringia, now flooded and known as the Bering Strait, was exposed intermittently during the great Pleistocene glaciations. Exposure to air of this strip of land connecting Asia and North America, 1300 miles across at maximum width separating the Bering and Chukchi seas, is easy to visualize if one is familiar with the shallow, flat bathymetry of the surrounding sea floor. The Bering Strait does not exceed 50 m depth. This was an important fact during the last glacial maximum, because as more ice was incorporated into the glaciers, the less water filled the ocean basins. During maximum glaciation at 21,000–25,000 years ago, sea level fell by 125 m (Lambeck et al. 2014).

Although one might think that Beringia, being close to the north polar regions, would have been covered by an impassible ice sheet, paleoclimatological evidence indicates that there was a zone of warm air over the Bering and Arctic seas (and thus over Beringia). A zone of warm air was required to acquire the moisture that fed ice and snow to the continental glaciers. Therefore, although there may have been thousands of square kilometers of glaciated surface between Alaska and the New England, at least there was a passable land route from Siberia to Alaska. This Bering Bridge, which has fluctuated between being dry, wooded land and being completely submerged scores of times during the past 15 million years, was used extensively if unwittingly by Asiatic and American flora and fauna to cross from one continent to the other, either by an eastward or westward route. Horses, bison, mastodons, mammoths and even muskrats slowly extended their ranges across the Beringia bridge onto a new continent. The direction of introduction was usually from Asia to America, but occasionally the migration took place in the other direction, as for example when horses first reached Asia via Beringia. Horses later went extinct in America until reintroduced by Spanish conquistadors.

Our picture of the Pleistocene ecology of North America has improved considerably due to the Ziegler Reservoir fossil site in the Rocky Mountains near Snowmass Village, Colorado, USA. This site, which includes parts of Marine Oxygen Isotope Stages 3–6, exceeds 100,000 years in age (Johnson et al. 2014). The site, known as "The Snowmastodon Project," has produced salamanders, birds, fish, snakes, lizards, frogs, beavers, rabbits, minks, muskrats, otters, big horn sheep, coyotes, black bears, bison, deer, camels, horses, ground sloths, mastodons and mammoths, all living in an alpine Pleistocene ecosystem at an elevation of 2705 m (Johnson et al. 2014). Exceptional preservation has produced beetle elytra (forewings) that were still iridescent, and willow and sedge leaves that were still green after having been buried for 100,000 years.

The topography of Beringia was low relief, allowing large animals to move freely across during both glacial and interglacial intervals. This of course poses two key questions for archaeology. When did humans first cross the Beringia bridge? Were they the first humans to reach North America?

Following on earlier prejudices regarding the earliest occupation of North America, for many decades it was archaeological orthodoxy that people first arrived in the Americas only 5000 years ago (Martin 1973). A series of important discoveries, however, including the Lehner and Naco mammoth kill sites in Arizona (USA), pushed the Beringia-crossing date back to 11,000–13,000 years ago (Haury et al. 1959; Antevs 1959). There the date 'fossilized,' and was not seriously reconsidered until the late 1960s and 1970s.

In an article entitled "The Discovery of America," Paul S. Martin (1973) defended the 11,000 year position. Martin (1973) maintained that the first significant human population reached America only 12,000 years ago, and then proceeded to decimate its Pleistocene megafauna (large mammals such as mammoths, mastodons, and ground sloths) within a mere 1000 years. Martin postulated that these hunters swept down the continent in an arcuate front that moved 16 km a year, executing overkill on the game animals in any region within a decade, and then moving on to 'happier hunting grounds.' This expanding front of faunal destruction swept down through all of North and South America, and within a few hundred years had completely wiped out the megafauna. After the mammoths, mastodons, giant ground sloths (*Megatherium*), horses, and camels had disappeared, Martin (1973) said that "the hunters [were] forced, by necessity, to learn more botany" and develop an agricultural subsistence basis.

Martin's (1973) hypothesis hinged on a group of well-dated New World mammoth kill sites that are all approximately 11,000 years old (e.g., Haury et al. 1959), and also on the observation that the initial human colonization of islands leads to extermination of endemic vertebrate species (Grayson and Meltzer 2003). Martin (1973) insisted that the discovery of the Americas triggered a human population explosion that shortsightedly destroyed its primary subsistence base. Martin (1973) argued that the extinction chronology of the Pleistocene megafauna can be used to map the spread of *Homo sapiens* though the Americas. The only conceivable circumstance of man discovering this continent and failing to exploit and decimate the megafauna in 1000 years, which had very limited time to adjust to human predation, was in Martin's (1973) opinion "that sometime before 12,000 years ago, the earliest early man came over the Bering Strait without early woman."

Martin's (1973) paper was convincing to many scientists (Grayson and Meltzer 2003), especially as it seemed in accord with other hypotheses that dealt with American prehistory. Beringia seemed to have been passable by people at least once between 25,000 and 10,000 years ago. Twelve thousand years was thought to be long enough for the first hunting and gathering humans to give rise to the Paleoindians (considered culturally equivalent to the European Paleolithic), who were responsible for first the Clovis, and then the Folsom biface assemblages (Cremin 2007) found across North America. This dating has since been confirmed. The remains of an infant (*Anzik-1*) buried with Clovis artifacts, dated to approximately 12,632 years old, was discovered in Montana in 2014 and yielded the most ancient genome of an American. Genetic sequencing of *Anzik-1's* genome demonstrates that the Clovis people are most closely related to the ancient inhabitants of Siberia (Rasmussen et al. 2014).

The age of *Anzik-1* would seem to be close to that of a submerged skeleton from Tulúm, México on the Yucatán Peninsula. The skeleton was on the floor of the Chan Hol cave, adjacent to a submerged stalagmite that projected upward like a tombstone. The skeleton is estimated to be approximately 13,000 years old, but dating has been complicated by the fact that, 1 month after discovery, most of the skeleton was looted by divers who raided the Chan Hol cave (Stinnesbeck et al. 2017) in a theft that may be compared to the loss of Peking Man. Fortunately, photographs of the undisturbed site allow Stinnesbeck et al. (2017) to infer that the individual died in the cave before submersion, and that he was male.

A skeleton named 'Naia' (for naiad or water spirit), a teenage female skeleton discovered from another submerged Yucatán cave (called Hoyo Negro, and evidently not looted in this case, although some of the bones of Hoyo Negro have been disturbed), has yielded a radiometric age of 12,000–13,000 years (Chatters et al. 2014; de Azvedo et al. 2015a, b).

Paleoindians such as these allegedly wiped out the megafauna, and in turn gave rise to the Archaic (ca. 8000–5000 years ago) assemblage. The Archaic inhabitants were equivalent to the European Mesolithic peoples, and went on to become the ancestors of the great Mesoamerican Mayan and Incan civilizations as well as the other Native American groups scattered throughout the Americas and discovered at various stages of cultural development by European invaders.

Martin's "overkill hypothesis" is periodically challenged in the archaeological community (Grayson and Meltzer 2003; Nogués-Bravo et al. 2008). In a lively critique of the overkill hypothesis Grayson and Meltzer (2003) take Paul Martin to task for his claim that the search for North American sites earlier than Clovis is "something less than serious science, akin to" searching for Big Foot or the Loch Ness Monster. And indeed, if there were people in the Americas thousands of years before the Clovis culture, that does complicate Martin's overkill hypothesis. Grayson and Meltzer (2003) attribute the popularity of the overkill hypothesis with contemporary concern for the environment, noting how well it accords with "the homily of ecological ruin." Grayson and Meltzer's (2003) main quibble with the overkill hypothesis, however, is supposed lack of evidence that it ever occurred, and Martin's assertion that "the overkill model predicts a lack of supporting evidence, thus turning the absence of empirical support into support for his beliefs. We suggest that this feature of the overkill position removes the hypothesis from the realm of science and places it squarely in the realm of faith. One may or may not believe in the overkill position, but one should not confuse it with a scientific hypothesis about the nature of the North American past."

This is a remarkable juncture in the archaeological sciences, with opponents on either side of the overkill debate accusing the other side of engaging in pseudoscience! A more reasonable approach in this case is to combine the merits of both views, in a search for coincidences and combined effects. For example, Nogués-Bravo et al.

(2008) conclude that "the disappearance of climatically suitable areas for woolly mammoths and the increase in anthropogenic impacts in the Holocene, the coup de grace, likely set the place and time for the extinction of the woolly mammoth."

Martin (1973) argued that any human artifact found in North America and dated as earlier than 15,000 years old, of which there were only a handful of examples in 1973, were misdated because of careless dating technique. He might therefore just barely be able to accept the Page-Ladson site from a karstic/sinkhole setting in the stream bed of the Aucilla River, Florida, that has been dated to 14,550 years ago (Halligan et al. 2016). The Page-Ladson is considered by some to be a challenge to Martin's overkill hypothesis (by suggesting that humans and mastodons lived together for an extended period), and in any case further highlights the geochronology problem, a problem that must be solved in order to learn when humans actually did first arrive in the Americas.

In nearly every archaeological project, the first and most import question about a site is "How old is it?" Reliable dating techniques are thus among the most valuable tools available to the archaeologist. Much of what we do know about the ages of Paleoindian sites is thanks to radiocarbon dating. At the Lehner mammoth kill site, where numerous Clovis assemblage bifaces were discovered with the remains of nine immature mammoths, charcoal from a campfire hearth yielded the first Clovis radiocarbon dates (Haury et al. 1959; Antevs 1959). The Lehner dates averaged 11,260 ± 360 years.

Unfortunately, radiocarbon dating has serious limitations as we have seen with the Shroud of Turin and other historical items that have been subjected to radiocarbon analysis. The limit of radiocarbon dating is about 70,000 years, and this is out of range of the more extreme estimates of the occupation of the Americas by *Homo sapiens*. Not subject to this limitation are the various methods of dating by Pleistocene geochronology. These methods involve correlating land forms, such as marine terraces, geomorphic scars left by glacial advances and retreats, and glacial lake sediments. All of these geomorphic features may be associated with archaeological materials.

When Louis S. B. Leakey (1903–1972), the renowned Kenyan archaeologist of Olduvai Gorge fame, coauthored an article in *Science* (Leakey et al. 1968) reporting knapped chert fragments in Pleistocene alluvial fan deposits dated to much older than 40,000 years, the archaeological community was shocked at these seemingly outlandish claims. Mary Leakey (1984) revealed that Calico Hills was "catastrophic to [Louis's] professional career and was largely responsible for the parting of our ways." The putative artifacts were discovered in 1958, approximately the same time that accurate radiocarbon dates were becoming available for the Lehner mammoth site.

The chert fragments from the Calico Mountains in California continue to attract heterodox archeological support (Budinger 1983), but the general consensus is that the supposed artifacts are actually 'geofacts' formed by natural processes of rock breakage (Haynes 1973). Nevertheless, Louis Leakey maintained to the end of his life that "any day, any moment, proof will be found" that humans arrived in North America many tens of thousands of years before the generally accepted date.

A comparable cautionary tale is associated with an American Professor of Geography at Texas A&M University, the late George Francis Carter (1912–2004). Carter had long been considered an archaeological heretic for insisting, like Louis Leakey, on much older dates for humanity's entrance into the Americas. Even worse than Leakey, Carter's academic reputation was compromised by his heterodox claims, so much so that the specimens Carter used to support his hypotheses were dubbed 'cartifacts' by his detractors. Nevertheless, Carter retired as professor emeritus after publishing *Earlier Than You Think: A Personal View of Man in America* in 1980. Not surprisingly, in this book Carter (1980) leaps to defend the Calico site: "The biggest myth in American archaeology is that nature breaks rocks by percussion and pressure with considerable frequency and that this breakage reproduces human work."

As a sophomore majoring in geology at Stanford University, I had become fascinated with the question of the antiquity of man in the Americas, so I wrote to Carter and he graciously replied as follows on May 11, 1977 (the physical letter is in the geological literature archives of MAS McMenamin):

Dear Mark

I don't know how to say just what evidence is most convincing. I think that the cross-checking system of geology, geomorphology, pedology, climatology that I reported in *Pleistocene Man at San Diego* (1957) is as convincing as anything can be. The radiometric systems simply verify the findings. I have said over and over that if I was wrong that it was on the conservative side and that now begins to appear possible as there are lines of evidence pointing to 200,000. One of the key things is the recognition of artifacts (cartifacts). No one last summer challenged them and this included Junius Bird, Scotty MacNeish and a flock of youngsters, some of them trained in lithic analysis.

Then there is the matter of the proliferation of the sites. We now have many sites with the blade and core work found at Texas Street. Many at San Diego, and numerous in the Imperial Valley. I saw much the same material from the Catskills, and it is also at the bottom of the Shequiandah Site in Canada (Thomas Lee's work). Both include [material] "beneath the glacial till." We now have this lithic assemblage in almost every kind of material, and from almost every kind of geomorphic situation (in and under glacial deposits, in and between riverine deposits, on old land surfaces where nothing more than sheet runoff from rain storms have moved in the past 500,000 years). We are left with only one constant: man at work. I lectured for the lithic analysis class at U. T. Austin last week. At the end the kids asked: 'Well, we accept these as artifacts, but we'd like to see some of the famous Cartifacts.' They couldn't believe it when I said: 'You are looking at them.'

There is one new bit of evidence at San Diego. The blade and core work cuts off abruptly at about 100 feet elevation. Something has erased all the evidence of this early stage below that level. Well what do we know about the coast? We know it to be a relatively stable block—see the Cal Tech Seismology Group reports on that. So, the eraser would be a sea level of that magnitude. So, see Evans (1971)—and you will find that he has taken the Milankovitch explanation for the glacials (it is fully supported by the dated ocean cores, the dated terraces, etc.) and on Evans' system the 100 foot sea stand dates to 200,000.

I am acutely uncomfortable with any one measure of time, and have always sought cross checking systems. We have them, and we are now cross checking them with radiometric systems. When we put man into that system it keeps coming out at 100,000 or 200,000; and do I kind [of] think maybe that just might be about what we are someday going to have to admit.

Yours

George Carter

This reply from Carter was interesting; he is clearly trying to drum up support with a fresh young recruit. So I sought a second opinion from Brian Reeves, who at the time was Associate Professor in the Department of Archaeology at the University of Calgary. On May 31, 1977, Reeves replied: "When did man arrive in the Americas? Probably during the Sangoman, i.e., *ca* 120,000. The profession, in general, refused to believe any of this, but that is typical." It is safe to say that by 1977, there was an active group of scientists interested in pushing the initial occupation of the Americas far back into time.

George Carter's tattered scientific reputation is such that he did not even receive a courtesy citation in a recent report on the Cerutti Mastodon site in San Diego County, California. The Cerutti site article arguably provides at least some vindication for the Carterian approach. The authors (Holen et al. 2017) were perhaps concerned that their report of a 130,000 year old archeological site in California was going to be controversial enough without invoking the memory of cartifacts. Holen et al. (2017) present a very interesting study, especially in light of the historical background in the search for evidence for humans in North America before 100,000 years ago. Its publication was long delayed.

The sedimentology in Holen et al. (2017) is unconvincing. For example, Holen et al. (2017) note cross-bedding immediately beneath the mammoth bones, but they do not provide paleocurrent direction(s). The concentrations of bones/stones in Holden et al.'s illustration (2017, their fig. 1a) form two or maybe three parallel bands, almost like the crests of very large ripples. Could a flash flood/tempestite/debris flow-like event have caused this, and created a geofact site as per Haynes (1973)? Nevertheless, the results of Holen et al. (2017) are not easily dismissed out of hand and must be added to the long list of sites suggestive of (but not proving) early occupation of the Americas.

In terms of conventional archaeology, coastal Southern California and its Channel Islands have an interesting and not fully understood history recorded largely by means of biface artifacts, some of which have unusual shapes. Due to a thin, distinctive white cortex on cobbles of what they refer to as the 'Tuqan' Monterey chert, Erlandson (2008) concluded that Paleocoastal people had utilized these distinctive cobbles on San Miguel Island for at least 10,000 years. Erlandson and Braje (2008) reported a distinctive artifact from San Miguel Island, also fashioned from Monterey Chert, known as a chipped stone crescent. The stone crescent is from a site known as Lotah's Wheel site, that preserves the remains of a 'medicine wheel' constructed during a National Park Service and University of California, Santa Barbara archaeological project by consultant Kate Lotah in 1982 (Erlandson and Braje 2008). Needless to say, building a medicine wheel seems to be an odd activity to pursue in while attempting to conduct government-sponsored archaeological research, as the stone structure could potentially complicate or contaminate future archaeological investigations.

Archaic (ca. 8000–5000 BP) "lozenge shape" spear points or knife blades from coastal California constitute an end member in a homologous series of coeval projectile points belonging to the Vandenberg Contracting Stem type. Five different point subtypes can be recognized in the series, with differences occurring in the

Fig. 11.1 Vandenberg
Contracting Stem type
knife blade (subtype
VCS-5) from near Coal Oil
Point, Isla Vista,
California. Scale bar in cm
(Photograph by Mark
McMenamin)

length and width of the contracting base, the presence or absence of barbs, and the position of the shoulder relative to the transverse midline of the point. Transition from one point type to the next in the series required only slight modification of the knapping method. The homologous series thus provides five different subtypes (VCS 1-5) based on variations to a fundamental ground plan for point manufacture. Subtype primary functions are inferred here as follows: small game projectile point (VCS-1 to VCS-3), larger game projectile point (VCS-4), and knife blade (VCS-5). A single knapping paradigm thus provided an extensive toolkit for ancient inhabitants of coastal Southern California.

Heavy rainfall on the Southern California coast in 1982 exposed a diamond or lozenge-shaped bifacial tool (Fig. 11.1) of red-mottled, green Franciscan Chert. The point was collected as a surface find near Coal Oil Point, Isla Vista, California (Fig. 11.2). The point is attributed here to the Vandenberg Contracting Stem type (Overstreet 2007). Bifacial points belonging to this type are found in coastal and island areas of Southern California. Typically fashioned from brown Monterrey Chert or greenish Franciscan Chert, these points have traditionally been attributed to the Canalino/Chumash (Miles 1963; Hudson and Blackburn 1982) peoples of coastal Southern California. More recent research, however, indicates an Archaic period age for the points, dating back to 5000–8000 years BP (Overstreet 2007).

Vandenberg Contracting Stem points may be organized into a series ranging from barbed points to lozenge-shaped points (Fig. 11.3). As is the case with some Paleoindian points (McMenamin 2011), Vandenberg Contracting Stem examples may show evidence of resharpening by means of an arcuate scar along the cutting edge of the point.

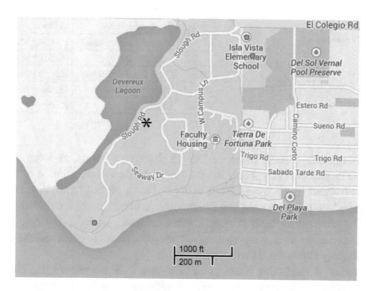

Fig. 11.2 Locality map for bifacial artifact shown in previous image. Collecting site indicated with asterisk (Cartography: Mark McMenamin)

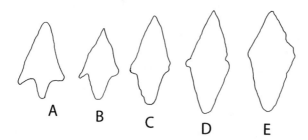

Fig. 11.3 Homologous series of the Vandenberg Contracting Stem projectile point type, showing a diagrammatic profile for each subtype. (**a**) subtype VCS-1; (**b**), subtype VCS-2; (**c**) subtype VCS-3; (**d**) subtype VCS-4; (**e**) subtype VCS-5. The relative sizes of the subtypes are arbitrary, and the diagram is not drawn to scale

Subtypes are referred to here as Vandenberg Contracting Stem types 1-5 (VCS 1-5). Subtypes consist of point morphologies that occur among points of various sizes, thus the subtype distribution seems to be roughly independent of tool size. Stems are usually ground (smoothed) in all five subtypes. Points with unbarbed, reduced or contracting shoulders have larger and wider stems.

VCS-1 is a barbed point with one barb more pointed and elongate than the other (second image from right in Fig. 1.144 of Miles [1963]; Fig. 11.3a). The shoulder is in the lower third of the point. VCS-2 (right image in Fig. 1.144 of Miles [1963]; Fig. 11.3b) is the second point in the series. The barbs are similar to VCS-1 but the shoulder is somewhat higher on the point and the stem is proportionately larger. In VCS-3 (left image in Fig. 1.144 of Miles [1963]; Fig. 11.3c), the shoulders are

Table 11.1 Vandenberg contracting stem point subtypes

	Characteristics	Inferred function
VCS-1	Barbed, low shoulder	Hunting small game
VCS-2	Larger stem	Hunting small game
VCS-3	Contracted shoulders	Hunting small game
VCS-4	Very large stem	Hunting large game
VCS-5	Largest stem	Knife

much contracted in comparison to the previous two points, and the barbs are reduced. The shoulder is a short distance below the transverse midline of the point. In VCS-4 (second image from the left in Fig. 1.144 of Miles [1963]; Fig. 11.3d), the barbs are reduced to mere nubs and the shoulder is at the approximate midline of the point. The contracting stem is considerably enlarged. In VCS-5 (Figs. 11.1 and 11.3e), the contracting stem is now larger than the rest of the point and the shoulder is at or slightly above the transverse midline.

Vandenberg Contracting Stem point subtypes VCS 1-5 are all variations on a common theme. As such they represent a significant technological advance over the earlier Clovis-type lithic technology that was also indigenous to coastal California (Hyland and De La Luz Gutiérrez 1995). A broken Clovis-type point was discovered on Isla Cedros off the coast of Baja California at the Arce-Meza site (Des Lauriers 2008).

Each VCS subtype appears to have been fashioned for a particular suite of functions. VCS-1 occupies a classic arrowhead function, with a large barb that might be particularly effective for bow fishing or otherwise spearing fish. VCS-2, particularly in its smaller variants, is well suited to serve as a bird point. VCS-3, with its contracted shoulders, is more massive and might be better suited for bringing down small mammals. Charles Miles (1963) considered the lozenge shapes of VCS 3-5 to be most useful as spear points or as knife blades. A knife blade function for VCS-5 seems likely. Indeed, the large ground stem in VCS-5 would provide a secure handle for the knife blade. Hypothesized subtype functions are summarized in Table 11.1.

Thus a single, generalized mode of manufacture was adjusted to produce stone tools that partitioned the multifaceted hunting and field dressing needs of the ancient peoples of coastal Southern California who were exploiting a stable and reliable resource (cf. Jones et al. 2016). Each tool is a variation on a common theme and yet nicely adapted to a particular inferred function. For example, perhaps the best interpretation of the San Miguel Island stone crescent is as a transverse bifacial projectile point for bringing down shorebirds (Erlandson and Braje 2008).

The Connecticut Valley region of Connecticut and Massachusetts has produced evidence for human occupation for over 12,000 years (Hoffman 1992). Fluted points are known from Chicopee Falls, Hamden, Greenfield, Montague (Fowler 1973), Hadley, Deerfield (the Sugarloaf or Dedic site; Gramly 1994; Fogelman 1998), and Agawam, in Massachusetts, as well as Glastonbury and Washington, in Connecticut (Moeller 1980). As is often the case with Paleoindian projectile points, many of the points are made of materials exotic to the region (Wray 1948). For example, a fluted point found at Westover Air Force Base, Chicopee, Massachusetts in 1939 was knapped from New York chert. This biface was on display as of 2009 at the Springfield (Massachusetts) Museum of Science. New Hampshire has produced fluted points as well (Boisvert 1999), including fluted Paleoindian projectile points associated with a rhyolite mineshaft at Mt. Jasper, New Hampshire (Thompson et al. 2009).

This question of initial colonization draws interest; for example, the above-the-fold story on the front page of the *Daily Hampshire Gazette* is entitled "Stone Age finds: Discovery of 13,000-year-old arrowhead in Meadows spurs dig" (LaFerney 2017). The report discusses ongoing excavations associated with the surface find of a Clovis point discovered in the Northampton Meadows area by Jason Lovett. The biface is associated with quartz debitage and Hudson River Valley flint, indicative that the area was used as a hunting area over an extended period. Other Paleoindian sites nearby include an excavation in 2013 at the Sugarloaf site between the Deerfield, Massachusetts Industrial Park and the Whatley town line (Curtis 2014).

Chet and Butch Yazwinski are asking the University of Massachusetts Archaeological Field School to return the artifacts taken from their land in the Pine Hill area of Old Deerfield, Massachusetts (McKiernan 2014). As Kathleen McKiernan (2014) explains: "Much of the conflict…is a result of a contradiction in state laws regarding archaeology. The state has a different set of regulations for archaeology digs on private and public lands. Under state General Law Chapter 9, Section 27C, a state body—such as the UMass field school—is required to get a state permit from the Massachusetts Historical Commission to work on public land. Artifacts found on public land by a state body with a state permit belong to the state, according to the law. The state law, however, does not address artifacts found on private property. Federal law, on the other hand, states if the land is privately owned, the artifacts belong to the private landowner." Clearly, this situation has implications for the access to important archaeological discoveries. Private land owners should be encouraged to ensure that their important artifacts eventually end up properly catalogued in institutions where they can be curated and studied by all interested parties.

The recent discovery of a fluted Clovis point (approximately 9.5 cm long) in the Meadows of Northampton represents an exciting addition to knowledge of Northampton's prehistory (LaFerney 2017). Fluting refers to the channel flake and corresponding flute removed near the base of the projectile point (Fogelman and Lantz 2006). Removal of this flake by the point manufacturer forms a channel or flute parallel to the long axis of the projectile point. This groove allowed for easier and more secure hafting onto a shaft, and is diagnostic of the tool kit of early New Englanders. But were these fluted points the tool kit of the *earliest* New Englanders?

Study of the earliest Americans will surely involve analysis of stone tools as variations on a common theme, and also the reuse and recycling of tools by subsequent cultures. A residential relandscaping project, triggered by a municipal road widening project in South Hadley, Massachusetts produced a specimen of a small polycrystalline quartz bifacial lanceolate projectile point (Figs. 11.4, 11.5 and 11.6). The point is provisionally assigned here to a pre-Clovis Paleoindian Phase

Fig. 11.4 Bifacial projectile point from South Hadley, Massachusetts, resharpened (vitreous surface) by a subsequent (much later) edge bevel. This side of the biface shows a shallow flute as well as the resharpening bevel on the upper left edge. Gray-translucent faintly banded polycrystalline quartz, sample 1 of 7/30/2009. Projectile point is 2.3 cm in length (Photograph by Mark McMenamin)

Fig. 11.5 Bifacial projectile point from South Hadley, Massachusetts, sample 1 of 7/30/2009. Note the projecting platform or fluting nipple at the base. The tan patch near the tip is a relict sample of a light colored loess soil horizon (the rest of which was evidently lost to erosion) in which the biface was originally buried during the late Pleistocene. Projectile point is 2.3 cm in length (Photograph by Mark McMenamin)

(18,000 years ago), a time interval that spans the end of glacial Lake Hitchcock times in the Connecticut Valley. The projectile point was given a percussion blow edge bevel retouch for reuse by a later cultural phase. The artifact shows a very shallow flute on one side of the point, and retains a projecting platform or flute-precursor nipple on the other side. This point represents the earliest evidence for Paleoindian occupation of South Hadley, Massachusetts, and indicates a high level of knapping skill with a challenging lithic type (polycrystalline vein quartz).

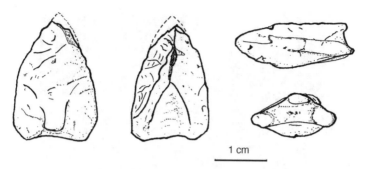

Fig. 11.6 Bifacial projectile point from South Hadley, Massachusetts, line art sketches from various perspectives (both faces, one side, and view looking at the base of the point). The top left of the center image shows the scar left by the re-sharpening bevel. Scale bar 1 cm

The residential landscaping project in South Hadley, Massachusetts (N42°15.7′ W72°33.8′) uncovered in 2009 at a depth of 79–85 cm on private property, a small Paleoindian projectile point fashioned from locally-derived gray-translucent faintly banded polycrystalline quartz (sample 1 of 7/30/09; Figs. 11.4, 11.5 and 11.6). The projectile point is 2.3 cm in length, 17.4 mm wide and 1 cm thick. The entire surface of the point appears to have been smoothed or ground, and appears to have been subsequently slightly etched by aqueous dissolution of the quartz. The point has been shortened and resharpened (with a roughly planar edge bevel) by an indigenous representative of a culture subsequent to the Paleoindian phase.

Polycrystalline quartz with large crystals as seen in this example is a difficult lithic type to fashion by knapping, because of the tendency of the preform to fracture along irregular quartz grain boundaries (Fig. 11.7). The original point in this case was thus fashioned with considerable skill, what Ellis (2008) calls "excellence in manufacture." The tip of the original point, plus several millimeters of the original point length, were removed by a much later resharpening that used one main percussion blow to resharpen the tip and remove a tapered prism-shaped wedge of quartz, creating a new edge by means of a roughly planar edge bevel.

The original point was approximately 2.5 cm in total length. The point has a superficial resemblance to the Squibnocket Triangle, a point type characteristic for New England that ranges in age from 6000 to 1000 BP with a most typical age of 4000 BP (Boudreau 2008). The new point nevertheless differs from the Squibnocket Triangle type by lacking a basal corner that bends inward (Overstreet 2007), and by the presence of a fluting nipple projecting platform. The point shows truncated auricles of essentially the same shape on either side. It thus represents an unusual typology otherwise unknown from the New England region.

The point is fluted on one side, and as just noted, retains a projecting platform or flute-precursor nipple on the other side. The nipple is quite prominent and imparts a roughly triangular shape to the base of the point as viewed from its end opposite the tip (Fig. 11.6). Its relative age is very roughly indicated by the fact that the point was resharpened and reused at least once by members of a subsequent (Paleoindian to Woodland) cultural phase. The more freshly knapped region seems to represent

Fig. 11.7 Polycrystalline quartz (vein quartz) as seen in petrographic thin section and viewed in cross-polarized light, from a glacial drift cobble from the same area in South Hadley, Massachusetts, that produced the projectile point described here. Note the interlocking crystal grain boundaries, rendering this a particularly difficult rock type for knapping into projectile points or other tools. Sample 3 of 12/15/09. Width of image approximately 1 mm

sharpening by a member of a profoundly different lithic culture, one that specialized in crude fracturing of quartz, and thus provides a confident relative indicator for the two lithic traditions.

Arguments against a Paleoindian attribution to this point include the relative thickness of the point (1 cm), which would seem to contrast with the typical razor-thin profile of many Paleoindian points. The coarse polycrystalline quartz that constitutes this piece, however, requires a correspondingly greater thickness for strength and structural integrity. This point nevertheless required more fashioning skill than would be required for a texturally homogenous rock type such as chert. The early American who fabricated the point was well aware of the constraints of working with polycrystalline quartz, and the resulting point was in fact sturdy enough to remain intact until sharpened and reused by a subsequent cultural phase.

Richard Boisvert (personal communication 2009; Bradley et al. 2008) notes that projecting platform or flute-precursor nipples of the type somewhat similar to that seen on the South Hadley point are virtually "restricted to the middle part of [Paleoindian phase] (in the Northeast)." If the date assigned here (18,000 years old) is correct, this point would represent an extension of the range of projecting platform morphology back to pre-Paleoindian (pre-Clovis equivalent) times. This age is estimated by comparison with the McCary Blade Point from Virginia and a biface from the Delmarva Peninsula in Maryland.

The South Hadley point has a point morphology similar, in terms of the medial ridge and short flute, to that of the McCary Blade Point of the Atlantic Coastal Plain (southern Virginia), a type with a ground base that has been proposed as a pre-Clovis point type, a likely precursor to the development of Clovis technology. The

Table 11.2 Earliest Eastern North American projectile point typometric comparison

	South Hadley biface	Miles Point biface
Point type	Bifacial lanceolate projectile point	Bifacial lanceolate projectile point
Estimated age	'Pre-Clovis' phase, *c.* 18,000 BP	'Pre-Clovis' phase, *c.* 18,000–25,000 BP
Sample number	1 of 7/30/09	18TA365
Recovery depth	79–85 cm	98 cm
Morphology	Short lanceolate	Short lanceolate
Length	2.3 cm	3.5 cm
Width	1.74 cm	2.4 cm
Thickness	1 cm	0.75 cm
Point ratio (width/length)	0.76	0.69
Base form	Concave	Straight to slightly concave
Tip angle	79° (as resharpened)	105°
Blade curvature (cord height of curve from shoulder to tip, unretouched side)	3 mm	3.3 mm
Angle between side and base (unretouched side)	90°	89°

McCary Blade Point has been given an estimated age of 17,000 BP (Hranicky 2007, p. 43). The fact that the fluting nipple is retained on the South Hadley point suggests that it (aside from the much later resharpening) represents the original point rather than a point worked down by expention, that is, in the sense of an expended larger bifacial point (Hranicky 2007).

The biface most similar to the South Hadley point is a very ancient point described by Lowery et al. (2010) from the western Delmarva Peninsula, eastern Maryland, USA at Miles Point. At approximately 3.5 cm length, the Miles Point biface is somewhat larger than the South Hadley point, but is still relatively small in comparison to Clovis points that occur stratigraphically upsection. This point is from the 2Btxb horizon of the Miles Point Loess, and occurs directly beneath the 2ABtxb soil horizon (Tilghman Soil). The Miles Point biface is dated by Lowery et al. (2010; their Figs. 5 and 6) to 18,000–25,000 years ago. This point has a relatively straight base, a short auricle on the left basal corner that resembles the South Hadley point, gently curved sides that come to a point that resembles that of the South Hadley point (especially before it was resharpened by a subsequent lithic culture), and a gently cuspate knapped edge that strongly resembles the edge of the South Hadley point (Figs. 11.4, 11.5 and 11.6).

A typometric comparison of the South Hadley and Miles Point lanceolate bifaces is shown in Table 11.2. I consider these two points to represent a pre-Clovis typology dating to 18,000–25,000 years ago, a lithic technology that is represented in both Maryland and Massachusetts. The South Hadley point may be dated to precisely 18,000 years, because that is the earliest that the glacier ice would have been melted away from this part of southern New England, thus encouraging human

Fig. 11.8 A relict patch of South Hadley Loess adhering (probably due to silicification) to the South Hadley biface. This patch of light soil adhered to the biface before the resharpening event. Width of view 16 mm (Photograph by Mark McMenamin)

occupation. The great age for the Miles Point biface is confirmed by the fact that it was recovered stratigraphically downsection from a Clovis point (dated to c. 13,000 years ago) at the Btxg horizon of the Paw Paw Loess just above an unconformity and stratigraphically separated by a paleosol (Tilghman Soil) from the Miles Point biface (Lowery et al. 2010). This relationship of stratigraphic superposition strongly supports the argument made here that the South Hadley point is of comparable antiquity. Further support comes from the observation that both bifaces are associated with loess.

The resharpening facet is brightly vitreous, but the remaining surfaces of the South Hadley point have a dull luster. Etched/ravined surfaces indicate, on microtribological grounds, fairly extensive etching of the South Hadley biface due to extended burial time in an eluvial soil. Also, the surfaces may have been smoothed or ground before natural etching of the quartz.

Two types of soil matrix adhere to the South Hadley biface. The first is a dark gray modern soil (Ninigret fine sandy loam; 10YR 6/1) common to the recovery site. The second, occurring in sheltered pockets, is silica-indurated and older—a light-colored (10YR 8/2-3) indurated loess. This sequal soil was evidently lost to deflation, and may have been completely lost to erosion apart from a light-colored silicified remnant that adheres to the biface (Figs. 11.8, 11.9 and 11.10). The South Hadley loess, comparable to the Miles Point Loess of the Delmarva Peninsula, Maryland, provides evidence that the South Hadley point is associated with the immediate aftermath of deglaciation in the southern Connecticut Valley region. As such it represents the oldest evidence for human occupation of New England.

The middle-late Paleoindian Phase coincides with a transition from grassland/tundra to a terrain of trees, shrubs and other taller vegetation. Pleistocene megafauna was still extant at this time, but the large animals, presumably including mammoths and possibly caballoid horses, became extinct after another thousand years. Large fluted points are known from this interval (Boudreau 2008). Before this interval, the evidence provided by the smaller Miles Point and South Hadley lanceolate

Fig. 11.9 Enlargement of the patch of light soil (South Hadley Loess) shown in the previous image. The dark oval item at the upper right of the light soil patch is unidentified fossil material, possibly a plant fragment or tiny fish scale. Width of view 7 mm (Photograph by Mark McMenamin)

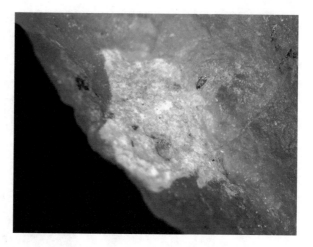

Fig. 11.10 SEM photomicrograph of enigmatic oval microfossil that adheres to the South Hadley biface. It may possibly represent a fragment of plant material (seed case) or broken edge of a fish scale. The fossil is nestled into the silicified patch of South Hadley Loess and is evidently the same age as the light soil. Scale bar 100 microns

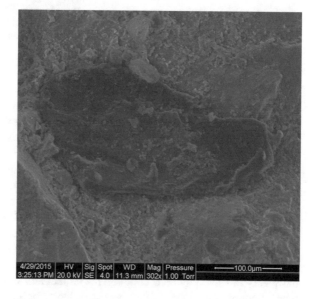

bifaces indicate that smaller artiodactyl ungulate game animals (white tail deer, caribou, elk, and moose) and even smaller animals were of primary importance for earliest human inhabitants of eastern North America. This realization led to David Starbuck's (2006) remark that "the First Americans were much more likely to eat small game than mammoths," a reflection of Roger W. Moeller's (1980; see also Meltzer, 1988) comment that man "cannot live by mastodon alone, nor would he have wanted to depend upon any single food source."

Fig. 11.11 Cumberland-
Barnes projectile point
from Norwich and Preston
region, southeastern
Connecticut, USA
(McMenamin 2011).
Surface find. Sample
MM-10-1. Scale bar in cm
(Photograph by Mark
McMenamin)

Relatively small 'pre-Clovis' points of distinctive morphology may therefore be characteristic for immediately post-glacial southern New England at 18,000 years ago. This represents an important datum for our species, as it represents one of or perhaps the very last major continental region (New England) to be colonized by *Homo sapiens*. Thus it took our species (setting our origination date at approximately 180,000 years ago, presumably in Africa) a mere 162,000 years to reach the 'ends of the earth'. Researchers who advocate for a much more ancient human occupation of the Americas, say in the 100,000–200,000 year range (which might imply that America was initially colonized by a species of *Homo* that was not *H. sapiens*), have yet to present convincing archaeological and paleontological data to support their case. Observe that this skepticism is voiced by someone who was initially very intrigued by their claims. This lamentable situation could change at any time, of course. For example, the earliest evidence for human occupation in Australia has just been pushed back to 65,000 years ago by excavations carried out at a rock shelter site (Madjedbebe) in northern Australia (Clarkson et al. 2017)

As a final note, I can say that the New Englander reputation for frugality, recycling and reuse has a long history, considering that the most ancient projectile points made of local materials, the pre-Clovis South Hadley biface and the small Cumberland-Barnes biface (McMenamin 2011) projectile point from Connecticut (Figs. 11.11, 11.12 and 11.13), both show evidence of resharpening and reuse.

Fig. 11.12 Cumberland-Barnes projectile point from Norwich and Preston region, southeastern Connecticut, USA (McMenamin 2011). This face of the projectile point has a shallow flute, anterior to the basal thinning scar, that is formed by a natural curvature in the rock cleavage. Scale bar in cm (Photograph by Mark McMenamin)

Fig. 11.13 Cumberland-Barnes projectile point from Norwich and Preston region, southeastern Connecticut, line art sketch of both faces. Note assymetrical base. Scale bar 1 cm

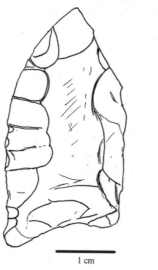
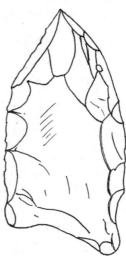

1 cm

References

Adovasio JM, Page J (2002) The first Americans: in pursuit of archaeology's greatest mystery. Random House, New York

Antevs E (1959) Geological age of the Lehner mammoth site. Am Antiq 25(1):31–42

Boisvert RA (1999) Paleoindian occupation of the White Mountains, Hampshire. Géog Phys Quatern 53:159–174

Boudreau J (2008) A New England typology of native American projectile points. Freedom Digital, Ashland

Bradley JW et al (2008) What's the point? Modal forms and attributes of Paleoindian bifaces in the New England-Maritimes region. Archaeol East N Am 36:119–172

Budinger FE (1983) The Calico early man site. Calif Geol 36(4):75–82

Carter GF (1957) Pleistocene man at San Diego. Johns Hopkins University Press, Baltimore

Carter GF (1980) Earlier than you think—a personal view of man in America. Texas A & M University Press, College Station

Chatters JC et al (2014) Late Pleistocene human skeleton and mtDNA link Paleoamericans and modern native Americans. Science 344:750–754

Clarkson C et al (2017) Human occupation of northern Australia by 65,000 years ago. Nature 547:306–310

Cremin A (2007) The world encyclopedia of archaeology. Firefly Books, Buffalo

Curtis C (2014) Deerfield showcases artifacts found as part of archaeological dig at site in Sugarloaf. Daily Hampshire Gazette (Northampton, Massachusetts) 229(195):B3

de Azvedo S et al (2015a) Ancient remains and the first peopling of the Americas: reassessing the Hoyo Negro skull. Am J Phys Anthropol 148:514–521

de Azvedo S et al (2015b) The first human settlement of the New World: a closer look at craniofacial variation and evolution of early and late Holocene Native American groups. Quat Int 431:152–167

Des Lauriers MR (2008) A paleoindian fluted point from Isla Cedros, Baja California. J Isl Coast Archaeol 3:271–227

Ellis C (2008) The fluted point tradition and the Arctic small tool tradition: What's the connection? J Anthropol Archaeol 27(3):298–314

Erlandson JM (2008) Tuqan chert: a "mainland" Monterey Chert source on San Miguel Island, California. Pac Coast Archaeol Soc Q 40(1):23–34

Erlandson JM, Braje TJ (2008) A chipped stone crescent from CA-SMI-681, San Miguel Island, California. J Calif Gt Basin Anthropol 28(2):184–188

Evans P (1971) Towards a Pleistocene time-scale. Part 2 of the Phanerozoic time-scale—a supplement. Spec Publ Geol Soc Lond 5:123–356

Fogelman GL (1998) Sugarloaf site, mass. Indian Artifact Mag 17(4):65

Fogelman GL, Lantz SW (2006) Pennsylvania fluted point survey. Fogelman Publishing, Turbotville

Fowler WS (1973) Bull Brook: a paleo complex site—Massachusetts. Archaeol Soc Bull 34(1&2):1–6

Gramly RM (1994) A member's suggestion: pre-printed site recording forms and an ASAA archive: Dedic Site—Deerfield, MA. Amateur Archaeol 1:78–81

Grayson DK, Meltzer DJ (2003) A requiem for North American overkill. J Archaeol Sci 30:585–593

Gregor AJ (2017) Why the West abandoned standard rules of conduct. New Oxf Rev 84:24–27

Halligan JJ et al (2016) Pre-Clovis occupation 14,550 years ago at the Page-Ladson site, Florida, and the peopling of the Americas. Sci Adv 2(5):e1600375. https://doi.org/10.1126/sciadv.1600375

Haury EW et al (1959) The Lehner mammoth site, southeastern Arizona. Am Antiq 25(1):1–30

Haynes CV (1973) The Calico site: artifacts or geofacts? Science 181:305–310

Hoffman C (1992) A handbook of Indian artifacts from southern New England. Massachusetts Archaeological Society Special Publication No. 4, Middleborough

Holen SR et al (2017) A 130,000-year-old archaeological site in southern California, USA. Nature 544(479). https://doi.org/10.1038/nature22065

Hranicky WJ (2007) Prehistoric projectile points found along the Atlantic coastal plain, 2nd edn. Universal Publishers, Boca Raton

Hudson T, Blackburn TC (1982) The material culture of the Chumash interaction sphere. Volume I. Food procurement and transportation. Ballena Press, Los Altos

Hyland JR, De La Luz Gutiérrez M (1995) An obsidian fluted point from central Baja California. J Calif Gt Basin Anthropol 17(1):126–128

Johnson KR et al (2014) The snowmastodon project. Quat Res 82(3):473–476

Jones TL et al (2016) The archaeology of fish and fishing on the central coast of California: the case for an under-exploited resource. J Anthropol Archaeol 41:88–109

LaFerney D (2017) Stone Age finds: discovery of 13,000-year-old arrowhead in Meadows spurs dig. Daily Hampshire Gazette (Northampton, Massachusetts) 231(225):A1–A7. http://www.gazettenet.com/An-archaeological-dig-in-the-meadows-in-Northampton-9887462

Lambeck K et al (2014) Sea level and global ice volumes from the last glacial maximum to the Holocene. Proc Natl Acad Sci (USA) 111(43):15296–15303

Leakey M (1984) Disclosing the past—an autobiography. Doubleday, New York

Leakey LSB et al (1968) Archaeological excavations in the Calico Mountains, California. Preliminary report. Science 160:1022–1023

Lowery DL et al (2010) Late Pleistocene upland stratigraphy of the western Delmarva Peninsula, USA. Quat Sci Rev. https://doi.org/10.1016/j.quascirev.2010.03.007

Martin PS (1973) The discovery of America. Science 179:969–974

McKiernan K (2014) Deerfield farmer wants Indian artifacts back. Daily Hampshire Gazette (Northampton, Massachusetts) 228(255):A1–B3

McMenamin MAS (2011) A recycled small Cumberland-Barnes biface projectile point from southeastern Connecticut. Bull Mass Archaeol Soc 72(2):70–73

Meltzer DJ (1988) Late Pleistocene human adaptations in eastern North America. J World Prehist 2(1):1–52

Miles C (1963) Indian and Eskimo artifacts of North America. Bonanza Books, New York

Moeller RW (1980) 6LF21: a paleo-indian site in western Connecticut. American Indian Archaeological Institute, Washington

Nogués-Bravo D et al (2008) Climate change, humans, and the extinction of the woolly mammoth. PLoS Biol 6(4):e79. https://doi.org/10.1371/journal.pbio.0060079

Overstreet RM (2007) Official Overstreet identification and price guide to Indian arrowheads. Gemstone Publishing, New York

Rasmussen M et al (2014) The genome of a Late Pleistocene human from a Clovis burial site in western Montana. Nature 506:225–229

Starbuck DR (2006) Archeology of New Hampshire: exploring 10,000 years in the granite state. University of New Hampshire Press, Lebanon

Stinnesbeck W et al (2017) The earliest settlers of Mesoamerica date back to the late Pleistocene. PLoS One 12(8):e0183345. https://doi.org/10.1371/journal.pone.0183345

Thompson WB et al (2009) Glacial geology, climate history, and late-glacial archaeology of the northern White Mountains, New Hampshire (Part 2). In: Westerman DS, Lathrop AS (eds) Guidebook for field trips in the northeast kingdom of Vermont and adjacent regions, New England Intercollegiate Geological Conference, 101st Annual Meeting. Lyndon State College, Lyndonville, Vermont, pp 225–242

Wray CF (1948) Varieties and sources of flint found in New York State. Pa Archaeol 18(1&2):25–45

Chapter 12
Feldspar Point

Abstract A small stemmed projectile point from western Massachusetts possibly dating to the Late Woodland Phase (1100–300 years ago) but of unknown age (but in any case less than about 18,000 years old), was fabricated from a locally available crystal of orthoclase feldspar derived from granite or granodiorite. Fabrication of the point demonstrates a fairly sophisticated appreciation of the sculpting potential inherent in a potassium feldspar megacryst. Use of intact single crystals found in the environment by organisms has a long history on earth, dating back at least to the Cambrian Explosion. Between Phase I of crystal utilization (Proterozoic-Early Cambrian) and Phase II of crystal utilization (*Homo sapiens* crystal usage) is what is called here the Crystal Gap, a long stretch of geological time during which there is no evidence for use of discreet types of crystals by animals that have intentionally selected minerals of particular compositions from their environment.

Keywords Feldspar · Projectile point · Connecticut Valley · Feldspar cleavage · Mineral evolution · Evolution of mineral use

In my Paleontology and Stratigraphy course at Mount Holyoke College, near the end of the clastic sediments part of the course I present the students with an unknown object (Fig. 12.1). The clast is a smooth orange rock sample that fits nicely in one's pocket like small hard objects that are sometimes marketed as 'worry stones'. The question to the students reads as follows: "Mystery clast—An exercise in geo-sleuthing! Examine this clast and answer the questions below. A sketch or two would be an asset to your responses." I then ask the students to describe the clast's sphericity, angularity and smoothness. I ask if it is siliciclastic and inquire if they can say anything further about its composition. I ask them to cite any unusual features of the clast. And finally, I ask them to describe the rock type or rock formation from which they believe this clast was most likely derived.

The clast is locally derived, and it represents a single megacryst of locally-derived potassium feldspar (orthoclase). The flat surfaces on either side of the piece represent feldspar cleavage planes, and preferential fractures along these planes account for the flattened shape (with rounded edges) of this piece. It was probably part of the Mesozoic Sugarloaf Arkose (Walsh 2008), a distinctive local sedimentary rock that is characterized by abundant orange-colored potassium feldspar

M.A.S. McMenamin, *Deep Time Analysis*, Springer Geology, https://doi.org/10.1007/978-3-319-74256-4_12

Fig. 12.1 Orthoclase clast consisting of a single crystal of pink-orange potassium feldspar, derived from crystalline basement rocks that form the boundaries of the Connecticut Valley in western Massachusetts, USA. This rounded, flattened feldspar crystal is probably a clast derived from the Mesozoic Sugarloaf Arkose that was eroded out of the arkose to resume transport as an isolated clast. Scale in cm (Photograph by Mark McMenamin)

clasts, before being eroded from that conglomeratic rock and transported as a stream cobble (Fig. 12.1).

In this chapter I will describe an intriguing artifact that was discovered in South Hadley, Massachusetts. It is composed entirely of a single feldspar crystal that has been shaped into a small biface for hafting to a narrow shaft. It is an important artifact because it represents the directed and intentional acquisition and utilization of a particular type of crystal by a member of our species, an action that is reminiscent of the acquisition of tourmaline trigonal crystals by the crystal creatures of Proterozoic Sonora discussed above.

A residential landscaping project in South Hadley, Massachusetts revealed an unusual small projectile point fashioned from a single crystal of pink feldspar (orthoclase or potassium feldspar; sample 4 of 6/23/09). The projectile point is 1.335 cm in length, 6.32 mm wide and 4 mm thick (Figs. 12.2, 12.3, 12.4, 12.5, 12.6, 12.7 and 12.8). The manufacturer of the point cleverly exploited the pinacoid crystal form and cleavage of the feldspar crystal to form the projectile point (*cf.* Flenniken and Raymond 1986). Note how feldspar cleavage has been used to control the knapping fracture, forming a rectangular notch that sharpens the point (Figs. 12.4, 12.7 and 12.8).

One side of the projectile point is a translucent light pink color, whereas the other side is an opaque pink. This suggests that the crystal used for the point may have been derived (possibly by glacial transport) from the Early Devonian Kinsman granodiorite of southern New Hampshire or related pluton. The Kinsman granodiorite is known for its white to light pink potassium feldspar megacrysts that reach 15 cm in length (Dorais 2003).

Fig. 12.2 Feldspar
projectile point, top
surface. Note sharpening
notch at tip. Field sample 4
of 6/23/09. Length of point
15 mm (Photograph by
Mark McMenamin)

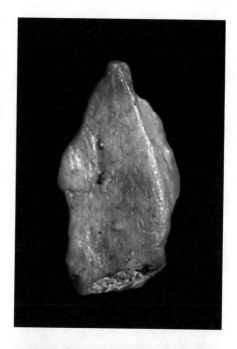

Fig. 12.3 Feldspar
projectile point, underside
surface. Length of
projectile point 15 mm
(Photograph by Mark
McMenamin)

Fig. 12.4 Feldspar
projectile point, oblique
view of left side of point.
Note how feldspar
cleavage has been used to
control the knapping
fracture, forming a
rectangular notch that
sharpens the point. Length
of projectile point 15 mm
(Photograph by Mark
McMenamin)

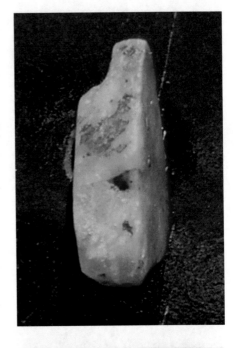

Fig. 12.5 Feldspar
projectile point, top surface
view. Scale in cm
(Photograph by Mark
McMenamin)

The pinacoid crystal form of the projectile point (see Table 12.1) has two flat sides determined by basal cleavage along the [001] or c face. The "bottom" or "underside" side as defined (Figs. 12.3 and 12.6) here shows evidence of sharpening at the tip of the projectile point by a groove that removed a roughly prismatic chunk of crystal at the tip, reducing the thickness of the point to 2.5 mm in this region. The vertical part of the sharpening scar (that is, perpendicular to the underside surface)

Fig. 12.6 Feldspar
projectile point, underside
view. Scale in cm
(Photograph by Mark
McMenamin)

Fig. 12.7 Feldspar
projectile point, left side.
Note how feldspar
cleavage has been used to
control the knapping
fracture, forming a
rectangular notch that
sharpens the point. Scale in
cm (Photograph by Mark
McMenamin)

represents cleavage along the [010] or b face, whereas the horizontal part of the sharpening scar represents another cleavage along the [001] or c face.

The [010] or b cleavage is also expressed on the shoulder of the blade. The edges of the blade, although roughly planar and at approximate 60–67° to the face of the point, do not represent true cleavage planes but rather broken or carved surfaces. These inclined blade edge surfaces nevertheless exploit the imperfect cleavage of the [110] or m face, steepened by grinding to give a tip angle of 41°.

The blade edges cannot represent two cleavage plane surfaces. In orthoclase feldspar, the (001) and (010) cleavages intersect at 90°, as do the (001) and (110) cleavage planes. The (110) and (010) cleavages meet at an angle of approximately 57°. By trigonometry, none of the cleavage planes can intersect any of the two others 43°, the tip angle of the projectile point.

Fig. 12.8 Feldspar
projectile point, right side.
Note how feldspar
cleavage has been used to
control the knapping
fracture, forming a
rectangular notch that
sharpens the point. Scale in
cm (Photograph by Mark
McMenamin)

Table 12.1 South Hadley
feldspar projectile point
typometric comparison

Point type: feldspar crystal point
Associated date: possibly 1100–300 years, may be as old as 18,000 years
Sample number: 4 of 6/23/09
Recovery depth: 52–57 cm
Morphology: Small stemmed
Length: 13.3 mm
Width: 6.3 mm
Point ratio (width/length): 0.47
Length of stem: 5.0 mm
Width of stem: 5.5 mm
Stem length/point length ratio: 0.38
Stem width/point width ratio: 0.87
Stem ratio (stem length/stem width): 0.91, proportionate type point
Stem form: parallel
Base form (subjective): slanted
Tip angle: 41°
Blade ratio ([length point- length stem]/ width): 1.32, isosceles type point
Blade curvature (cord height of curve from shoulder to tip): straight

This Late Woodland Phase feldspar crystal projectile point shows creative local use of an abundant local lithic resource, in an area where easily knapped rocks such as chert are rare or absent (any chert would have to be a raw material imported overland from a source to the west; Wray 1948; Hoffman 1992). Fabrication of the point demonstrates a fairly sophisticated utilization of the sculpting potential inherent in

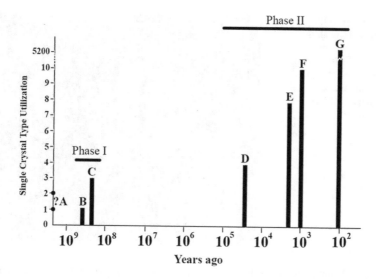

Fig. 12.9 Histogram of occurrences of the use by organisms (or 'proto-organisms') of found crystals of single mineralogy dating back to the origin of life. *Phase I* represents single crystal type acquisition and utilization by animals in the vicinity of the Proterozoic-Cambrian boundary; *Phase II* represents single crystal type acquisition and utilization by *Homo sapiens*. Letters denote the following: **A**, putative utilization of quartz or clay minerals as template during biopoesis (n = 1–2); **B**, black tourmaline use by the Proterozoic crystal creature (n = 1); **C**, Early Cambrian crystal use (n = 3; *Onuphionella* [muscovite mica]; *Campitius* [ilmenite]; *Salterella* [anatase]); **D**, hematite, pyrolusite, kaolin and muscovite mica sought by early man as pigments (red, black, white and white, respectively; n = 4); **E**, add use of native gold, native copper, gypsum (alabaster), and steatite (talc) by ancient Egyptians (n = 8); **F**, add feldspar (as a unique projectile point type [date a conservative estimate here]) and calcite (as a sunstone; n = 10); and **G**, modern technological recognition of 5200+ mineral types. Halite (rock salt) and galena could be added to Phase II but their time of first use by humans is uncertain

a potassium feldspar megacryst. As such, this projectile point represents the earliest known "mineralogical study" conducted on American soil.

The use of intact single crystals found in the ambient environment by organisms has a long history on earth, dating back at least to the Cambrian Explosion. Figure 12.9 shows a plot of the use of single found crystals by organisms dating back to the origin of life. Each of the data points displayed on this plot are worthy of individual discussion. The earliest foraminifera (Cambrian) formed agglutinated tests, and later agglutinizing forams may show preferences for particular sizes and colors of grains, but these have not been included in Fig. 12.9.

The first point, provided here as a starting point to address question, is the use of the quartz crystals and/or clay crystals as templates during the processes that led to the origin of life, in other words, led to biopoesis. This was the primary thesis of Cairns-Smith's (1982) book on genetic takeover and the mineral origins of life (see also Cairns-Smith and Hartman 1986). This is not an implausible hypothesis, as experiments by Pawlikowski (2012) have shown that "vibrating quartz may alter the internal structure of selected amino acids," as indicated by changes in infrared peak positions.

Fig. 12.10 Acicular
tourmaline developing
crystal fans. Mount
Holyoke College Geology
Collection Sample No.
4332. Scale in cm
(Photograph by Mark
McMenamin)

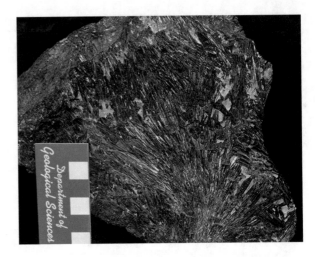

The next use of crystals of a selected, particular type is by the crystal creature described in this book that used black tourmaline crystals to form what was apparently a dorsal covering of small tourmaline crystals derived from crystals found in sands in a Proterozoic shallow marine sandy environment. The crystal creatures, from the Clemente Formation of northwestern Sonora, México, utilized acicular tourmaline crystals that were derived from crystal fans as seen in this specimen (Fig. 12.10) from the Mount Holyoke College mineral collection (locality unknown, Sample No. 4332). Thus the cyclosilicate mineral tourmaline was the earliest known discrete mineral type to have been actively selected, sought out and utilized by an organism of any kind.

The next organisms to select particular types of crystals from the environment were the Cambrian animals that built an agglutinated 'scleritome': *Onuphionella*, *Campitius titanius* and *Salterella*. *Campitius* and *Salterella* used the minerals ilmenite and anatase, respectively, to construct their agglutinated armor. The worm or worm-like creature *Onuphionella* used muscovite mica to construct its tube. All three of these genera are Early Cambrian in age, hence they constitute further testimony to the degree of innovation that occurred in the marine biosphere across the Proterozoic-Cambrian boundary. In contrast to many of the innovations that first appear at the onset of the Cambrian, however, crystal selection has a clear Proterozoic precedent in the crystal creature of the Clemente Formation. Selection of a particular type of crystal from marine sands would seem to require a certain degree of perception, sensory acuity and intelligence on the part of the crystal-collecting animal. If this is indeed the case, then it seems clear that these capabilities were already part of the animalian behavioral tool kit by the late Proterozoic. Gastroliths are not included in this analysis, because they can be composed of any of a number of different rock lithologies, and there is no evidence that crystals of a particular mineralogy were ever preferentially selected to serve as gastroliths.

These Cambrian animals, who turned the Cambrian sea floor into a rudimentary maker space, constitute Phase I of crystal utilization (Fig. 12.9). There is a gap of approximately one half billion years before the onset of Phase II of crystal utilization (Fig. 12.9). This is of course utilization of discrete mineral types, sought out as such, by members of our species *Homo sapiens*, sometimes informally referred to as "Homo faber," Man who Makes. Between Phases I and II is what we might call the Crystal Gap, a long stretch of geological time during which there is no evidence for use of discreet types of crystals by organisms that have intentionally selected said crystals from their environment. A search is underway for examples of crystal utilization during the times of the Crystal Gap. One possible example are foraminifera that use grains of the green authigenic mineral glauconite to form their agglutinated tests. In his book on *Seismosaurus*, Hallet (1994) notes:

> [Some paleontologists] have suggested that sauropods selected only quartz-rich rocks for gastroliths. Again, I disagree. The fact that Sam's gastroliths (the only sauropod gastroliths thoroughly documented) are all quartz does not lead to that conclusion; instead, these may have been the only stones to survive the vigorous action of the digestive tract.

Curiously, according to currently available archaeological data, selection of discrete minerals by members of *Homo sapiens* is a relatively late development. Selection of rock types suitable for knapping has a long history going back over a million years, but selection of particular minerals as raw materials first occurs only fairly recently, perhaps extending back approximately 50,000 years in Europe with selection of hematite (red pigment) pyrolusite (black pigment), kaolin (white pigment) and white mica (another white pigment) to compose the color pallet of ancient cave painters.

Ancient Egyptians made use of gold as far back as 5000 BC. The probable next example, obviously at a less advanced cultural stage but later in time, is represented by the feldspar projectile point described in this chapter. Next is the putative Viking use of optical calcite to make sunstones, used to aid navigation on cloudy days. The search continues for an actual ancient Viking example of a sunstone; however, one was found on a 1592 (Elizabethan) shipwreck discovered in the English Channel. By the 1850s, larger sheets of mica were used as isinglass to form windows on coal and wood-fired stoves. The mica sheets could also be fashioned into translucent curtains, as for example the isinglass curtains mentioned in the song "The Surrey with the Fringe on the Top" from the 1943 musical production *Oklahoma*.

Muscovite or white mica, no doubt due to the high reflectivity of its perfect plane of cleavage, was probably one of the first single mineral type to be collected and utilized by early man. By the present day, according to the International Mineralogical Association, 5200+ minerals have been described by geologists, and many are sought by miners, collectors and mineral enthusiasts for a variety of purposes. Crystal utilization is currently at an all-time high. Interestingly, human activity on the planet has led to the spontaneous synthesis of more than 200 new mineral types as a product of disturbances to the surface of the planet (such as mining) caused by humans (Hazen et al. 2008, 2017).

In defining the ten stages of mineral evolution, Hazen et al. (2008) make an interesting observation on the use of the word 'evolution,' which they say might seem inappropriate to use in the context of mineralogy:

At the most basic level, evolution is simply defined as 'change over time,' and in this respect there can be no question that mineral evolution has occurred. However, we suggest that mineral evolution implies something more, as it arises in part from a sequence of deterministic, irreversible processes that lead from the mineralogical parsimony of the pre-solar era to progressively more diverse and complex phase assemblages. In this sense, mineral evolution is a fascinating specific example of the more general process of cosmic evolution.

Hazen et al. (2008, 2017) define 11 partially overlapping stages of mineral evolution. The successive stages are as follows: (1) primary chondrite minerals; (2) planetesimal alteration/differentiation; (3) igneous rock evolution; (4) granite formation; (5) plate tectonics; (6) anoxic biological world; (7) Paleoproterozoic atmospheric changes; (8) intermediate ocean; (9) Neoproterozoic biogeochemical changes; (10) Phanerozoic Era, and (11) Anthropocene mineralogy. In the system proposed here of crystal utilization, we see a crystal utilization Phase I at the end of Hazen et al.'s (2008) Phase 10 and a crystal utilization Phase II at the beginning of Hazen's et al.'s (2017) Phase 11. Both crystal utilization phases are associated with tremendous leaps in the cognitive abilities of the 'crystal creatures.'

The origin of life may have taken place during Hazen et al.'s (2008) Phase 3. It may very well be the case that this momentous event was associated with discrete mineral selection, in some sense comparable to crystal utilization of Phases I-II. The nature of the minerals involved, and the agency(ies) involved in directing the mineral selections, constitute enormous puzzle pieces in our search for the origin of life.

References

Cairns-Smith AG (1982) Genetic takeover and the mineral origins of life. Cambridge University Press, Cambridge

Cairns-Smith AG, Hartman H (1986) Clay minerals and the origin of life. Cambridge University Press, Cambridge

Dorais MJ (2003) The petrogenesis and emplacement of the New Hampshire plutonic suite. Am J Sci 303:447–487

Flenniken JJ, Raymond AW (1986) Morphological projectile point typology: replication experimentation and technological analysis. Am Antiq 51:603–614

Hallet M (1994) *Seismosaurus*: the earth shaker. Columbia University Press, New York

Hazen RM et al (2008) Mineral evolution. Am Mineral 93(11–12):1693–1720

Hazen RM et al (2017) On the mineralogy of the 'Anthropocene Epoch'. Am Mineral 102(3). https://doi.org/10.2138/am-2017-5875

Hoffman C (1992) A handbook of Indian artifacts from southern New England, Massachusetts Archaeological Society special publication #4. Massachusetts Archaeological Society, Middleborough

Pawlikowski M (2012) Atomic structural templates of the earliest life on earth: vibration and lightning experiments with quartz and amino acids. In: Seckbach J (ed) Genesis—In the beginning. Cellular origin, life in extreme habitats and astrobiology, vol 22. Springer, Dordrecht

Walsh MP (2008) Petrology and provenance of the Triassic Sugarloaf Arkose, Deerfield Basin, Massachusetts. M.S. Thesis, University of Massachusetts at Amherst

Wray CF (1948) Varieties and sources of flint found in New York State. Pa Archaeol 18(1&2):25–45

Systematics

Repository abbreviations are as follows: Institute of Geology Museum (IGM), Departmento de Paleontología, Instituto de Geología, Cuidad Universitaria, Delegacíon de Coyoacán, 04510, México, D. F.; Mount Holyoke College Paleontology Collection (MHC), Department of Geology and Geography, Mount Holyoke College, South Hadley, Massachusetts, USA.

Systematic Paleontology

Zirabagtaria n. gen.

Diagnosis: A small solzid kimberellomorph with an 'oval' body shape (*sensu* Schwabe 2010). The holotype is 11 mm long and 7 mm wide. The anterior 3.5 mm consists of a largely smooth head region. The thorax shows a faint, slightly curved axial lobe and is covered by numerous epidermal papillae or weakly-mineralized sclerites that are in many cases elongated in a direction perpendicular to the main body axis. The dorsal surface of the organism posterior of the head region bears what appears to be a scleritome consisting of tiny rectangular tubercles or weakly mineralized sclerites.

The remainder of the body lacks a dorsal keel but has a very faint, slightly curved low-relief axial lobe and is covered by numerous dorsal epidermal papillae or weakly mineralized sclerites that are in many cases elongated in a direction perpendicular to the main body axis. The papillae resemble tiny rectangular tubercles. Many of the papillae show transverse elongation, and are organized in longitudinal columns, especially in the axial part of the organism. There are approximately ten longitudinal columns of mostly transversely-elongated papillae in the axial lobe region, and roughly 8–10 irregular columns on either side of the axial lobe in the 'pleural' fields. This gives a rough maximum of up to 30 longitudinal rows of sclerites. The 'glabella' region is covered in tiny circular papillae.

© Springer International Publishing AG 2018
M.A.S. McMenamin, *Deep Time Analysis*, Springer Geology,
https://doi.org/10.1007/978-3-319-74256-4

Zirabagtaria ovata n. gen. n. sp. M. A. S. McMenamin
Figures 3.26, 3.27, 3.46 and 3.47
Holotype: IGM 4995.
Diagnosis. As for genus.
Material: One specimen.

Discussion: The dorsal surface of the organism posterior of the head region bears what appears to be a scleritome consisting of tiny rectangular tubercles or weakly mineralized sclerites. The array is somewhat similar to the tubercles described by Ivantsov (2007, his Pl. 2, Figs. 1–3) of *Lossinia lissetskii*, although unlike *Lossinia*, tubercles on the cephalic region of the solzid are sparse and the scleritome is mostly restricted to the post-cranial region. *Zirabagtaria* differs from *Kimberella* by lacking a "scalloped margin to the dorsal covering" (Ivantsov 2017).

The papillae/sclerites were preserved by a thin parting of clay that settled on the dorsal surface of the *Zirabagtaria* before it was buried in sand. The presumably aragonitic sclerites (if they were mineralized at all) were subsequently dissolved out and replaced by the same siliceous cement that lithified the sandstone.

Interestingly, a lopsided cephalic region, reminiscent of and perhaps homologous (McMenamin 2003) to the effaced cephalon of an agraulid trilobite (McMenamin 2004), is apparent on both *Lossinia lissetskii* Ivantsov (2007, his Pl. 2, Fig. 3) and *Zirabagtaria ovata* n. gen. n. sp. (Figs. 3.26 and 3.27). A similar morphology of lopsided cephalic area is seen in the nektaspid trilobite-like Ediacaran *Keretsa brutoni* Ivantsov 2017 from the Arkhangelsk region of the Winter Coast, White Sea, Zimnie Gory, Russia.

The papillate scleritome of *Zirabagtaria ovata* n. gen. n. sp. shows clear evidence for strong influence by a morphogenetic field in accordance with the Second Law of Morphogenetic Evolution. This evidence is twofold. First, the individual papillae are in many cases elongated in a transverse direction, and thus appear to be following the latitudinal field lines of the toroidal metazoan morphogenetic field. Second, the sclerites themselves are largely organized into longitudinal columns, and thus track the longitudinal field lines. The papillae at the posterior tip of *Zirabagtaria ovata* n. gen. n. sp. are of approximately the same size as the rest of the sclerites that constitute the scleritome. This suggests that the *Zirabagtaria* anus is not directly at the posterior tip of the animal, otherwise the posteriormost sclerites would presumably diminish in size as the longitudinal field lines converge at the posterior pole of the torus. This is precisely the position of the cloaca in solenogastres, and quite possibly *Zirabagtaria* had a similar morphology.

Locality: Field sample 6 of 3/16/95; GPS coordinates of site are N30°24.041′, W111°57.141′ (average of seven measurements), altitude 526 m (average of six measurements). The fossil occurrence is approximately 5–10 m below the Clemente oolite, in unit 4 of the Clemente Formation.

Palankiras n. gen.
Diagnosis: A praecambridiid bilateralomorph (approximately 18 mm long and 11 mm wide) with a wide cephalic region, narrow body and 'Y' shaped posterior region. A prominent and well preserved anterior axial lobe ('glabella'; Figs. 3.29

and 3.30) occupies the anterior part of the cephalic region. A long genal spine arches downward almost to the transverse midline of the organism (Fig. 3.29). A linear depression at the proximal-posterior edge of the genal spine is interpreted here as a rudimentary facial suture. The posterior of the trunk of the creature is split into two slightly flared projections. Six faint tubercles occur along the midline of the trunk or thorax region; the anteriormost of these is largest.

Palankiras palmeri n. gen. n. sp. M. A. S. McMenamin
Figures 3.28, 3.29, 3.30 and 3.31
Holotype: IGM 4997.
Diagnosis. As for genus.
Material: One specimen.

Discussion: The specimen is preserved as a convex epirelief. The incomplete nature of the fossil suggests the possibility that it represents a shed molt, which if so would make it the oldest known such structure in the fossil record. No appendages are preserved on the holotype. It is possible, however, that appendages did occur on either side of the narrow middle thoracic region of *Palankiras palmeri* n. gen. n. sp.

An illustration of *Praecambridium* in Glaessner and Wade (1971, their Figure 1C) shows certain similarities to *Palankiras* as shown in Figs. 3.28, 3.29, 3.30 and 3.31, particularly as regards the first 'metamere' or isomere partition that in both *Praecambridium sigillum* and *Palankiras palmeri* n. gen. n. sp. is drawn out into an arching spine or spine-like structure.

Several key features link *Palankiras* to both *Praecambridium* and trilobites. These are: presence of a genal spine, presence of a prominent axial lobe (glabella), and rudimentary facial sutures. The genal spine is clearly visible in Figs. 3.28 and 3.29. A linear depression at the proximal-posterior edge of the genal spine is interpreted here as a rudimentary facial suture. This interpretation strongly supports the concept that *Palankiras* and *Spriggina* were both ecdysozoans that were required to molt. The prominent ovoid-spherical anterior lobe in the new praecambridiid is very reminiscent of the anterior lobe in *Praecambridium sigillum* as reconstructed by Glaessner and Wade (1971, their Fig. 3).

The overall body form of *Palankiras palmeri* n. gen. n. sp. (Fig. 3.31) is intriguing. Several trilobite orders have produced strange looking forms with what has been called a 'classic fish skeleton' appearance: large spherical glabella ("eye of fish"), long curved genal spines ("skull of fish"), separated pleural segments ("fish ribs") and two prominent spines of the pygidium ("fish tail"). These trilobites include the Silurian cheirurid *Deiphon forbesi* (Fig. 3.32) and the Devonian giant lichid trilobite (>60 cm long) *Terataspis grandis*. Complete specimens of *Terataspis grandis* are unknown but the overall body shape of the trilobite may be accurately inferred from isolated pieces. The similar and odd body form of *Deiphon* and *Terataspis*, which by the way are not closely related trilobites, strongly suggests that their joint similarity to *Palankiras palmeri* n. gen. n. sp. is a result of an evolutionary atavism. *Palankiras palmeri* n. gen. n. sp. shares the entire "fish skeleton" outline, lacking only the 'ribs' (separated pleural segments). Curiously, this "fish skeleton" trilobite form is currently unknown from Cambrian strata. The evolutionarily-convergent atavisms thus appear to be post-Cambrian Explosion phenomena.

Locality: Field sample 6 of 3/16/95; GPS coordinates of site are N30°24.041′, W111°57.141′ (average of seven measurements), altitude 526 m (average of six measurements). The fossil occurrence is approximately 5–10 m below the Clemente oolite, in unit 4 of the Clemente Formation.

Vendamonia n. gen.

Diagnosis: A praecambridiid bilateralomorph (23 mm wide and 21 mm long) with three pairs of parapodia-like structures, a small anterior lobe ('glabella'), and very wide, straight posterior margin that does not taper to form a pygidium or other 'tail' structure. The organism consists of a bar-shaped structure (the parapodial component of the great cirri or cirri base; Meyer 1926), a domal anterior axial lobe (buccal cavity), and two metamers (neuropodial or parapodial pairs) posterior to the anterior axial lobe (Fig. 3.34) A bifurcate distal appendage extends from the left parapodium of the first postcephalic pair.

Vendamonia truncata n. gen. n. sp. M. A. S. McMenamin

Figures 3.33, 3.34, 3.35 and 3.36
Holotype: IGM 4996.
Diagnosis. As for genus.
Material: Two specimens.

Discussion: The specimen is preserved as a convex epirelief. In addition to the holotype, a second specimen of *Vendamonia truncata* n. gen. n. sp., an apparent juvenile, was discovered in June 2017 (Figs. 3.35 and 3.36). This fossil is 4.3 mm in width and 5 mm in length. The original specimen creature was originally identified as the anterior portion of a *Tomopteris*-like worm, with the posterior of the animal excised to form a straight posterior margin. However, discovery of the second specimen indicates that the anvil-shaped body and flat posterior margin (bottom of the anvil) in fact represents the entire outline of the body of the creature.

There could nevertheless still be a relationship between *Vendamonia truncata* n. gen. n. sp. and the bioluminescent polychaete worm *Tomopteris helgolandicus*. In living members of the Tomopteridae, the bioluminescence is associated with certain parapodial glands (Phillips Dales 1971), raising the intriguing possibility that if *Vendamonia truncata* n. gen. n. sp. is indeed a tomopteroid, it may provide evidence for Proterozoic animalian bioluminescence.

Living tomopterids are nectonic marine predators. They are usually classified as a unique group of polychaete worms, characterized by an elongate, flattened metameric body, paddle-shaped parapodia, and long anterior cirri. Glaessner's (1958) suggestion of a phylogenetic link between the Ediacaran bilateralomorph *Spriggina* and modern tomopterids was challenged by subsequent researchers (Briggs and Clarkson 1987). *Vendamonia truncata* n. gen. n. sp. is remarkable both for its great antiquity, its size, and for the support that it provides to Glaessner's (1958, 1959) original suggestions concerning the link between *Spriggina* and tomopterids.

Shortly after I located *Vendamonia truncata* n. gen. n. sp. in the field, Dave Evans nicknamed it "the anvil" on account of its being shaped like an old fashioned iron anvil. Note the appearance of a very similar anvil pattern in the anterior portion of *Praecambridium sigillum* as shown in Glaessner and Wade (1971), their Figure 1C. The anteriormost first two "segments" in this image appear larger than the other isomers and separate from the posterior part of the fossil (Glaessner and Wade 1971); this part ends up looking like an anvil that has been so heavily hammered that its horns are bent. I propose that *Vendamonia* had a similar morphology, with a smaller and fragile posterior segment series that separated from the more robust first two in the Sonoran fossil.

Locality: Field sample 4 of 3/16/95; GPS coordinates of site are N30°24.041′, W111°57.141′ (average of seven measurements), altitude 526 m (average of six measurements). The fossil occurrence is approximately 5–10 m below the Clemente oolite, in unit 4 of the Clemente Formation.

Crown group Aculifera
Discussion: Aculifera is now widely accepted as monophyletic (Sigwart 2017).
Total group ?Polyplacophora
Order unknown
Family unknown

Korifogrammia n. gen.
Diagnosis: An aculiferan shaped like a swollen circular disc 9 mm long and 9 mm wide, with a lunate anterior or cephalic field and a dorsal midline keel divided into numerous valves. A dorsal medial keel consists of eight carinated trapezoidal valves (*sensu* Schwabe 2010). Paired lateral riblets are preserved on the right sides of the anteriormost two or three plates. Faint possible additional valves (bringing the total number to ten) may occur anterior and posterior to the eight more clearly preserved valves. The trapezoidal valves are beaked, and the anteriormost edge of each valve (top of the trapezoid) forms a broad 'W' shape. The posterior edge of each valve flares out laterally. Paired lateral riblets are preserved on the right sides of the anteriormost three plates. A lunate field is present at the anterior end of the creature. The lunate anterior or cephalic field is covered in numerous stellate tubercles. Faint 'segmental' divisions are visible along the post-cephalic margin of the organism.

Korifogrammia clementensis n. gen. n. sp. M. A. S. McMenamin
Figures 3.22. 3.38, 3.39, 3.40, 3.41, 3.42 and 3.43
Holotype: IGM 4998.
Diagnosis. As for genus.
Material: Two specimens, one complete and one fragmentary, 5 mm apart on IGM 4998.

Discussion: A fragmentary specimen 750 microns long and 600 microns wide, consisting of three trapezoidal valves (valves IV-VI), is preserved to the left of the holotype (at a distance of 4.5 mm). Valve morphology is more distinct on the fragmentary specimen, and the 'W' shape anterior edge of the trapezoid is clearly visible (Figs. 3.42 and 3.43). As for *Clementechiton sonorensis*, the other aculiferan from the Clemente Formation, original valve composition in *Korifogrammia clementensis* n. gen. n. sp. was probably aragonitic (Lowenstam 1967).

Close inspection of the dorsal median keel shows that it is divided into eight small carinated mucro-anterior plates with beaked anterior margins (Figs. 3.38 and 3.39). These form a series of eight keeled impressions that constitute the dorsal medial ridge or crest. On the holotype, a slightly curved sagittal groove divides each plate roughly in half. Paired lateral riblets are preserved on the right sides of the anteriormost two or three plates. The lunate head region is covered with numerous stellate tubercles. These are best preserved on the left side of the specimen, but apparently covered the entire cephalic region (Fig. 3.39). The edge of the abdomen is marked by faint curving partitions that, adjacent to the edge, appear to break up into fine marginal spines. This apparent segmentation may have a relationship to the "metameric pattern" inferred for *Kimberella* by Fedonkin et al. (2007). Very faint triangular impressions, probably tubercles, are seen in the pleural field between the valves and the curved sculpturing at the margin.

The stellate tubercles of the head region of *Korifogrammia clementensis* n. gen. n. sp. are reminiscent of similar structures seen in *Onega* and *Lossinia* from the White Sea Vendian biota (Ivantsov 2007; Ivantsov and Leonov 2008). The spacing and arrangement of the stellate tubercles in *Korifogrammia clementensis* n. gen. n. sp. is similar to that of a "retracted" specimen of *Kimberella quadrata* from the White Sea biota (Ivantsov 2009). However, the tubercles in the Russian specimen cover the body of the animal and are not merely restricted to the head region. Thus *Korifogrammia clementensis* n. gen. n. sp., as it bears eight aculiferan (chiton) plates, appears to represent an intermediate form between mollusc-like Ediacarans such as *Kimberella* and the more arthropod- or annelid-like "Proarticulata" forms, and thus shows an intriguing mosaic of aculiferan and proarticulate features. *Korifogrammia clementensis* n. gen. n. sp. may potentially represent a stem form to both mollusks and arthropods.

Locality: Field sample 7 of 3/16/95; GPS coordinates of site are N30°24.041′, W111°57.141′ (average of seven measurements), altitude 526 m (average of six measurements). The fossil occurrence is approximately 5–10 m below the Clemente oolite, in unit 4 of the Clemente Formation.

References

Briggs DEG, Clarkson ENK (1987) The first tomopterid, a polychaete from the Carboniferous of Scotland. Lethaia 20(3):257–262

Fedonkin MA, Simonetta A, Ivantsov AY (2007) New data on Kimberella, the Vendian mollusc-like organism (White Sea region, Russia): palaeoecological and evolutionary implications. Geol Soc Lond Spec Publ 286(1):157–179

Glaessner MF (1958) New fossils from the base of the Cambrian in South Australia. Trans R Soc S Aust 81:185–188

Glaessner MF (1959) The oldest fossil faunas in South Australia. Geol Rundsch 47(2):522–531

Glaessner MF, Wade M (1971) *Praecambridium*-a primitive arthropod. Lethaia 4:71–77

Ivantsov AY (2007) Small Vendian transversely articulate fossils. Paleontol J 41(2):113–122

Ivantsov AY (2009) A new reconstruction of *Kimberella*, a problematic Vendian metazoan. Paleontol J 43(6):601–611

Ivantsov AY, Leonov MV (2008) Otpechatki vendskix zhivotnykh-unikal'nye paleontologicheskie ob'eky Archangel'skoi oblasti. Arkhangel'sk, Russia

Ivantsov AY (2017) The most probable Eumetazoa among late Precambrian macrofossils. Invertebr Zool 14(2):127–133

Lowenstam HA (1967) Lepidocrocite, an apatite mineral, and magnetite in teeth of chitons (Polyplacophora). Science 156(3780):1373–1375

McMenamin MAS (2003) *Spriggina* is a trilobitoid ecdysozoan. Geol Soc Am Abstr Programs 35(6):105

McMenamin MAS (2004) The ptychoparioid trilobite *Skehanos* gen. nov. from the Middle Cambrian of Avalonian Massachusetts and the Carolina Slate Belt, USA. Northeast Geol Environ Sci 24(4):276–281

Meyer A (1926) Die Segmentalorgane von Tomopteris catharina (Gosse) nebst Bemerkungenueber das Nervensystem, die rosetten-formigen Organe und die Colombewimperung. Zeitschrift fur Wissenchaftliche Zoologie 127:297–402

Phillips Dales R (1971) Bioluminescence in Pelagic Polychaetes. J Fish Res Board Can 28(10):1487–1489

Schwabe E (2010) Illustrated summary of chiton terminology (Mollusca, Polyplacophora). Spixiana 33(2):171–194

Sigwart JD (2017) Zoology: molluscs all beneath the sun, one shell, two shells, more or none. Curr Biol 27:R702–R719

Index

© Springer International Publishing AG 2018
M.A.S. McMenamin, *Deep Time Analysis*, Springer Geology,
https://doi.org/10.1007/978-3-319-74256-4

Printed in the United States
By Bookmasters